AF740086

Perspective of Carbon Nanotubes

Edited by Hosam El-Din Saleh and Said Moawad Mohamed El-Sheikh

Published in London, United Kingdom

IntechOpen

Supporting open minds since 2005

Perspective of Carbon Nanotubes
http://dx.doi.org/10.5772/intechopen.77444
Edited by Hosam El-Din Saleh and Said Moawad Mohamed El-Sheikh

Contributors
Narsimha Mamidi, Khalid Parwez, Suman Budihal, Neeraja Turagam, Durga Prasad Mudrakola, GuiPing Dai, Muzamal Hussain, Dr. Muhammad Nawaz Naeem, Mohamed R. Berber, Inas Hafez, Mohamed Mustafa, Norihiro Shimoi, S Karthikeyan, V. S. Angulakshmi Lakshmi, S Mageswari, Oleg Il'in, Marina Il'ina, Alexandr Fedotov, Nikolay Rudyk, Oleg Alexeevich Ageev, Arti Rushi, Kunal Datta, Prasanta Ghosh, Mahendra Shirsat, Firas Abdulrazzak, Marianna Kharlamova, Dominik Eder, Vladimir Samuilov, Nikolai A. Poklonski, Hosam El-Din M. Saleh

© The Editor(s) and the Author(s) 2019
The rights of the editor(s) and the author(s) have been asserted in accordance with the Copyright, Designs and Patents Act 1988. All rights to the book as a whole are reserved by INTECHOPEN LIMITED. The book as a whole (compilation) cannot be reproduced, distributed or used for commercial or non-commercial purposes without INTECHOPEN LIMITED's written permission. Enquiries concerning the use of the book should be directed to INTECHOPEN LIMITED rights and permissions department (permissions@intechopen.com).
Violations are liable to prosecution under the governing Copyright Law.

(cc) BY

Individual chapters of this publication are distributed under the terms of the Creative Commons Attribution 3.0 Unported License which permits commercial use, distribution and reproduction of the individual chapters, provided the original author(s) and source publication are appropriately acknowledged. If so indicated, certain images may not be included under the Creative Commons license. In such cases users will need to obtain permission from the license holder to reproduce the material. More details and guidelines concerning content reuse and adaptation can be found at http://www.intechopen.com/copyright-policy.html.

Notice
Statements and opinions expressed in the chapters are these of the individual contributors and not necessarily those of the editors or publisher. No responsibility is accepted for the accuracy of information contained in the published chapters. The publisher assumes no responsibility for any damage or injury to persons or property arising out of the use of any materials, instructions, methods or ideas contained in the book.

First published in London, United Kingdom, 2019 by IntechOpen
IntechOpen is the global imprint of INTECHOPEN LIMITED, registered in England and Wales, registration number: 11086078, 7th floor, 10 Lower Thames Street, London, EC3R 6AF, United Kingdom
Printed in Croatia

British Library Cataloguing-in-Publication Data
A catalogue record for this book is available from the British Library

Additional hard and PDF copies can be obtained from orders@intechopen.com

Perspective of Carbon Nanotubes
Edited by Hosam El-Din Saleh and Said Moawad Mohamed El-Sheikh
p. cm.
Print ISBN 978-1-78984-401-6
Online ISBN 978-1-78984-402-3
eBook (PDF) ISBN 978-1-83968-026-7

For EU product safety concerns:
IN TECH d.o.o., Prolaz Marije Krucifikse Kozulić 3, 51000 Rijeka, Croatia, info@intechopen.com or visit our website at intechopen.com.

We are IntechOpen, the world's leading publisher of Open Access books

Built by scientists, for scientists

4,500+
Open access books available

118,000+
International authors and editors

130M+
Downloads

151
Countries delivered to

Our authors are among the
Top 1%
most cited scientists

12.2%
Contributors from top 500 universities

WEB OF SCIENCE™

Selection of our books indexed in the Book Citation Index in Web of Science™ Core Collection (BKCI)

Interested in publishing with us?
Contact book.department@intechopen.com

Numbers displayed above are based on latest data collected.
For more information visit www.intechopen.com

Meet the editors

Hosam El-Din Saleh is a professor of radioactive waste management in the Radioisotope Department, Atomic Energy Authority, Egypt. He has been awarded MSc and PhD degrees in Physical Chemistry from Cairo University. His area of interest is innovative economic and environment-friendly techniques for the management of hazardous and radioactive wastes. He has authored many peer-reviewed scientific papers and book chapters, and is the editor of several books from by international publishers.

Prof. El-Sheikh was awarded his PhD from Cairo University in 2000. He is currently the Head of the Nanomaterials Department, Central Metallurgical R&D Institute. His research interests are nanomaterials, wastewater treatment, conversion of CO2 to renewable fuel, and environmental pollution control. He has published 110 papers in international journals and more than 60 presentations at national and international conferences.

Contents

Preface

This book provides guidance on the synthesis and characterization of carbon nanotubes and includes a number of applications. The authors collaborating in this project have summarized their experience and present advances in fields related to the synthesis and characterization of carbon nanotubes. The book contains 14 chapters, organized in four sections that cover important research aspects in modern technologies of carbon nanotubes. The first section consists of an introductory chapter prepared by the editor Hosam Saleh with Martin Koller aiming to present a brief background on the synthesis, characterization, and applications of carbon nanotubes.

The second section presents the synthesis and production of carbon nanotubes; it comprises four chapters of the following titles: Chapter 2 "Synthesis of Carbon Nanotubes by Catalytic Chemical Vapor Deposition, prepared by Xiaodi Wang, K. Vinodgopal, and Gui-Ping Dai. Chapter 3, "Synthesis and Properties of Single-Walled Carbon Nanotubes Filled with Metal Halogenides and Metallocenes," is presented by Marianna V. Kharlamova and Dominik Eder. Chapter 4, "Carbon Nanotube from Unconventional Precursor-Optimization of Synthesis Parameters" is prepared by S. Karthikeyan and V. S. Angulakshmi. Chapter 5, "Vertically Aligned Carbon Nanotubes Production by PECVD," is submitted by Oleg I. Il'in, Marina V. Il'ina, Nikolay N. Rudyk, Alexandr A. Fedotov, and Oleg A. Ageev.

The third section presents the characterization of carbon nanotubes, which includes six chapters conducting "Functionalized Carbon Nanotubes for Detection of Volatile Organic Pollutant", Chapter 6 written by Arti Rushi, Kunal Datta, Prasanta Ghosh, Ashok Mulchandani, and Mahendra Shirsat. Chapter 7, "Behavior of X-Ray Analysis of Carbon Nanotubes," is by Firas H. Abdulrazzak, Ayad F. Alkaim, and Falah H. Hussein. Chapter 8, "Conductive Effect of Increased Crystallinity of Single-Walled Carbon Nanotubes as Field Emitter," is by Norihiro Shimoi. Chapter 9, "Electron Transport in the Assemblies of Multi-Wall Carbon Nanotubes," is by Vladimir Samuilov, Jean Galibert, and Nikolaj Poklonski. Chapter 10, "Surface Functionalization of Carbon Nanotubes for Energy Applications," is by Mohamed R. Berber, Inas H. Hafez, and Mohamed Y. Mustafa. Chapter 11, "Vibration Characteristics of Single-Walled Carbon Nanotubes Based on Nonlocal Elasticity Theory Using Wave Propagation Approach (WPA) Including Chirality," is by Muzamal Hussain and Muhammad Nawaz Naeem.

The fourth section describes the risk management and limitation of carbon nanotubes, and includes three chapters discussing the "Cytotoxicity Evaluation of Carbon Nanotubes for Biomedical and Tissue Engineering Applications", Chapter 12 written by Narsimha Mamidi. Chapter 13, "Quality Control and Risk Management of Carbon Nanomaterials," is by Khalid Parwez and Suman V. Budihal. Chapter 14, "Advantages and Limitations of CNT-Polymer Composites in Medicine and Dentistry," is by Neeraja Turagam and Durga Prasad Mudrakola.

The editor wishes to express his thanks to all the participants of this book for their valuable contributions and to Ms. Ivana Barac for her assistance in finalizing the work. Acknowledgment also goes to the IntechOpen staff members responsible for the completion of this book and other publications for free visible knowledge.

Hosam El-Din Mostafa Saleh
Atomic Energy Authority of Egypt,
Cairo, Egypt

Said M. El-Sheikh
Central Metallurgical Research and Development Institute,
Helwan, Egypt

Section 1

Introduction

Chapter 1

Introductory Chapter: Carbon Nanotubes

Hosam M. Saleh and Martin Koller

1. Background

As a family of rather new nanomaterials, carbon nanotubes (CNTs) are emerging since about two decades. However, their origin dates back almost 70 years ago, when they were observed and described in 1952 for the first time by Radushkevich and Lukyanovich; in 1976, Oberlin and colleagues described the microscopic observation of single- or double-walled carbon nanotubes. In 1991, Iijima demonstrated for the first time a process for preparation of multi-walled carbon nanotubes (MWNTs); this discovery occurred rather fortuitously during testing a new method for arc evaporation to fabricate C_{60} carbon molecules. Soon later, two seminal studies by the groups of Iijima and Bethune provided mechanistic descriptions of the growth process involved in the formation of single-walled carbon nanotubes [1]. Structurally, such single-walled carbon nanotubes (SWNT) can be conceived as one atom-thick sheets of graphite ("graphene"), which are rolled up (wrapped) to form tubes, as illustrated in **Figure 1**.

Since their discovery in 1991, CNTs have experienced considerable investigative efforts, especially regarding potential smart applications. Those structures first reported in 1991 were MWNTs with a broad range of dimensions. These were basically distant relatives of highly defective carbon nanofibers grown via catalytic chemical vapor deposition.

Real molecular nanotubes *sensu stricto* only came up when they were by chance detected while a catalyst (Fe and Co) material was inserted in the anode during electric-arc discharge synthesis. For the first time, it became possible to synthesize molecular fibers exclusively based on carbon; one can imagine that the excitement was tremendous, since many physical properties of such a fiber had already been theoretically predicted [6].

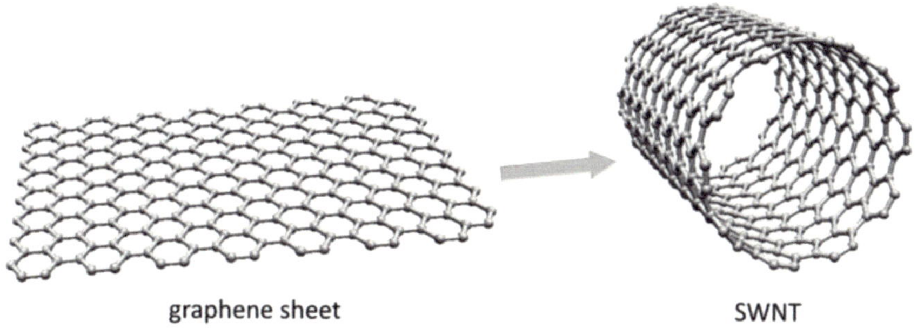

Figure 1.
Wrapping of graphene sheet to form single-walled carbon nanotubes [1].

2. Preparation of carbon nanotubes

High temperature preparation techniques such as arc discharge or laser ablation were the first techniques applied to produce CNTs [1]. In order to examine the properties of nanotubular structures, tubes of well-defined morphology, thickness, length, and a number of concentric shells are needed; however, established carbon-arc synthesis and other established methods generate a range of differently structured tubes [2]. Nowadays, these methods have been substituted by low temperature chemical vapor deposition (CVD) techniques (<800°C), which offer the possibility to fine tune orientation, alignment, length and diameter, purity and density of CNTs in a precise way. The most frequently applied methods and some additional non-standard techniques such as liquid pyrolysis and bottom-up organic approaches are introduced below. Most of these techniques need auxiliary gases and *in vacuo* operation, although CNT growth has been already reported also at atmospheric pressure. However, gas-phase methods are volumetric, which makes them suitable for diverse new applications such as development of novel composite materials, which require high amounts of nanotubes and industrial-scale synthesis methods in order to make their production feasible in economic terms. Yet, there are some drawbacks when resorting to gas-phase synthesis methods, namely short catalyst lifetime, low catalyst yields by only a low percentage of catalysts actually forming nanotubes, and low catalyst number density. Independent on which CNT preparation method is applied, CNT production is always accompanied with formation of certain impurities that qualitatively and quantitatively depends on the technique applied. Most of above-mentioned methods generate powders, which contain not only a small fraction of CNTs, but also other carbonaceous particles like nanocrystalline graphite, amorphous carbon, fullerenes, and different heavy metals such as Fe, Co, Mo or Ni, which were needed as catalysts for CNT synthesis. Such contaminations negatively affect the desired properties of CNTs, and cause a serious obstacle for detailed characterization and application of CNTs. Therefore, the development of efficient and convenient purification methods is among the most central challenges in CNT research. Most common purification methods are based on acid treatment of prepared CNTs.

3. Nanotube growth and characterization

Both single-walled carbon nanotubes (SWNTs) and multi-walled carbon nanotubes (MWNTs) have turned out as excellent electron emitters due to their large aspect ratio, high chemical stability, and small end radius [3]. However, there has been no successful alignment of single-wall individual or bundle nanotubes reported. Auspiciously, growth of large arrays of well-aligned MWNTs in an area up to the dimension of inches has been accomplished on glass, nickel, and silicon by plasma-enhanced chemical vapor deposition (PECVD), and on silica, porous silicon, and porous aluminum by chemical vapor deposition. Yet, the emission property of these arrays is not acceptable, probably because of the high site density (>109/cm^2), which leads to small electrical enhancement at the tips. In this context, it has been calculated that a site density of about 10^7/cm^2 constitutes the adequate number for optimal electron emission characteristics in terms of both emission site and current density. The growth of arrays of well-aligned unconnected MWNTs with sharp tips and controlled dimensions and site spacing has been realized by using PECVD. This technique offers a new strategy to obtain optimal emitter density of CNTs. For aligned MWNTs grown on a uniform Ni layer, a catalyst particle at the tip of each nanotube was detected. Moreover, this metal particle had a preferential orientation relative to the catalytically active surface, which was explained by a model showing that Ni surfaces except (220) get quickly covered, with the catalytically active (220) surface remaining the only exposed surface.

4. Carbon nanotube metrology

Broad commercialization of CNTs requires bulk material measurements, which are fast, inexpensive, reliable, and traceable to rigorous and quantitative methods. Information such as the volume fraction of semiconducting to metallic bulk material is needed to optimize manufacturing processes, thus enabling successful end-user applications like development of multi-function composites. Chirality and diameter of individual CNTs are crucial for matching species with high-end applications in (micro)electronics and optoelectronics. Metrology at the scale of individual CNTs involves complex optical techniques and imaging, and further launches the fundamental traceability of bulk measurements. Moreover, this introductory chapter describes NIST standards to establish recommended practice guides and reference materials based on industry demand, and shows that collaboration of researchers, stakeholders such as CNT manufacturers, NASA, and other end users is needed for ultimate success of CNTs on the market.

Single-walled carbon nanotubes (SWCNTs) offer multiple opportunities for applications in the fields of electronics, optoelectronics, photonics, or photovoltaics. SWCNTs are definitely almost ideal models of infinite π orbital conjugations, where the sp^2 lattice provides exceptional charge carrier properties. In addition, their real one-dimensional character enables well-defined optical transitions from excitonic states. While their Raman and photoluminescence characteristics have already been comprehensively studied, the basic absorption properties of SWCNT are still scarcely investigated on a quantitative basis, although especially these properties are of major significance for applications. In fact, described absorption measurements on SWCNT assemblies suffer from the problem of sample heterogeneity characteristic for standard synthesis methods.

The huge multiplicity of nanotube chiralities and dimensions, together with the occurrence of impurities from the synthesis process encumber the assessment of the exact SWCNT chirality contributing to an absorption signature [4].

5. Electrochemistry of carbon nanotubes

CNTs have attracted attention because of their exceptional morphology, nanosized dimensions, unique physico-chemical properties, and moreover, their multifarious fields of applications [5]. As illustrated in **Figure 1**, CNTs could be visualized as wrapped graphene sheets (sp^2 carbon arranged in a honeycomb-like grid). As mentioned above, two groups of carbon nanotubes are distinguished, namely multi-walled carbon nanotubes (MWNTs) and single-walled carbon nanotubes (SWNTs). In the case of MWNTs, these can be visualized as concentric and closed graphite tubules with multiple graphene sheet layers defining a hole of typically 2–25 nm, separated by about 0.34 nm. SWNTs consist of a simple single-rolled graphite sheet, forming a cylinder of 1–2 nm diameter. Depending on the structure, mainly on the diameter and helicity, CNTs can behave like metals or semiconductors. Electrically, MWNTs display superconductivity with a relatively high transition temperature. Electron transport in nanotubes is ballistic; hence, the nanotubes' resistance does not depend on its length, because the mean free path λm is longer than the length of the nanotube itself. For example, Berger *et al.* reported the mean free path with λm = 30 μm, which was considerably longer than the nanotube used for the study. Recent measurements of the magnetic properties of nanotubes evidence that SWNT might be the heavily desired materials of choice to be used as room-temperature superconductors.

6. Applications of carbon nanotubes

Various potential applications have been proposed in literature for CNTs, encompassing conductive and high-strength composites, sensors, energy storage and energy conversion devices, hydrogen storage media, field emission displays and radiation sources, and nanometer-sized semiconductor devices, probes, and interconnects [6]. Some of these applications were already successfully realized in products, while others were only shown in lab-scale to prototype devices. Cost for nanotube preparation, polydispersity in a given type of nanotubes, and immature processing and assembly methods are still important barriers for some applications, especially in the case of SWNTs [7].

6.1 Potential application of CNTs in vacuum microelectronics

Field emission, a quantum effect, is a smart source to generate electrons in comparison to thermionic emission. When subject to a sufficiently high electric field, electrons near the Fermi level can overcome the energy barrier to get released to the vacuum level. The basic physics of electron emission is already well understood. The emission current from a metal surface is determined by the Fowler–Nordheim equation: $I = aV\,2 \exp(-b\varphi 3/2/\beta V)$, where I, V, φ, β are the current, applied voltage, work function, and field enhancement factor, respectively. Electron field emission materials have been broadly investigated for different technological uses, for example, for flat panel displays, microwave amplifiers, or electron guns in electron microscopes. For technological applications, electron emissive materials should have low threshold emission fields and should be stable at high current density. A current density of 1–10 mA/cm^2 is required for displays and >500 mA/cm^2 for a microwave amplifier.

In order to minimize the electron emission threshold field, emitters with a low work function and a large field enhancement factor are favorable. The work function is an intrinsic material property. The field enhancement factor mainly depends on the emitter's geometry, and can be approximated by: $\beta = 1/5r$, where r is the radius of the emitter tip. Various processing techniques to generate different emitter types have been established like Spindt-type emitters, with a sub-micron tip radius. Yet, this process is expensive, and the emitters have only limited lifespan. Malfunction is often caused by ion bombardment from the residual gas species that dull the emission tips.

6.2 Energy storage

CNTs are also considered for energy production and storage. Graphite, carbonaceous materials, and carbon fiber electrodes are used since decades in fuel cells, batteries, and various other electrochemical applications. Nanotubes are unique due to their small dimensions, the smooth surface topology, and perfect surface specificity, which is based on the fact that only the basal graphite planes are exposed in their structure. The electron transfer rate at carbon electrodes finally governs the performance of fuel cells, which depends on different factors, such as arrangement and morphology of the carbon material used in the electrodes. Several experiments have demonstrated that, compared to traditional carbon electrodes, electron transfer kinetics are fastest on nanotubes, following ideal Nernstian behavior. Nanotube microelectrodes have been built using a binder; such devices have been successfully tested for bioelectrochemical reactions (e.g., oxidation of dopamine). They outperformed the established carbon electrodes in terms of reaction rates and reversibility. Pure MWNTs and MWNTs deposited with noble metal catalysts (Pd, Pt and Ag) have been applied for electro-catalysis of oxygen reduction reactions, which are pivotal for fuel cells. Several studies demonstrate that nanotubes constitute excellent alternatives for traditional carbon-based electrodes. Likewise, the enhanced selectivity of nanotube-based catalysts in heterogeneous catalysis was confirmed. In this context, Ru-supported nanotubes outperformed the same metal on graphite and on other carbon materials in the liquid phase hydrogenation reaction of cinnamaldehyde. The properties of catalytically grown carbon nanofibers (which can principally be conceived as defective nanotubes) turned out to be appropriate for high power electrochemical capacitors.

6.3 Filled composites

The mechanical characteristics of CNTs are stimulating since they are considered the "ultimate" carbon fiber ever produced. Compared with steel, traditional carbon fibers have about 50 times the specific strength (strength/density), and constitute outstanding load-bearing reinforcements when incorporated in composites. Consequently, nanotubes should be the candidates of choice for structural applications. In fact, carbon fibers have successfully been applied as reinforcements in high strength, low density, and high performance composites. They are typically found in a variety of products ranging from expensive tennis rackets to aircraft and spacecraft body parts. In this context, NASA has recently made large investments in developing novel CNTs-based composites to be applied for, for example, the futuristic Mars mission. In addition, carbon nanotubes were also applied as filler materials to fine-tune the mechanical properties of films consisting of biodegradable, microbial bioplastics [8].

6.4 Nanoprobes and sensors

The tiny and constant dimensions of nanotubes pave the way for some highly intriguing applications. With extremely small sizes, high mechanical strength and elasticity, and high conductivity, nanotubes have the potential to ultimately become essential nanoprobes for technological use. Such probes can be conceived as being applied for various applications, such as high resolution imaging, nanoelectrodes, nanolithography, drug delivery systems, sensors, or field emitters. The option of developing nanotube-based field emitting devices has already been discussed above in this chapter. Moreover, also using a single MWNT attached to the end of a scanning probe microscope tip for imaging has already been described. Since MWNT tips are electrically conductive, they can be applied in STM and AFM devices as well as in other scanning probe instruments, such as in electrostatic force microscopes. The benefit of nanotube tips is their thinness and the possibility to image particular structures (such as tiny, deep surface cracks), which are very difficult to probe with larger, blunter-etched Si or metal tips. Compared to conventional STM tips, also large biomolecules, such as DNA or proteins, can be imaged with excellent resolution using nanotube tips. MWNT and SWNT tips were successfully used in a tapping mode to image biomolecules like amyloid-b-protofibrils (related to Alzheimer's disease) with unprecedented resolution. Moreover, based on the high elasticity of the nanotubes, the tips are not damaged by contact with the substrates. Impacts merely cause buckling of the nanotube, which usually is reversible on retracting the tip from the substrate.

6.5 Templates

Since nanotubes have quite straight and thin channels in their cores, it was discussed from the very beginning that it might be possible to fill different materials into these cavities in order to generate one-dimensional nanowires. Early assumptions proposed that pronounced capillary forces exist in nanotubes, sufficiently strong to hold back gases and liquids inside the cavities. In 1993, this was for the first time proofed experimentally by filling and solidifying molten lead inside MWNT channels. By this technique, wires only 1.2 nm in diameter were successfully manufactured inside nanotubes. Nowadays, a large number of studies dealing with this topic exist, many of them describing the filling of nanotubes with metallic and ceramic materials. Thus, nanotubes constitute proficient templates to create nanowires of different structure and composition.

Author details

Hosam M. Saleh[1]* and Martin Koller[2]

1 Radioisotope Department, Nuclear Research Center, Egyptian Atomic Energy Authority, Egypt

2 University of Graz, Office of Research Management and Service, c/o Institute of Chemistry, NAWI Graz, Graz, Austria

*Address all correspondence to: hosamsaleh70@yahoo.com

IntechOpen

© 2019 The Author(s). Licensee IntechOpen. This chapter is distributed under the terms of the Creative Commons Attribution License (http://creativecommons.org/licenses/by/3.0), which permits unrestricted use, distribution, and reproduction in any medium, provided the original work is properly cited.

References

[1] Prasek J, Drbohlavova J, Chomoucka J, Hubalek J, Jasek O, Adam V, et al. Methods for carbon nanotubes synthesis. Journal of Materials Chemistry. 2011;**21**:15872-15884

[2] Iijima S, Ichihashi T. Single-shell carbon nanotubes of 1-nm diameter. Nature. 1993;**363**:603

[3] Wen JG, Huang ZP, Wang DZ, Chen JH, Yang SX, Ren ZF. Growth and characterization of aligned carbon nanotubes from patterned nickel nanodots and uniform thin films. Journal of Materials Research. 2001;**16**:3246-3253

[4] Oudjedi L, Parra-vasquez ANG, Godin A, Oudjedi L, Parra-vasquez ANG, Godin A, et al. Metrological investigation of the (6,5) carbon nanotube absorption cross section. The Journal of Physical Chemistry Letters - American Chemical Society. 2013;**4**:1460

[5] Elrouby M. Electrochemical applications of carbon nanotube. Journal of Nanotechnology & Advanced Materials. 2013;**1**:23-38

[6] Ajayan PM, Zhou OZ. Applications of carbon nanotubes. In: Dresselhaus MS, Dresselhaus G, Avouris P, editors. Carbon Nanotubes. Berlin/Heidelberg: Springer-Verlag; 2001. pp. 391-425

[7] Baughman RH, Zakhidov AA, De Heer WA. Carbon nanotubes—The route toward applications. Science. 2002;**297**:787-792

[8] Yun SI, Gadd GE, Latella BA, Lo V, Russell RA, Holden PJ. Mechanical properties of biodegradable polyhydroxyalkanoates/single wall carbon nanotube nanocomposite films. Polymer Bulletin. 2008;**61**:267-275

Section 2

Synthesis and Production of Carbon Nanotubes

Chapter 2

Synthesis of Carbon Nanotubes by Catalytic Chemical Vapor Deposition

Xiao-Di Wang, K. Vinodgopal and Gui-Ping Dai

Abstract

As a new carbon material in the twenty-first century, carbon nanotubes (CNTs) have excellent optical, electrical, magnetic, thermal, chemical, and mechanical properties. There are many synthesis methods to produce CNTs. Compared with other methods, chemical vapor deposition (CVD) is the most effective method that has broad prospects for large-scale control of CNTs in recent years due to its simple equipment, simple operation, and lower cost. In order to gain a comprehensive understanding of the controlling parameters about the formation of CNTs, this chapter reviews the latest progress in the preparation of CNTs by CVD from three of the most important influencing factors: carbon sources, catalysts, and substrates. Among them, the catalyst is the most influential factor for the morphology, structure, and properties of CNTs. It should be pointed out that many growth factors can control the particle size distribution, composition, and structure of the catalysts, such as catalyst substrate, metal transition components added, calcination temperature, etc.

Keywords: synthesis, CVD, carbon nanotubes

1. Introduction

In order to meet the challenge of the increase of global energy consumption and the augment of global financial markets and population in the past time, people have opened up new fields in materials science such as nanotechnology. Carbon-based nanomaterials have caused great concern in the light of their unique structure and extraordinary physical properties (**Table 1**).

Generally speaking, carbon is mainly retained as solid carbon including graphite and diamond. With the rapid development of science and technology, more and more carbon nanostructured materials are discovered. In 1975, Oberlin et al. [1] successfully prepared carbon nanofibers by CVD, which has aroused the upsurge in the research of carbon nanomaterials. In 1985, fullerene [2], a zero-dimensional carbon nanomaterial, was discovered. Later, Iijima [3] accidentally discovered what is now known as the CNT needle (**Figure 1**) when he used TEM to observe the products by the arc discharge process. The investigation of CNTs was heating up since their discovery. The preparation and application of CNTs are one of the examples for the current industry technology. In 2004, Geim and Novoselov [4] mechanically peeled off the graphite repeatedly until it formed a plane of a single atom and found another allotrope of carbon named graphene.

Properties		SWCNTs	MWCNTs	Comparison
Mechanical properties	Young's modulus	~1 TPa	~1–1.2 TPa	About 5 times stronger than steel
	Tensile strength	~60 GPa	~0.15 TPa	About 100 times stronger than steel
Electronic properties	Bandgap	When n-m is divisible by 3 (0 eV, metallic)	~ 0 eV (non-semiconducting)	
		When n-m is not divisible by 3 (0.4–2 eV, semiconducting)		
Thermal properties	Thermal conductivity at room temperature	1750–5800 W/mk	>3000 W/mk	About 3 times better than diamond
Electrical properties	Typical resistivity	10^{-6} Ω m		
	Typical maximum current density	10^{7}–10^{9} A cm^{-2}		
	Typical quantized conductance (measured)	12.9 $k\Omega^{-1}$		

Table 1.
Some excellent properties of CNTs.

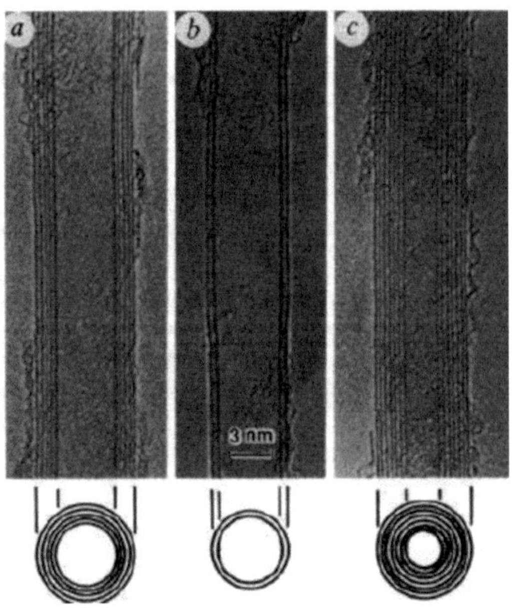

Figure 1.
CNTs firstly discovered by Iijima in 1991. Reproduced with permission of Ref. [3].

CNTs are tubes that are rolled up by a layer or several layers of graphite, or that can be coaxially wrapped together by different diameters of microtubules, which have only a few nanometers in diameter. The spacing between the tube walls is about the size of graphite layer spacing. There are two ways of hybridizing carbon

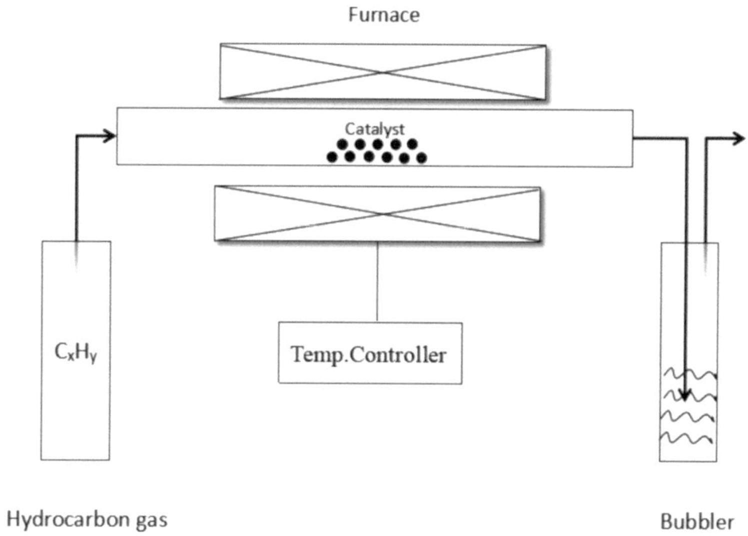

Figure 2.
Simple form of CVD setup.

atoms in CNTs, mainly SP^2 hybridization, and specific SP^3 bonds can also be formed between hexagonal carbon atoms.

CNTs are classified as single-walled CNTs and multi-wall CNTs according to the layer number of graphene sheets. Because there is only one layer of graphene in SWCNTs, the diameter range of SWCNTs is smaller than that of MWCNTs, which is only 0.4–3 nm. Therefore, the application of single-walled CNTs in nanomaterials field is more promising. Multi-walled CNTs have graphene layers ranging from 2 and more. SWCNTs were firstly discovered by Iijima and Ichihashi [5] in 1993, CNTs formed by the coaxial crimping of two or more layers of graphene are called MWCNTs [3]. Besides, SWCNTs can be classified into three types according to their structural characteristics (depending on graphene sheet in a rolled manner as shown in **Figure 2**): armchair form, zigzag form, and chiral form. This difference is due to the fact that in the case of single-walled CNTs, the graphene sheets can be curled at any angle and therefore have different chiral structures. The chirality index of CNTs is directly related to the helicity and electrical properties of CNTs.

In addition to above, there is another fibrous carbon nanostructure which is rolled by multi-layer graphene called carbon nanofibers. Carbon nanofibers, which are fibrous carbon nanomaterials made from multilayer graphene sheets, are generally 10–500 nm in diameter and 0.5–100 nm in length, with high orientation, conductivity, and thermal conductivity.

CNTs are characterized by high mechanical strength, high specific surface, high conductivity and strong interfacial effect. They are widely used in energy storage devices, electrode materials, field emission displays, semiconductor materials, catalyst carriers, composite reinforcement, etc. The scale of all these applications mainly depends on quality, output, and cost of CNTs.

CNTs are strictly studied by scientists and engineers because of their special properties. The range of applications of CNTs covers almost every field of materials science; a wide variety of applications have also been widely reported such as electromagnetic and microwave absorbing coatings, sensors, radiation sources, and nanometer-sized semiconductor devices and thermal interface materials. CNT arrays can provide good thermal interface conductance for ionic and electronic transport devices such as batteries [6]. Dalton et al. [7] discovered that the low density, high aspect ratio, and extraordinary mechanical properties of CNTs make fibers super-tough particularly.

Method	Chemical vapor deposition	Arc discharge	Laser ablation
Condition	Low pressure inter gas (argon)	Argon or nitrogen gas at 500 Torr	High temperatures about 500–1000 °C at high energy laser beam
Yield	High (60–90%)	Low (20–100%)	Low (up to 70%)
Purity	Medium to high	Medium	Low
Temperature	500–1200°C	~4000°C	25–1000°C
Product	SWCNTs: long tubes with diameters ranging from 0.6 to 4 nm	SWCNTs: short tubes with diameters of 0.6–1.4 nm	SWCNTs: long bundles of tubes (5–20 μm) with individual diameter from 1 to 2 nm
	MWCNTs: long tubes with diameter ranging from 10 to 240 nm	MWCNTs: short tubes with inner diameter of 1–3 nm and outer diameter of approximately 10 nm	MWCNTs: not very much interest in this technique

Table 2.
Comprehensive comparison of CVD and other preparation methods.

Directional carbon nanotube arrays can be used in photonic crystals, optical antennas, optical waveguides, solar cells, and so on [8–12]. Carbon nanotube arrays used in optoelectronic devices are three-dimensional ordered structure and are usually fabricated by PECVD. Carbon nanotubes are considered to be one of the best electron-emitting materials because of their high aspect ratio, which induces very high local electric fields at the top of carbon nanotubes [13]. The field emission characteristics of vertically oriented CNTs can be used in flat panel displays, high-strength electron sources, X-ray generators, and microwave devices [14–16].

If CNTs cannot be achieved as desired scale and controllable parameters according to the pre-designed structural model, all valuable applications disappear. Therefore, it is one of the key technologies for the development of CNTs with specific characteristics such as height, diameter, aspect ratio, morphology, large area, and scale preparation.

There are three most common preparation methods (**Table 2**) for CNTs: chemical vapor deposition, arc discharge, and laser ablation [6]. Among them, CVD is the most widely used method because of the continuous mass production and low cost.

Despite tremendous advances in carbon nanotube research in the past years, we have been unable to mass produce well-performing CNTs with cost-effective technologies. This chapter focused on chemical vapor deposition of CNTs. In terms of reaction materials, we discussed three aspects in detail: carbon source, catalyst and substrate, and growth process in CNTs. For growth parameters, we mainly discussed the effects of temperature on the morphology and properties of CNTs. In industrial production, we emphasized high yield, large yield, and high purity. Finally, we look forward to the future prospects of CNTs.

2. Chemical vapor deposition

It is important that large quality and high purity CNTs require highly reliable synthesis techniques. Understanding the influencing factors and control conditions of specific carbon nanotube synthesis is what we need.

At present, the preparation process of multi-walled CNTs is quite mature, and industrial production has been realized by the CVD. The production cost of

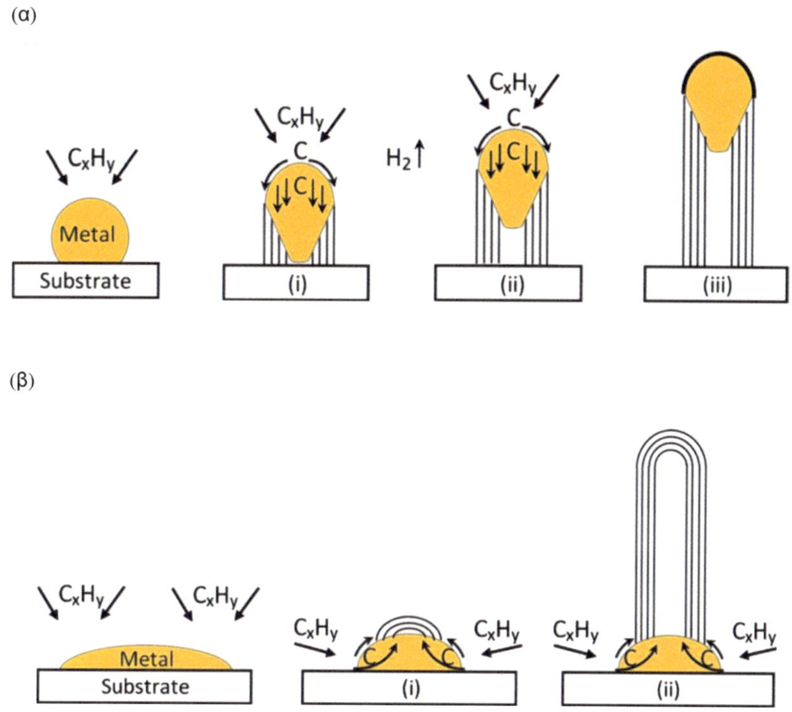

Figure 3.
Two growth model diagrams for CNTs: (α) tip-growth mechanism and (β) root-growth mechanism. Reproduced with permission of Ref. [21].

single-walled CNTs is still quite high, and the preparation of macroscopic arrays of some oriented single-walled CNTs has not been realized.

The chemical vapor deposition method is to cleave a carbon atom-containing gas continuously flowing through the catalyst nanoparticle to generate carbon atoms and then generate CNTs on the surface of the catalyst or the substrate. The synthesis process is to let catalyst decompose carbon source (usually hydrocarbon gas) at a sufficiently high temperature in a tubular reactor [6]. **Figure 3** is a schematic diagram of the most common carbon nanotube production process by CVD.

The history of chemical vapor deposition can be traced back to the nineteenth century. In 1890, French scientists observed carbon filaments in experiments in which cyanobacteria spread on hot porcelain [17]. This material was formed by the interaction of carbon-containing gases with the surface of molten iron, which was first reported in 1896, but its structural details and mechanical properties are the result of recent advances in electron microscopy. CVD may be the first in the name of filaments and fiber-planted CNTs. But in 1996, CVD became a method of mass production and synthesis of CNTs [18].

Compared to other two methods, in terms of crystallinity, although the crystallinity of CVD-grown MWCNTs is low, the crystallinity of CVD-grown SWCNTs is close to that by arc or laser. However, CVD is superior to arc and laser methods in terms of yield and purity. CVD is the only viable method for structural control or growth parameter control.

Although the growth mechanism of CNTs has been controversial, there are currently two widely accepted mechanisms, and they can be summarized as follows. When a hydrocarbon vapor is contacted with heated metal nanoparticles, it is first decomposed into carbon and hydrogen. Hydrogen leaves with the passing carrier gas or reducing gas, and carbon dissolves in the metal catalyst. When the temperature reaches the carbon solubility limit of the metal, the decomposed carbon particles precipitate and crystallize to form CNTs. The decomposition of hydrocarbons is an exothermic process, carbon crystallization is an endothermic process, and the thermal gradient continues this process. The reason for the two growth mechanisms is that the interaction between the catalyst and the substrate is different:

1. When the catalyst interacts weakly with the substrate, carbon decomposed from the hydrocarbon diffuses from the metal catalyst to the bottom of the metal catalyst and precipitates between the substrate and the metal catalyst, thereby promoting the growth of the entire metal catalyst nanoparticles. When the metal particle is entirely covered by excess carbon, growth stops, which is called tip-growth [19].

2. When the catalyst interacts strongly with the substrate, the carbon precipitates without pushing up the metal particles, so it is forced to precipitate from the top of the metal, which is called the "basic growth model," also called root growth. [20]

During the synthesis of CNTs, many parameters affect the final morphology and properties of CNTs, such as carbon source, catalyst, reactor temperature, system pressure, flow rate of carrier gas, deposition time, reactor type, the geometry of reactor, catalyst support, active metal components in catalyst, and so on.

3. Carbon source

Precursors in CVD can be of any state, whether solid, liquid or gas. In the case of a liquid carbon source, it is only necessary to heat the liquid to become steam, and then enter the tubular reactor along with the inert gas or the reducing gas. If a solid hydrocarbon is used as the carbon source, it can be stored directly in the quartz boat together with the catalyst and placed together in the tubular reactor. In the case of volatile materials (camphor, naphthalene, etc.), they can be converted directly from solids to steam and then reacted on the catalyst as the gas enters the high temperature zone.

Also, molecular structure of the carbon source, air pressure, and so on have a great impact on the formation of CNT morphology.

In addition, the concentration of carbon source gas has also an impact on the mentality of CNTs. If the concentration is too high, too many CNTs can be wrapped by amorphous carbon and carbon nanoparticles, resulting in rough product surface [22]. This is mainly due to the excessive supply of carbon during the growth process. Therefore, proper control of carbon source gas concentration can effectively reduce or even remove excess substances on the CNT surface. Secondly, the purity of gas can also affect the generation of CNTs, as reported [23]. The oxygen content can significantly change the growth kinetics of CNTs.

The most common carbon sources are carbon monoxide, ethane, ethylene, acetylene, benzene, and xylene [21].

If the carbon source is a linear hydrocarbon such as methane, ethylene or acetylene, they are generally decomposed into linear dimers of carbon atoms or carbon at high temperatures, and synthesis by CVD can produce straight carbon nanotubes, and if it is benzene, two cyclic hydrocarbons such as toluene, fullerene, and cyclohexene can produce relatively curved carbon nanotubes [24, 25].

SWCNTs had higher formation energy than MWCNTs because SWCNTs have the characteristics of small diameter, large curvature, and high strain energy. This is why most MWCNTs grow more easily than SWCNTs. In fact, MWCNTs is easily generated in a low temperature of 600–900°C, and usually at a higher temperature of 900–1200°C SWCNTs can be generated. SWCNTs must be grown from certain hydrocarbons, such as methane and carbon monoxide that maintain thermal stability at high temperatures. In contrast, the effective carbon sources for the synthesis of MWCNTs cannot be stable at high temperature, which can lead to a large amount of amorphous carbon impurities deposited outside the required products.

José-Yacamán et al. [26] produced helical MWCNTs at 700°C on iron nanoparticles as a catalyst using acetylene by the CVD method. And in an early research, Endo et al. [27] also developed MWCNTs using the same catalyst by the CVD method, except that the carbon source was changed to benzene, and the reaction temperature rose to 1100°C. Subsequently, cyclohexane [28] and fullerenes [24] were also found to be carbon sources for the synthesis of MWCNTs.

In 2010, Plata et al. [29] added small amounts of hydrocarbons, ethylene, and hydrogen to the reactor. The results show that the addition of alkanes can promote the synthesis of MWCNTs. Therefore, they demonstrate a unique mechanism for the formation of carbon nanotubes, which is quite different from the previously known; they believe that the carbon source may not decompose during C—C bond formation.

SWCNTs can be synthesized from toluene, methane, benzene, phenylacetylene, fullerene, and cyclohexane [6] as carbon sources.

By selecting a suitable carbon source and its gas pressure, the yield, quality, and purity of the carbon nanotubes can be simultaneously improved.

In 2002, Maruyama et al. [30] presented a report on the synthesis of high-purity SWCNTs from ethanol on a double-impregnated zeolite matrix at low temperatures. The advantage of ethanol as a carbon source is that the CNTs produced have almost no amorphous carbon. Since that, ethanol has become a commonly used carbon source for CVD synthesis of CNTs. The low price of ethanol can greatly reduce the cost, making the production of carbon nanotubes almost free. Subsequently, vertically aligned SWCNTs were also found to grow on silica and silicon substrates under the action of molybdenum and cobalt bimetallic catalysts [31, 32]. Recently, Maruyama's research team found that intermittent supply of acetylene can significantly help ethanol to maintain catalyst activity, thereby increasing the growth rate of carbon nanotubes [33].

In 2004, Hata et al. [34] reported that impurity-free SWCNTs could be efficiently synthesized on Si substrates using water assisted ethylene CVD. They proposed that the amorphous carbon can be selectively removed without damaging the CNTs in growth by controlling the water vapor supply to the CVD reactor as a weak oxidant. Balancing the relative content of ethylene and water is very important for prolonging the service life of the catalyst. Joshi et al. [35] produced ultra-long CNTs in 2010 using ethylene as a carbon source, assisted by hydrogen and water vapor. This study found that in order to generate super-long and high-quality CNTs on catalysts composed of complex surfaces, the hydrogen/water ratio should be very accurate.

In addition to the above conventional carbon sources, some organic compounds can also be used as carbon sources for the growth of carbon nanotubes, especially polymers, carbonization of poly-acrylonitrile, poly-furfuryl-alcohol, amino-dichloro-s-triazine [6], and carbonization of polymers.

Tripropylamine was pyrolyzed in the AFI nanochannel (0.73 nm) to obtain the nanotubes with the smallest diameter (0.4 nm) [36]. In 2009, Han [37] obtained the required MWCNT diameter by loading some polymer carbon source in a well-defined pore size on the alumina template carbonization with a temperature range of 400–600°C.

In recent years, metallic organic compounds, such as metallocene complex (cobalt diene, ferrocene, and nickelocene) [38] and nickel phthalocyanine [39], have been used as carbon sources and catalysts at the same time. Although CNTs grown have high metal encapsulation, their yields are very low.

Zdrejek et al. [40] demonstrated that CVD can be used to synthesize high quality SWCNTs and MWCNTs at a range of temperatures using pure propane as a carbon source in 2015. In their research report, it is proposed that adjusting the growth temperature can control the volume of carbon nanotubes. Cui et al. [41] synthesized a thin-walled, open-type N-doped CNT array by floating CVD, in which the carbon source was acetonitrile. The product obtained has a large thin-wall index, which is defined as the ratio of the inner diameter of the CNTs to the wall thickness.

In addition, CNTs can be successfully synthesized from kerosene, liquefied petroleum gas, gas, natural gas, waste plastics, green grass [6], and other daily raw materials as carbon sources. In addition, Terrones et al. [42] loaded a cobalt catalyst on a silica substrate by laser etching and then produced an oriented carbon nanotube by pyrolyzing 2-amino-4,6-dichloro-pyrolysis. This study demonstrates that the use of patterned catalysts significantly promotes the formation of aligned nanotubes.

Kumar et al. [43] used camphor to grow carbon nanotubes on a nanoporous zeolite substrate and first proposed optimized conditions for growing SWCNTs and MWCNTs. Brazilian scientists [21] conducted thermal annealing and

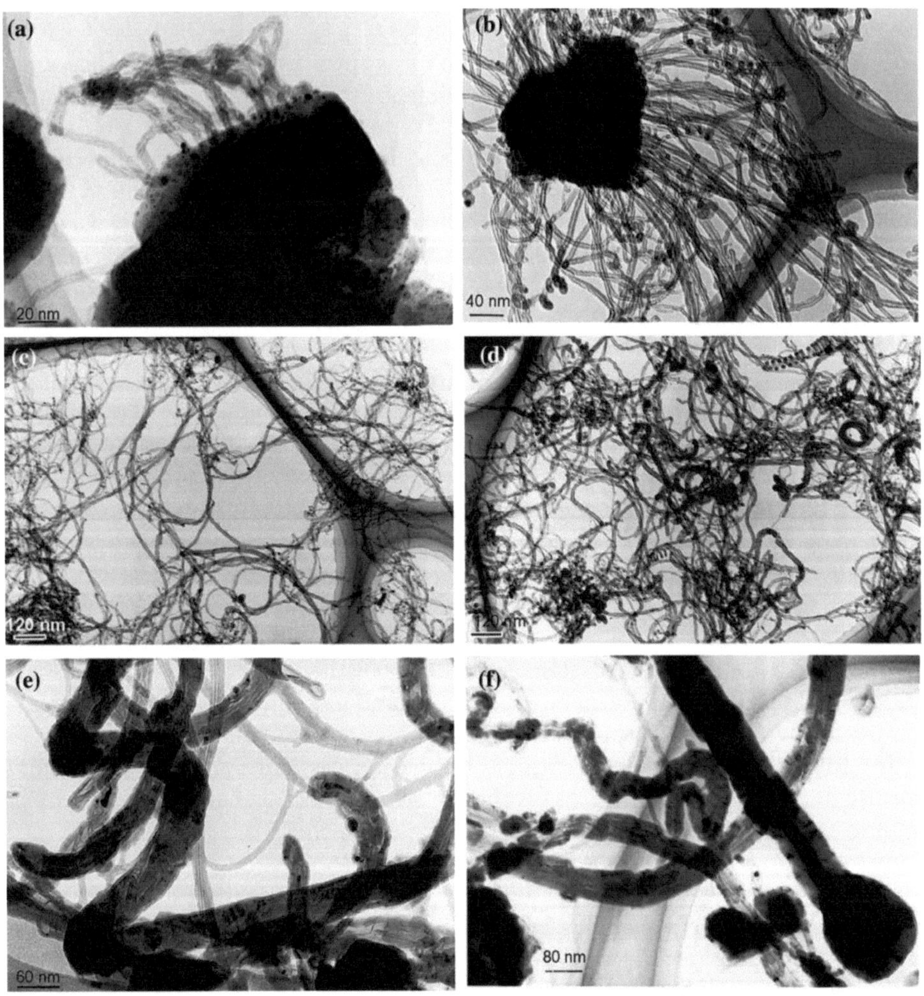

Figure 4.
Carbon nanotubes produced by camphor at different temperatures. Reproduced with permission of Ref. [43].

electrochemical purification of CNTs grown in camphor, and in China, Tang et al. [44] used ferrocene as a catalyst and camphor as a carbon source, in the argon atmosphere, prepared multi-branched dendritic carbon by CVD, and announced the effects of reaction temperature, carrier gas flow rate, and ferrocene density on the structure of CNTs. Subsequently, they published hydrogen storage analyses for different types of CNTs [45]. Therefore, camphor is a promising and efficient new carbon source. There is no doubt that this is an energy-saving way. As an agricultural product, camphor is definitely a renewable raw material; therefore, there is no danger of depletion of natural resources. Therefore, the synthesis of carbon nanotubes based on camphor largely conforms to the principles of green chemistry. This work caught the attention of industrial ecologists (**Figure 4**).

Subsequently, cheap raw materials such as turpentine [46–48] and eucalyptus oil [49] have also been reported for growth of CNTs.

In nanoelectronics, SWCNTs are more popular than MWCNTs because of their properties and bandgap displacement due to small changes in the diameter and wrapping angle of the nanotubes, but this property of SWCNTs can only be fully exploited if they are structurally pure.

Therefore, in the past 20 years, the biggest challenge facing the scientific community is how to control the generation process of CNTs and obtain SWCNTs with a single structure. Researchers are currently working on how to control their chirality when generating SWCNTs.

In 2013, He et al. [50] reported the use of carbon monoxide as a carbon source, giving priority to under 500°C to generate very high density of CNTs. Yang et al. [51] stated that chiral SWCNTs were directly synthesized from solid alloy catalysts with ethanol. Both experiments and simulations have shown that the highly selective growth of SWCNTs is that the catalyst has a high melting point, so that they can maintain the crystal structure during chemical vapor deposition.

In fact, under certain experimental conditions, any containing carbon can be used as a carbon source. It is necessary to identify cheap carbon sources that can control chiral growth of CNTs under CVD in the future.

4. Catalyst

The catalyst is indispensable in the process of preparing CNTs by the CVD method. It can reduce the decomposition temperature of the carbon source and promote the nucleation of CNTs, which is the most important influencing factor for the preparation of CNTs. Specific carbon nanotube synthesis requires different catalysts. Synthetic SWCNTs usually require nanosized particles as catalysts, and MWCNTS can also be produced without catalysts. At present, the preparation process of multi-walled CNTs is quite mature, and industrial production has been realized by the CVD method. The production cost of single-walled CNTs is still quite high, and the preparation of macroscopic array single-walled CNTs has not been realized.

Catalyst is the carbon source decomposition active center and graphite carbon deposition center in the growth process of CNTs. The selection, preparation, and carrier selection of CNTs will also be very important for nucleation, growth rate, density, separation, and purification of CNTs. This large influence can lead to the different morphology and structure of CNTs, which is significant for the preparation of CNTs. The results showed [52, 53] that the effective dispersion of catalyst particles is one of the key factors in the synthesis of CNTs, and the size of catalyst particles directly determines the diameter of CNTs.

Therefore, the preparation of a catalyst having a specific particle diameter with high catalytic activity, high selectivity, and good dispersibility is the key to whether the CVD method can mass produce high-purity CNTs. The catalysts prepared by different methods have different dispersibilities and large difference in catalytic activity, which results in different carbon nanotube content and tube diameter. The composite catalyst prepared by the same method has a closely related relationship between the catalytic activity and the ratio of the effective component and the carrier, and the optimum ratio.

Some popular transition metal nanoparticles are usually used as catalysts in CVD synthesis of CNTs [54]. It is reported that Fe, Co, Ni, Pd, Pt, Au, Mn, W, Ti, Co, Mg, Al, In, Na, K, Cs, and other metals reported can be used in the preparation of MWCNT in recent years, and the diameter of these catalyst nanoparticles is usually larger than 3 nm. Single-walled CNTs are usually produced when the active components of Fe, Co, Ni, Pd, Pt, Ru, Cu, Ag, Au, Mn, Cr, Mo, and other metals act as catalysts, and the particle size of the generated nanoparticles is often less than 3 nm.

Due to the high carbon solubility of Fe, Co, and Ni, transition metal carbides can be formed. With a high diffusion rate in these transition metals, nucleation and growth of carbon atoms can be carried out [55]. Cu, Au, Ag, Pt, and Pd can also be used to synthesize CNTs by catalyzing appropriate carbon sources [56]. Among the catalysts, transition metals (Fe, Co, and Ni) are the most widely studied and widely used catalyst for the nucleation and growth of CNTs due to their relatively high carbon solubility and carbon diffusion coefficient. They are more effective in forming SWCNTs [21]. In addition, due to the high melting point of these metals and the low equilibrium vapor pressure, this provides a wide range of CVD temperatures for various carbon sources and can be used to grow carbon nanotubes at different temperatures. It is worth noting that transition metals are not only effective catalysts in CVD, but are also effective catalysts in other preparation processes.

Solid organometallic metallocene compound (ferrocene, cobalt, nickel–ene) is widely used as carbon nanotube catalysts for the reason that they can release metal nanoparticles, resulting in more effectively catalyzing the decomposition of hydrocarbons. Chaisitsak et al. [57] have also prepared single-walled CNTs by floating catalytic method by using ferrocene as a catalyst and ethanol as a carbon source. Other nanoparticles can also be used as catalysts for the synthesis of CNTs, such as diamond and semiconductor nanoparticles (Si and Ge). Chemical vapor deposition is employed to grow single-walled CNTs on sapphire substrates [58]. In addition, CNTs were grown on scratched silicon wafers [59] without any other catalyst. The growth of CNTs does not require metal at all, and the nonmetallic matrix itself can be used as a catalyst. The synthesis of metal-free CNTs (MWCNTs) is undoubtedly a breakthrough in the research of CNTs and opens up a new approach to nanotechnology.

SWCNTs have many excellent properties, and their atomic scale changes resulted diversity. They are still limited by large-scale, high-purity, ordered, and specific chiral controllable preparations. Single-walled CNTs undergo nucleation growth and structural regulation on the catalyst.

In 1996, Dai et al. [60] realized the synthesis of single-walled CNTs in the chemical vapor deposition (CVD) using Mo or Ni–Co catalysts. Bachilo et al. [61] obtained single-walled CNTs using Co–Mo alloy catalyst in 2003. In 2004, Hata et al. [34] prepared single-walled carbon nanotube vertical arrays using Fe as a catalyst. In conclusion, the catalysts for the synthesis of single-walled CNTs are generally Fe, Co, Ni, or bi-metallic compound catalysts.

Shin et al. [62] used Ni-based and Fe-based catalytic systems, and high-yield synthesis of SWCNTs was demonstrated. It is more difficult to synthesize SWNTs with a low temperature in nickel-based catalyst system than in iron-based catalyst system. It is likely related to the diffusion rate and carbon solubility of the catalyst metal. Jafarpour et al. [63] found that the double-layer catalyst for preparing Co–Mo: 0.5–0.1 nm at low deposition rate has a high prospect of application, because in this special structure, not only high quality but also long vertical SWNTs can be obtained. SWCNT thin films with high semiconductor content can also be obtained. Mendoza et al. [64] declared the influence of molybdenum for CVD synthesis of MWCNTs. The addition of a molybdenum-based catalyst based on magnesium oxide to a cobalt catalyst has a significant effect on the production of carbon nanotubes.

In 2002, Harutyunyan et al. [65] used methane to grow CNTs on alumina substrates using two catalysts, respectively, Fe and Fe/Mo two-component catalysts. It revealed that molybdenum had obvious synergistic effect, decreased the growth temperature, and it also eliminated the process of catalyst activation in hydrogen. Molybdenum is widely used as a cocatalyst in metal supported catalysts because it can increase the yield of CNTs. In addition, molybdenum is a significant promoter added to the iron-based catalyst, which increases the activity of the catalyst at lower temperatures.

In early reports, bi-metallic and tri-metallic compound catalysts can solve many problems of CNT productions. Iron plays a higher catalytic role in the decomposition of hydrocarbons, which leads to the increase of CNT deposition, but the graphitization degree of these CNTs descends. On the other hand, cobalt catalyst makes carbon nanotube graphitization superior, but the yield decreases. Therefore, the attempt to mix the advantages of two metals is successful, and the multicomponent catalyst is considered to be a better catalyst than the pure metal.

In addition to the composition of the catalyst, the preparation conditions of the different catalysts also have an effect on the growth of the CNTs. Flahaut et al. [66] used urea or citric acid as raw material to prepare catalysts by combustion. Studies have found that milder combustion conditions can limit the formation of carbon nanofibers, thereby increasing the selectivity to SWCNTs. Chang et al. [67] took Fe and Ni as an example, and they found that the composition and size of the bicomponent metal catalyst is the key to the large-scale preparation of single-walled CNTs.

In addition to the catalyst materials, the concentration of catalyst is also vital in the growth of CNTs. Lower concentration of catalyst (2.4–5%) showed that SWCNT increased (1–50% and above), while high concentration of Fe and Co was beneficial to the growth of MWCNTs. The yield of MWCNTs prepared with 40% Fe and Co catalyst concentration is the highest and the metal pollution is negligible [68]. These studies confirmed that the proper selection of catalyst materials and their concentrations can selectively culture SWCNTs or MWCNTs.

In the experiment of Zarabadi-Poor et al. [69], they first loaded the metal component iron on alumina, and then decomposed methane at 1000°C to prepare CNTs. Studies have shown that the diameter and length of the CNTs depend on two experimental conditions, one for the iron content of the catalyst and one for the reaction time. Thinner CNTs can be obtained at lower iron concentrations. In addition to affecting the diameter and length of CNTs, the concentration of iron also affects morphological control. As the concentration of iron in the catalyst increases, carbon tends to form carbon nanoribbons rather than carbon nanotube structures.

The proper calcination temperature can regulate the interaction between the active center and the support, make the active metal easy to reduce, and reduce the particle diameter of the catalyst, and ultimately affect the activity of the catalyst and the morphology and properties of the carbon product.

The general rule of thumb is that MWCNTs are easier to form at 600–900°C, and more favorable for SWCNT growth at 900–1200°C. Temperature has a great influence on the growth of CNTs: the higher the growth temperature, the better the degree of graphitization of CNTs, and the higher the yield of CNTs than at low temperature [70]. But it is easy to form carbon structure with five-member ring at too high temperature, which leads to defective CNTs with different shapes [71].

In addition to particle size and morphology, the influence of atmosphere on the catalyst may also change the properties of CNT by influencing the structure of the catalyst. Harutyunyan et al. [72] designed cleverly Ar/He/H_2/H_2O pretreatment gas effectively to increase the proportion of metallic SWCNTs. In 2014, Bayer et al. [73] used Co/SiO_2 to catalyze the growth of SWCNTs with concentrated chiral and pipe diameter distribution. The only explanation for the variation was the effect of ammonia pretreatment on the catalyst's valence state and particle size distribution.

In the process of CVD synthesis of CNTs, water vapor, oxygen (air), and ethanol are added into the growth atmosphere, and they can selectively remove amorphous carbon without damaging the nanotube and significantly improve the activity and life of the catalyst.

Hata [34] succeeded in production of SWNTs using a variety of catalysts including Fe nanoparticles from $FeCl_3$, silicon wafers, quartz, and metal foil sputtering metal films (Fe, Al/Fe, Al_2O_3/Co). The catalytic activity induced by water stimulation resulted in dense and vertically arranged SWNT forest.

Li et al. [74] prepared extremely long and compact CNT beams on ferrocene by air-assisted chemical vapor deposition in 2008. Air-assisted CVD can prolong the service life of the catalyst and rapidly synthesize long and well-arranged carbon nanotube bundles up to 1.5 cm in length.

Water-assisted chemical vapor deposition was used to synthesize hyper-beam single-walled CNTs by Zhao et al. [75] in 2015. Water vapor was introduced into the reactor and the yield of SWCNT was significantly increased from 40 to 206 wt.%. By adding water vapor, the average diameter of SWCNTs increases from 1.5 to 3.0 nm and the diameter distribution becomes broader.

In addition to the carrier gas and the reducing gas in the CVD process, it is also possible to add water vapor or ethanol in an auxiliary manner, which not only increases the yield of the carbon nanotubes, but also increases the activity of the catalyst.

5. CNT substrates

During the growth of CNTs, the catalyst needs to be supported on a suitable matrix to ensure the quality and yield of the CNTs, including the material, surface topography, and texture properties of the matrix. The various substrates used for CNT growth in CVD are silicon, silicon carbide, graphite, quartz, silica, alumina, magnesium oxide, calcium carbonate, zeolite, sodium chloride [6], and so on. The most commonly used catalyst substrates are silica, zeolite, and alumina. Among these catalyst-based materials, magnesium oxide is the most operative substance for producing SWCNTs.

Mattevi et al. stated [76] that alumina is a better catalyst carrier than silicon dioxide because of its strong metal-carrier interaction, high metal dispersion, and high catalytic site density. They used X-ray photoelectron spectroscopy to study the interaction of the catalyst with the catalyst support during the CVD process over a wide range of pressures. Due to the particularity of Fe/Al_2O_3 interface reaction, the addition of the Al_2O_3 support layer can increase the yield of CNTs. They observed that the carrier interaction of Al_2O_3 is much solider than that of SiO_2, which limits the surface fluidity of the iron.

In 2007, Noda et al. [77] found that there were significant differences between Al_2O_x and Al_2O_3 for the catalysts. When the particle size distribution of iron was relatively more than 0.6 nm, nanotube forests grew thicker in either catalyst. When the particle size distribution of iron was less than 0.6 nm, nanotube forests can only grow thicker under the catalyst of Fe/Al_2O_x catalyst. Al_2O_x catalyst carrier was more suitable than Al_2O_3 for the growth of SWCNTs.

Grain boundaries or grain grades of supporting materials play an irreplaceable role for synthesis of CNTs. Mattevi et al. [76] showed that the high-valence state of iron does not decrease during CNT synthesis, presumably related to the details of the interface between iron and alumina. The generation of a large amount of lattice strain in the metal particles is the root cause of this dynamic stability.

Some porous materials such as Al_2O_3, MgO, and zeolites are used as catalyst supports in order to obtain smaller and uniformly dispersed catalyst particles. In particular, some catalytic studies based on silica gel or zeolites support show that porous materials are excellent carriers for the production of CNTs. By making metal particles distribute uniformly in the support, the problems of agglomeration and enrichment of metal catalyst particles can be solved very well. Lee et al. [78] using porous SiO_2 as carrier first obtained a three-dimensional network of single-walled CNTs by chemical vapor deposition at 800°C.

However, a large amount of amorphous carbon tends to accumulate in the pores of porous materials, and the preparation of CNTs requires an extra step of purification, which may cause additional pollution or structural damage of CNTs. Couteau et al. [79] could overcome these difficulties by using $CaCO_3$ as a catalyst carrier. $CaCO_3$ is not a porous material, which decreases the production of amorphous carbon. Moreover, both metal particles and catalyst carriers can be dissolved in dilute inorganic acids, so CNTs with high yield and high purity can be obtained.

The catalyst substrate not only functions as a carrier, but also interactions between the substrate and the catalyst, chemical interactions, and physical interactions may occur, such as van der Waals forces and static forces. The physical interaction prevents the movement of the catalyst particles on the support material, and also reduces the thermal diffusion and sintering of the metal particles on the matrix material, which makes the particle size distribution of the catalyst tend to be stable during the synthesis of the carbon nanotubes. The chemical interaction also helps to maintain the particle size distribution of the catalyst particles.

Studies have found that some precious metals can also be used as catalyst active ingredients for growing CNTs. CNTs doped with heteroatoms (such as nitrogen) can enhance the bonding of metals to CNTs. Lv et al. [80] first synthesized high purity, open and well-aligned N-CNTs-graphenes by water-assisted CVD and then used them as support materials for noble metal nanoparticles. The results demonstrated that the electronic structure (N-doped) and the hierarchical structure (graphite-CNT hybridization) are improved.

6. Conclusion and future perspectives

To sum up, the formation of metal catalyst nanoparticles is extremely important for CVD synthesis of CNTs. Nanoparticles must be formed on the surface of a nonconductive substrate, which is conducive to the formation of these nanoparticles. The yield and quality of CNTs are greatly affected by the matrix material. The surface morphology and grain size of the matrix can greatly affect the quality and alignment of the carbon nanotubes. In addition, the substrate thickness of the catalyst has significant impact on the growth rate of the CNTs.

Chemical vapor deposition is an ideal and effective method for the preparation of large-scale chiral CNTs with low cost, high carbon source rate, high product purity, and easy control. Chiral growth control is the ultimate goal of CNTs. Its realization marks the controllable synthesis of CNTs in diameter, wall number, helicity, chiral angle, and structure. It can greatly promote the wide application of CNTs in micro-nanoelectronic devices, sensors, intelligent functional materials, optical devices, and other fields.

Catalyst type, catalyst particle size, carbon source gas type, and reaction temperature are the main factors affecting the yield and performance of CNTs. Finding suitable catalyst type and particle size, suitable carbon source gas, and reaction temperature is the prerequisite for large-scale preparation of high-quality CNTs. In the process of CNTs synthesis by the CVD method, there is still a lot of room for exploration.

1. The reason of catalytic activity of transition metal catalyst needs to be further analyzed.

2. To find a way to control the particle size of the catalyst and provide a possible method for preparing CNTs with specific layers and morphologies.

3. Chiral control is more challenging. The controlled preparation of single-walled CNTs with uniform structure, macroscopic length, single chirality, and excellent properties has always been a research focus.

4. How to selectively produce CNTs with specific morphology and properties?

5. How to improve the growth rate of CNTs, what is the most important step? What are the crucial steps in nucleation progress?

6. The growth of CNTs is currently highly dependent on high temperature environments. Finding a way to grow CNTs at room temperature is a big challenge.

7. When large quantities of CNTs are produced, they often contain many magazines, whether they are catalyst particles, substrates, or amorphous carbon, which require further processing. However, purification will greatly reduce the quality and final yield of CNTs, and more thinking is needed to achieve high purity growth.

Author details

Xiao-Di Wang[1], K. Vinodgopal[2] and Gui-Ping Dai[1*]

1 School of Resources, Environmental and Chemical Engineering, Nanchang University, Nanchang, Jiangxi, China

2 Department of Chemistry and Biochemistry, North Carolina Central University, Durham, NC, USA

*Address all correspondence to: nanodai@gmail.com

IntechOpen

© 2019 The Author(s). Licensee IntechOpen. This chapter is distributed under the terms of the Creative Commons Attribution License (http://creativecommons.org/licenses/by/3.0), which permits unrestricted use, distribution, and reproduction in any medium, provided the original work is properly cited. CC BY

References

[1] Oberlin A, Endo M, Koyama T. Filamentous growth of carbon through benzene decomposition. Journal of Crystal Growth. 1976;**32**(3):335-349

[2] Kroto HW, Heath JR, O'Brien SC, et al. C60: Buckminsterfullerene. Nature. 1985;**318**(6042):162

[3] Iijima S. Helical microtubules of graphitic carbon. Nature. 1991;**354**(6348):56

[4] Geim AK, Novoselov KS. The rise of graphene. Nanoscience and Technology: A Collection of Reviews from Nature Journals. 2010:11-19

[5] Iijima S, Ichihashi T. Single-shell carbon nanotubes of 1-nm diameter. Nature. 1993;**363**(6430):603

[6] Shah KA, Tali BA. Synthesis of carbon nanotubes by catalytic chemical vapour deposition: A review on carbon sources, catalysts and substrates. Materials Science in Semiconductor Processing. 2016;**41**:67-82

[7] Dalton AB, Collins S, Munoz E, et al. Super-tough carbon-nanotube fibres. Nature. 2003;**423**(6941):703

[8] Kempa K, Rybczynski J, Huang Z, et al. Carbon nanotubes as optical antennae. Advanced Materials. 2007;**19**(3):421-426

[9] Bissett MA, Shapter JG. Photocurrent response from vertically aligned single-walled carbon nanotube arrays. The Journal of Physical Chemistry C. 2010;**114**(14):6778-6783

[10] Bissett MA, Köper I, Quinton JS, et al. Dendron growth from vertically aligned single-walled carbon nanotube thin layer arrays for photovoltaic devices. Physical Chemistry Chemical Physics. 2011;**13**(13):6059-6064

[11] Paudel T, Rybczynski J, Gao YT, et al. Nanocoax solar cells based on aligned multiwalled carbon nanotube arrays. Physica Status Solidi (a). 2011;**208**(4):924-927

[12] Rybczynski J, Kempa K, Herczynski A, et al. Subwavelength waveguide for visible light. Applied Physics Letters. 2007;**90**(2):021104

[13] Zeng B, Xiong G, Chen S, et al. Enhancement of field emission of aligned carbon nanotubes by thermal oxidation. Applied Physics Letters. 2006;**89**(22):223119

[14] Saito Y, Uemura S. Field emission from carbon nanotubes and its application to electron sources[J]. Carbon. 2000;**38**(2):169-182

[15] Yue GZ, Qiu Q, Gao B, et al. Generation of continuous and pulsed diagnostic imaging X-ray radiation using a carbon-nanotube-based field-emission cathode. Applied Physics Letters. 2002;**81**(2):355-357

[16] Milne WI, Teo KBK, Minoux E, et al. Aligned carbon nanotubes/fibers for applications in vacuum microwave amplifiers. Journal of Vacuum Science & Technology B: Microelectronics and Nanometer Structures Processing, Measurement, and Phenomena. 2006;**24**(1):345-348

[17] Schultzenberger P, Schultzenberger L. Sur quelques faits relatifs a l'histoire du carbon. Comptes Rendus. Académie des Sciences. 1890;**111**:774

[18] Tempel H, Joshi R, Schneider JJ. Ink jet printing of ferritin as method for selective catalyst patterning and growth of multiwalled carbon nanotubes. Materials Chemistry and Physics. 2010;**121**(1-2):178-183

[19] Baker RTK, Barber MA, Harris PS, et al. Nucleation and growth of carbon deposits from the nickel catalyzed decomposition of acetylene. Journal of Catalysis. 1972;**26**(1):51-62

[20] Baker RTK, Waite RJ. Formation of carbonaceous deposits from the platinum-iron catalyzed decomposition of acetylene. Journal of Catalysis. 1975;**37**(1):101-105

[21] Kumar M, Ando Y. Chemical vapor deposition of carbon nanotubes: A review on growth mechanism and mass production. Journal of Nanoscience and Nanotechnology. 2010;**10**(6):3739-3758

[22] Huang ZP, Carnahan DL, Rybczynski J, et al. Growth of large periodic arrays of carbon nanotubes. Applied Physics Letters. 2003;**82**(3):460-462

[23] In JB, Grigoropoulos CP, Chernov AA, et al. Hidden role of trace gas impurities in chemical vapor deposition growth of vertically-aligned carbon nanotube arrays. Applied Physics Letters. 2011;**98**(15):153102

[24] Nerushev OA, Dittmar S, Morjan RE, et al. Particle size dependence and model for iron-catalyzed growth of carbon nanotubes by thermal chemical vapor deposition. Journal of Applied Physics. 2003;**93**(7):4185-4190

[25] Morjan RE, Nerushev OA, Sveningsson M, et al. Growth of carbon nanotubes from C60. Applied Physics A. 2004;**78**(3):253-261

[26] José-Yacamán M, Miki-Yoshida M, Rendon L, et al. Catalytic growth of carbon microtubules with fullerene structure. Applied Physics Letters. 1993;**62**(6):657-659

[27] Endo M, Takeuchi K, Igarashi S, et al. The production and structure of pyrolytic carbon nanotubes (Pcarbon nanotubes). Journal of Physics and Chemistry of Solids. 1993;**54**(12):1841-1848

[28] Liu ZJ, Che R, Xu Z, et al. Preparation of Fe-filled carbon nanotubes by catalytic decomposition of cyclohexane. Synthetic Metals. 2002;**128**(2):191-195

[29] Plata DL, Meshot ER, Reddy CM, et al. Multiple alkynes react with ethylene to enhance carbon nanotube synthesis, suggesting a polymerization-like formation mechanism. ACS Nano. 2010;**4**(12):7185-7192

[30] Maruyama S, Kojima R, Miyauchi Y, et al. Low-temperature synthesis of high-purity single-walled carbon nanotubes from alcohol. Chemical Physics Letters. 2002;**360**(3-4):229-234

[31] Murakami Y, Chiashi S, Miyauchi Y, et al. Growth of vertically aligned single-walled carbon nanotube films on quartz substrates and their optical anisotropy. Chemical Physics Letters. 2004;**385**(3-4):298-303

[32] Maruyama S, Einarsson E, Murakami Y, et al. Growth process of vertically aligned single-walled carbon nanotubes. Chemical Physics Letters. 2005;**403**(4-6):320-323

[33] Xiang R, Einarsson E, Okawa J, et al. Acetylene-accelerated alcohol catalytic chemical vapor deposition growth of vertically aligned single-walled carbon nanotubes. The Journal of Physical Chemistry C. 2009;**113**(18):7511-7515

[34] Hata K, Futaba DN, Mizuno K, et al. Water-assisted highly efficient synthesis of impurity-free single-walled carbon nanotubes. Science. 2004;**306**(5700):1362-1364

[35] Joshi R, Schneider JJ, Yilmazoglu O, et al. Patterned growth of ultra long carbon nanotubes. Properties and systematic investigation into their growth process. Journal of Materials Chemistry. 2010;**20**(9):1717-1721

[36] Tang ZK, Sun HD, Wang J, et al. Mono-sized single-wall carbon nanotubes formed in channels of $AlPO_4$-5 single crystal. Applied Physics Letters. 1998;**73**(16):2287-2289

[37] Kumar M, Ando Y. Chemical vapor deposition of carbon nanotubes: A review on growth mechanism and mass production[J]. Journal of Nanoscience and Nanotechnology, 2010;**10**(6):3739-3758

[38] Sen R, Govindaraj A, Rao CNR. Carbon nanotubes by the metallocene route. Chemical Physics Letters. 1997;**267**(3-4):276-280

[39] Yudasaka M, Kikuchi R, Ohki Y, et al. Nitrogen-containing carbon nanotube growth from Ni phthalocyanine by chemical vapor deposition. Carbon. 1997;**35**(2):195-201

[40] Zdrojek M, Sobieski J, Duzynska A, et al. Synthesis of carbon nanotubes from Propane. Chemical Vapor Deposition. 2015;**21**(4-5-6):94-98

[41] Cui T, Lv R, Kang F, et al. Synthesis and enhanced field-emission of thin-walled, open-ended, and well-aligned N-doped carbon nanotubes. Nanoscale Research Letters. 2010;**5**(6):941

[42] Terrones M, Grobert N, Olivares J, et al. Controlled production of aligned-nanotube bundles. Nature. 1997;**388**(6637):52

[43] Kumar M, Ando Y. Controlling the diameter distribution of carbon nanotubes grown from camphor on a zeolite support. Carbon. 2005;**43**(3):533-540

[44] Tang J, Jin GQ, Wang YY, et al. Tree-like carbon grown from camphor. Carbon. 2010;**48**(5):1545-1551

[45] Bianco S, Giorcelli M, Musso S, et al. Hydrogen adsorption in several types of carbon nanotubes. Journal of Nanoscience and Nanotechnology. 2009;**9**(12):6806-6812

[46] Chatterjee AK, Sharon M, Banerjee R, et al. CVD synthesis of carbon nanotubes using a finely dispersed cobalt catalyst and their use in double layer electrochemical capacitors. Electrochimica Acta. 2003;**48**(23):3439-3446

[47] Afre RA, Soga T, Jimbo T, et al. Growth of vertically aligned carbon nanotubes on silicon and quartz substrate by spray pyrolysis of a natural precursor: Turpentine oil. Chemical Physics Letters. 2005;**414**(1-3):6-10

[48] Afre RA, Soga T, Jimbo T, et al. Carbon nanotubes by spray pyrolysis of turpentine oil at different temperatures and their studies. Microporous and Mesoporous Materials. 2006;**96**(1-3):184-190

[49] Ghosh P, Soga T, Tanemura M, et al. Vertically aligned carbon nanotubes from natural precursors by spray pyrolysis method and their field electron emission properties. Applied Physics A. 2009;**94**(1):51-56

[50] He M, Jiang H, Liu B, et al. Chiral-selective growth of single-walled carbon nanotubes on lattice-mismatched epitaxial cobalt nanoparticles. Scientific Reports. 2013;**3**:1460

[51] Yang F, Wang X, Zhang D, et al. Chirality-specific growth of single-walled carbon nanotubes on solid alloy catalysts. Nature. 2014;**510**(7506):522

[52] Sun LF, Mao JM, Pan ZW, et al. Growth of straight nanotubes with a cobalt–nickel catalyst by chemical vapor deposition. Applied Physics Letters. 1999;**74**(5):644-646

[53] Li W, Qian L, QIAN S, et al. Structure and growth thermodynamics of carbon tubes. Science in China Series A-Mathematics. 1996;**39**(6):657-664

[54] Thess A, Lee R, Nikolaev P, et al. Crystalline ropes of metallic carbon nanotubes. Science. 1996;**273**(5274):483-487

[55] Zhong G, Hofmann S, Yan F, et al. Acetylene: A key growth precursor for single–walled carbon nanotube forests. Journal of Physical Chemistry C. 2009;**113**(40):17321-17325

[56] Ding F, Larsson P, Larsson JA, et al. The importance of strong carbon-metal adhesion for catalytic nucleation of single-walled carbon nanotubes. Nano Letters. 2008;**8**(2):463-468

[57] Han S, Liu X, Zhou C. Template-free directional growth of single-walled carbon nanotubes on a- and r-plane sapphire. Journal of the American Chemical Society. 2005;**127**(15):5294-5295

[58] Chaisitsak S, Nukeaw J, Tuantranont A. Parametric study of atmospheric-pressure single-walled carbon nanotubes growth by ferrocene–ethanol mist CVD. Diamond and Related Materials. 2007;**16**(11):1958-1966

[59] Huang S, Cai Q, Chen J, et al. Metal-catalyst-free growth of single-walled carbon nanotubes on substrates. Journal of the American Chemical Society. 2009;**131**(6):2094-2095

[60] Dai H, Rinzler AG, Nikolaev P, et al. Single-wall nanotubes produced by metal-catalyzed disproportionation of carbon monoxide. Chemical Physics Letters. 1996;**260**(3-4):471-475

[61] Bachilo SM, Balzano L, Herrera JE, et al. Narrow (n, m)-distribution of single-walled carbon nanotubes grown using a solid supported catalyst. Journal of the American Chemical Society. 2003;**125**(37):11186-11187

[62] Shin KY, Su HC, Tsai CH. In situ growth of single-walled carbon nanotubes by bimetallic technique with/without dielectric support for nanodevice applications. Journal of Vacuum Science & Technology B: Microelectronics and Nanometer Structures Processing, Measurement, and Phenomena. 2006;**24**(1):358-361

[63] Jafarpour SM, Kini M, Schulz SE, et al. Effects of catalyst configurations and process conditions on the formation of catalyst nanoparticles and growth of single-walled carbon nanotubes. Microelectronic Engineering. 2017;**167**:95-104

[64] Mendoza MP, Valles C, Maser WK, Martinez MT, Benito AM. Studies on carbon nanotubes synthesis via methane CVD process using Co catalyst on carbon supports. Nanotechnology. 2005;**16**:S224-S229

[65] Harutyunyan AR, Pradhan BK, Kim UJ, et al. CVD synthesis of single wall carbon nanotubes under "soft" conditions. Nano Letters. 2002;**2**(5):525-530

[66] Flahaut E, Laurent C, Peigney A. Catalytic CVD synthesis of double and triple-walled carbon nanotubes by the control of the catalyst preparation. Carbon. 2005;**43**(2):375-383

[67] Chiang WH, Sankaran RM. Linking catalyst composition to chirality distributions of as-grown single-walled carbon nanotubes by tuning Ni_xFe_{1-x} nanoparticles. Nature Materials. 2009;**8**(11):882

[68] Kumar M, Ando Y. Gigas growth of carbon nanotubes. Defence Science Journal. 2008;**58**(4):496-503

[69] Zarabadi-Poor P, Badiei A, Yousefi AA, et al. Catalytic chemical vapour deposition of carbon nanotubes using Fe-doped alumina catalysts. Catalysis Today. 2010;**150**(1-2):100-106

[70] Zhang H, Liang E, Ding P, et al. Layered growth of aligned carbon

nanotube arrays by pyrolysis. Physica B: Condensed Matter. 2003;**337**(1-4):10-16

[71] Jain SC, Willander M, Narayan J, et al. III–nitrides: Growth, characterization, and properties. Journal of Applied Physics. 2000;**87**(3):965-1006

[72] Harutyunyan AR, Chen G, Paronyan TM, et al. Preferential growth of single-walled carbon nanotubes with metallic conductivity. Science. 2009;**326**(5949):116-120

[73] Fouquet M, Bayer BC, Esconjauregui S, et al. Effect of catalyst pretreatment on chirality-selective growth of single-walled carbon nanotubes. The Journal of Physical Chemistry C. 2014;**118**(11):5773-5781

[74] Li X, Zhang X, Ci L, et al. Air-assisted growth of ultra-long carbon nanotube bundles. Nanotechnology. 2008;**19**(45):455609

[75] Zhao Y, Choi J, Kim P, et al. Large-scale synthesis and characterization of super-bundle single-walled carbon nanotubes by water-assisted chemical vapor deposition. RSC Advances. 2015;**5**(39):30564-30569

[76] Mattevi C, Wirth CT, Hofmann S, et al. In-situ X-ray photoelectron spectroscopy study of catalyst-support interactions and growth of carbon nanotube forests. The Journal of Physical Chemistry C. 2008;**112**(32):12207-12213

[77] Noda S, Hasegawa K, Sugime H, et al. Millimeter-thick single-walled carbon nanotube forests: Hidden role of catalyst support. Japanese Journal of Applied Physics. 2007;**46**(5L):L399

[78] Lee TJ, Seo J, Lee H, et al. Fabrication of single-walled carbon nanotube three-dimensional networks inside the pores of a porous silicon structure. Carbon. 2010;**48**(5):1473-1479

[79] Couteau E, Hernadi K, Seo JW, et al. CVD synthesis of high-purity multiwalled carbon nanotubes using $CaCO_3$ catalyst support for large-scale production. Chemical Physics Letters. 2003;**378**(1-2):9-17

[80] Lv R, Cui T, Jun MS, et al. Open-ended, N-doped carbon nanotube–graphene hybrid nanostructures as high-performance catalyst support. Advanced Functional Materials. 2011;**21**(5):999-1006

Chapter 3

Synthesis and Properties of Single-Walled Carbon Nanotubes Filled with Metal Halogenides and Metallocenes

Marianna V. Kharlamova and Dominik Eder

Abstract

This chapter reviews the current status of the research on the electronic properties of single-walled carbon nanotubes (SWCNTs) filled with metal halogenides and metallocenes and growth kinetics of inner SWCNTs inside metallocene-filled nanotubes. The chapter starts with the description of the peculiarities of the synthesis of metal halogenide-filled SWCNTs, comparison of different filling methods, their advantages, disadvantages, and restrictions. Then, we comprehensively summarize, compare, and critically discuss the recent studies on the electronic properties of metal halogenide-filled SWCNTs. After that, the synthesis methods of metallocene filled SWCNTs are described and the results of the investigation of the growth kinetics of inner SWCNTs inside the filled nanotubes are summarized. Then, the reports dedicated to the investigation of the electronic properties of metallocene-filled SWCNTs are reviewed. Finally, potentials for future research, development, and application of filled SWCNTs are highlighted.

Keywords: single-walled carbon nanotube, metal halogenide, metallocene, electronic properties, growth kinetics, optical absorption spectroscopy, Raman spectroscopy, X-ray photoelectron spectroscopy

1. Introduction

Carbon nanotube, a one-dimensional allotropic modification of carbon with sp^2-hybridization of atoms, can be represented as rolled-up graphene sheets. Depending on the number of graphene layers, their structure is classified into multi-, double- (DWCNTs), and single-walled nanotubes (SWCNTs). Thanks to their unique physical and chemical properties, SWCNTs can find applications in different fields, including next-generation nanoelectronic devices [1]. The electronic properties of SWCNTs depend on their atomic structures. Due to the lack of methodology and control, nanotubes produced via industrial synthesis methods, i.e., arc-discharge, laser ablation, and chemical vapor deposition (CVD), typically exhibit varying and mixed properties, thus limiting their applicability. Despite progress on synthesis [2–6] and sorting of SWCNTs [7–11] with defined atomic structures, new methods are required that allow controllable modification of the electronic properties of SWCNTs.

Recent research has been aimed at the modification of the electronic properties of SWCNTs by the covalent and noncovalent modification of their outer surface, substitution of carbon atoms by foreign atoms, intercalation of the bundles, and filling of the channels of nanotubes [12, 13]. The latter method is especially promising, because a variety of substances with different properties can be encapsulated inside SWCNTs. The filling of SWCNTs with fullerene C_{60} [14] and $RuCl_3$ [15] was firstly performed in 1998 and since then the topic has attracted increasing attention. SWCNTs were filled with different simple elemental substances, metals [16–22] and nonmetals [17, 23], chemical compounds, metal halogenides [24–29], metal chalcogenides [30–34] and metal oxides [35, 36] as well as molecules, fullerenes and their derivatives [37–41] and metallocenes [42, 43].

Metal halogenides are the largest group of introduced inorganic substances. Depending on the metal cation and halogen anion, they are semiconductors or insulators with different work functions. The filling of SWCNTs with these salts opens the way to stable doping of nanotubes and tailoring their doping level. This triggered extensive studies on the electronic properties of metal halogenide-filled SWCNTs.

Metallocenes are another popular group of encapsulated substances. In 2008, it was shown that the high vacuum annealing of metallocene-filled SWCNTs leads to the formation of inner carbonaceous tubes [44]. In this case, metallocene molecules served as carbon and catalyst source at the same time. In contrast to the typical CVD growth of nanotubes, where the growth process stops after tens of minutes due to the deactivation of catalyst, the growth inside metallocene-filled SWCNTs was considerably slower and could last up to tens of hours until the carbon source was consumed. This enabled not only a more in-depth investigation of the growth process of nanotubes, but also of their electronic properties, which continuously evolved by filling and annealing upon chemical transformation of metallocenes. Consequently, this has attracted considerable interest in the field.

This chapter reviews the current status of the research on the electronic properties of SWCNTs filled with metal halogenides and metallocenes and growth kinetics of SWCNTs inside metallocene-filled nanotubes. The first part of the chapter focuses on the synthesis and electronic properties of SWCNTs filled with metal halogenides. We review and compare the results of the studies on the electronic properties of the filled SWCNTs by state-of-the-art spectroscopic methods such as optical absorption spectroscopy (OAS), Raman spectroscopy, and X-ray photoelectron spectroscopy (XPS). In the second part of the chapter, the results of the investigation of the growth kinetics of inner SWCNTs inside metallocene-filled nanotubes by *in situ* Raman spectroscopy and the electronic properties of the filled nanotubes by XPS and ultraviolet photoelectron spectroscopy (UPS) are summarized.

2. SWCNTs filled with metal halogenides

2.1 Synthesis of metal halogenide-filled SWCNTs

Metal halogenides were encapsulated inside SWCNTs by several methods. Among them are the gas phase and liquid phase approaches. The latter includes the solution and melt techniques.

The gas phase approach implies the encapsulation of a substance in a gas state inside SWCNTs. In the filling process, SWCNTs and substances are heated up to the temperature that is higher than the boiling or sublimation point of the substance. During dwelling at the synthesis temperature, the vapor of the substance condenses and diffuses inside SWCNTs. Then, the system is cooled, which leads to the crystallization of the substance. The advantages of this method are that it is rather simple, and it allows obtaining high filling ratios of nanotubes. However, it has several restrictions. Firstly, the maximal boiling or sublimation point of substances cannot exceed 1000–1200°C, because at higher temperatures the destruction of nanotubes occurs. Secondly, there should be no decomposition of the substance during the evaporation or sublimation process. Thirdly, the substance should have a high vapor pressure at synthesis temperature [13]. These restrictions limit the list of substances that can be encapsulated inside SWCNTs by this approach.

The list of introduced substances can be widened by the use of the liquid phase approach. The parameters of the filling process can be broadly varied. This approach is the most popular filling technique of nanotubes.

The solution method implies the filling of SWCNTs with a chosen substance dissolved in a solvent. During the filling process, several parameters are important. For the successful filling, the solubility of the substance in the solvent should be high, whereas the surface tension coefficient of the solution and its viscosity should be low. The choice of appropriate solvents allows encapsulating different substances inside SWCNTs. However, the solution method has several disadvantages. The first of them is the contamination of filled SWCNTs with solvent molecules. This can be critical for further characterizations of filled SWCNTs. The second is low filling ratios of SWCNTs, which usually do not exceed 30%. The third is inhomogeneous morphology of introduced substances inside SWCNTs [13].

These disadvantages are absent in the melt method. This method implies the filling of SWCNTs with molten salt. In a typical experiment, opened SWCNTs are mixed with an excessive amount of salt in quartz ampoules. The ampoules are evacuated, sealed, and heated up to a temperature above the melting point of the salt. The liquid salt is pulled inside SWCNTs by capillary forces. After dwelling at high temperature for some time, the SWCNTs are saturated with filling ratios up to 90%. The final cooling rate is a way to control the crystallinity of the solidified salt inside SWCNTs. Very low cooling rates are required for obtaining one-dimensional nanocrystals. Several parameters are important for the successful filling. Firstly, the melting temperature of the salt should not exceed 1000–1200°C, because at high temperatures, the nanotubes start to degrade. Secondly, the surface tension coefficient of the melt of the salt and its viscosity should be low enough [13].

The melt method is the most popular method of the filling of SWCNTs with metal halogenides. It was applied for the filling of nanotubes in all papers dedicated to the investigation of the electronic properties of metal halogenide-filled SWCNTs.

2.2 Electronic properties of metal halogenide-filled SWCNTs

In the literature, there are the reports on the investigation of the electronic properties of SWCNTs filled with $MnCl_2$, $MnBr_2$ [45, 46], $FeCl_2$, $FeBr_2$, FeI_2 [47], $CoBr_2$ [48], $NiCl_2$ [49], $NiBr_2$ [49, 50], CuCl [51, 52], CuBr [51], CuI [51, 53, 54], $ZnCl_2$ [55, 56], $ZnBr_2$, ZnI_2 [55], RbI, $RbAg_4I_5$ [57], AgCl [58–61], AgBr, AgI [58], $CdCl_2$ [56, 62, 63], $CdBr_2$, CdI_2 [62], SnF_2 [64], $TbCl_3$ [56, 65, 66], $TbBr_3$, TbI_3 [66], $PrCl_3$ [65, 67], $ErCl_3$ [68], $TmCl_3$ [65, 69], and $HgCl_2$ [70]. The characterization of the electronic properties of the filled SWCNTs was performed by three main techniques: OAS, Raman spectroscopy, and XPS. These three methods are complementary, and they give a comprehensive picture of the modification of the electronic properties of SWCNTs upon their filling. OAS gives information about structure-dependent optical transitions of nanotubes. Raman spectroscopy allows studying the vibronic properties of SWCNTs. XPS investigates the Fermi level shift and bonding environment in filled SWCNTs.

Figure 1 shows the typical OAS spectrum of metal halogenide-filled SWCNTs by an example of the ZnI_2-filled nanotubes in comparison with the spectrum of the pristine SWCNTs [55]. The spectrum of the pristine SWCNTs includes the characteristic peaks corresponding to optical transitions between the first (E^S_{11}) and second (E^S_{22}) van Hove singularities (vHs) in the valence and conduction band of semiconducting SWCNTs and the first vHs (E^M_{11}) of metallic nanotubes. In the spectrum of the filled SWCNTs, there is the suppression of the E^S_{11} peak, which corresponds to the canceling of the optical transitions between the first vHs of semiconducting SWCNTs. This is a result of the shift of the Fermi level of SWCNTs below the first vHs in the valence band or above the first vHs in the conduction band of semiconducting SWCNTs. Thus, the OAS data testify to the presence of the charge transfer in the filled SWCNTs; however, the direction of the charge transfer cannot be determined. The same modifications were observed in the OAS spectra of SWCNTs filled with $FeCl_2$, $FeBr_2$, FeI_2 [47], $CoBr_2$ [48], $ZnCl_2$, $ZnBr_2$ [55], AgCl, AgBr, AgI [58], $CdCl_2$, $CdBr_2$, CdI_2 [62], CuCl, CuBr, CuI [51], CuCl [52], and $TbCl_3$ [56].

Raman spectroscopy allows obtaining further information about the charge transfer in the filled SWCNTs. A Raman spectrum of SWCNTs includes two main characteristic bands: radial breathing mode (RBM), which corresponds to radial vibrations of carbon atoms, and G-band, which belongs to longitudinal and tangential vibrations of carbon atoms [71]. **Figure 2** demonstrates the RBM and G-bands of Raman spectra of $CdCl_2$-filled SWCNTs in comparison with the spectra of the pristine nanotubes acquired at seven different laser wavelengths between 458 and 785 nm [63]. It is visible that all spectra of the filled SWCNTs are significantly modified as compared to the spectra of the pristine SWCNTs. The shift and change of the relative intensity of the peaks of the RBM band, the shift of the peaks of the G-band, and the change in the profile of the G-band are observed. These modifications are common for metal halogenide-filled SWCNTs. They were reported for SWCNTs filled with $MnCl_2$, $MnBr_2$ [45, 46], $FeCl_2$, $FeBr_2$, FeI_2 [47], $CoBr_2$ [48],

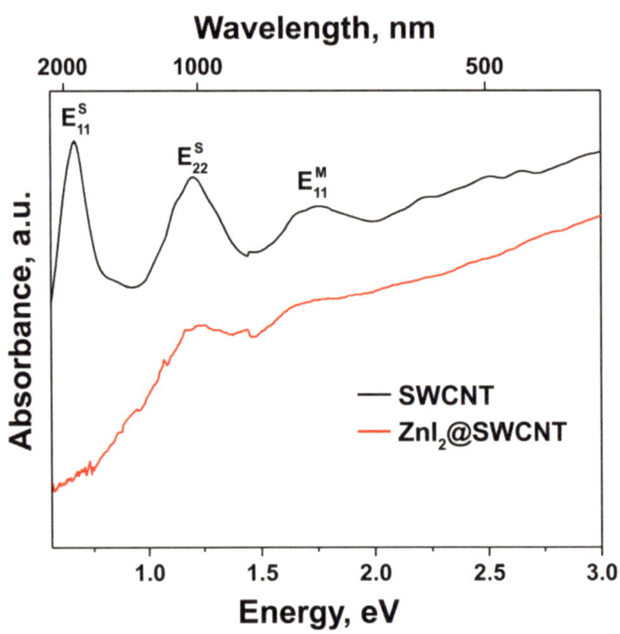

Figure 1.
The OAS spectra of the pristine and ZnI_2-filled SWCNTs. The peaks corresponding to optical transitions between the first (E^S_{11}) and second vHs (E^S_{22}) of semiconducting SWCNTs and the first vHs (E^M_{11}) of metallic SWCNTs are denoted. The data are replotted from ref. [55].

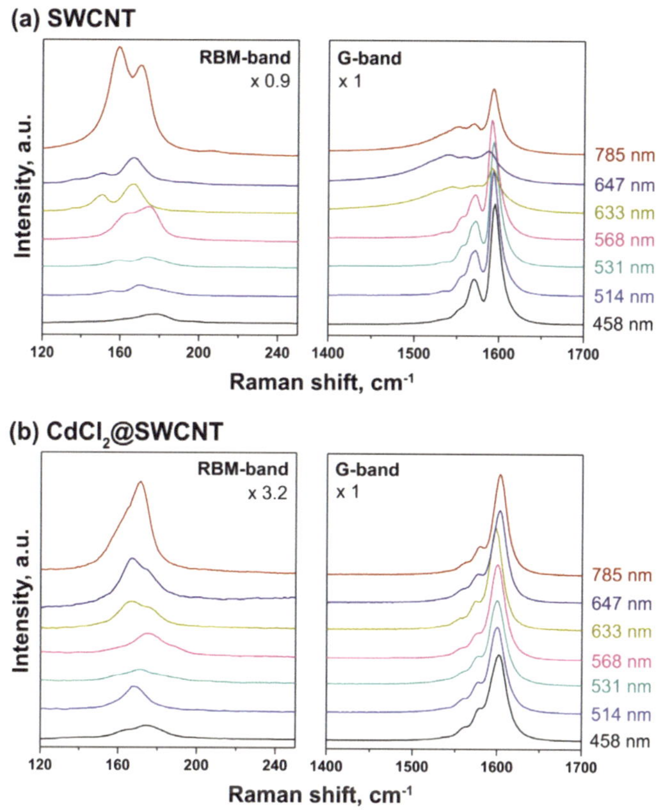

Figure 2.
The RBM and G-bands of Raman spectra of the pristine (a) and CdCl2-filled SWCNTs (b) acquired at different laser wavelengths. The spectra are normalized to the area intensity of the G-band and offset for clarity. The data are replotted from ref. [63].

$NiCl_2$, $NiBr_2$ [49], CuCl [51, 52], CuBr [51], CuI [51, 53, 54], $ZnCl_2$ [55, 56], $ZnBr_2$, ZnI_2 [55], $CdBr_2$, CdI_2 [62], SnF_2 [64], $RbAg_4I_5$ [57], $TbCl_3$ [56, 65, 66], $TbBr_3$, TbI_3 [66], $TmCl_3$ [65, 69], $PrCl_3$ [65, 67], AgCl [58–61], AgBr, AgI [58], and $HgCl_2$ [70].

The fitting of the RBM and G-bands of Raman spectra with individual components allows investigating in detail the observed modifications. **Figure 3** shows the fitting results of the spectrum of the $TbBr_3$-filled SWCNTs in comparison with the spectrum of pristine nanotubes acquired at laser wavelength of 633 nm [66]. The RBM-band of the pristine SWCNTs is fitted with two components at 156 and 172 cm^{-1}, which correspond to the nanotubes with diameters of 1.5 and 1.4 nm, respectively [72]. The G-band of the pristine SWCNTs is fitted with three components. The peak at 1540 cm^{-1} (G^-) belongs to longitudinal phonon in metallic SWCNTs, and the peaks at 1567 and 1591 cm^{-1} (G^+_{TO} and G^+_{LO}) are assigned to tangential and longitudinal phonons in semiconducting SWCNTs, respectively [73]. The RBM band of the filled SWCNTs is fitted with two components, as the spectrum of the pristine SWCNTs. However, the peak positions are shifted to 164 and 175 cm^{-1}, and also the ratio of the relative intensities of the peaks is changed from 0.32:0.68 to 0.53:0.47. These modifications are due to the changes in resonance conditions of the filled SWCNTs, which are caused by the charge transfer in the filled nanotubes. The G-band of the filled SWCNTs is fitted with three components, as the spectrum of the pristine nanotubes. However, their positions are upshifted to 1558, 1576, and 1602 cm^{-1}. This can be attributed to the p-doping of SWCNTs by the encapsulated compound. Additionally, there is the change of the profile

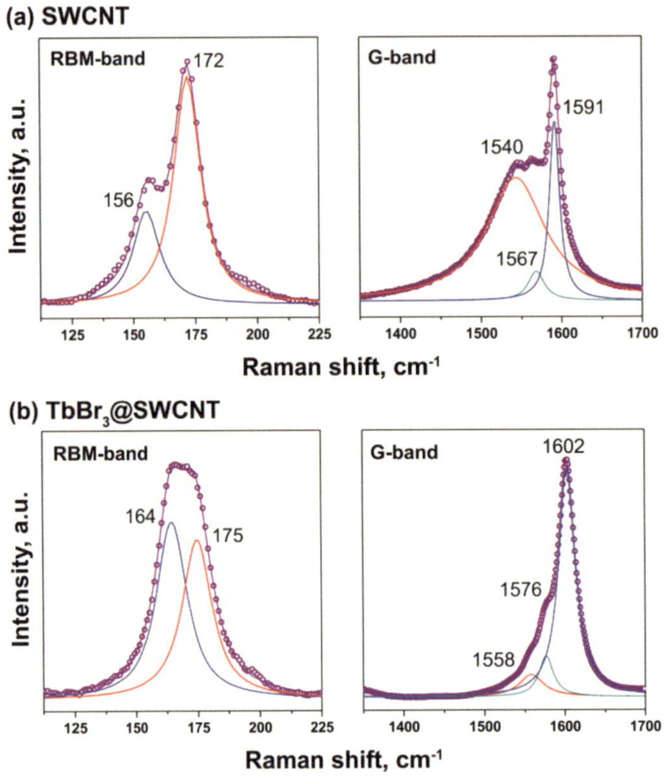

Figure 3.
The fitting of the RBM and G-bands of Raman spectra of the pristine (a) and $TbBr_3$-filled SWCNTs (b) acquired at laser wavelength of 633 nm with individual components. The peak positions are denoted. The data are replotted from ref. [66].

of the G-band of the filled SWCNTs from the Breit-Wigner-Fano shape, which is typical for metallic SWCNTs [71, 74] to the Lorentzian shape, which is a fingerprint of semiconducting SWCNTs [71, 74, 75]. This is probably a result of the filling-induced transition of metallic nanotubes into semiconducting state due to the opening of a band gap in their band structure.

Although similar modifications of Raman spectra of metal halogenide-filled SWCNTs testify to p-doping of nanotubes by the encapsulated salts, the doping level varies for different compounds. In Ref. [66], the doping level of SWCNTs by introduced $TbCl_3$, $TbBr_3$, and TbI_3 was compared. Authors analyzed the modifications of RBM and G-bands of Raman spectra of the filled SWCNTs acquired at laser wavelength of 633 nm. **Figure 4** shows the results of the analysis. In **Figure 4(a)**, the relative intensities of the RBM peaks of the pristine and filled SWCNTs are presented. In the case of the pristine SWCNTs, the second peak of the RBM band has the largest intensity and the ratio of relative intensities of two RBM peaks amounts to 0.32:0.68. In the case of the $TbCl_3$- and $TbBr_3$-filled SWCNTs, in contrast, the first peak has the largest intensity and the ratio is changed to 0.55:0.45 and 0.53:0.47, respectively. In the case of the TbI_3-filled SWCNTs, the ratio of the pristine SWCNTs (0.27:0.73) is recovered. Thus, the largest differences as compared to the pristine SWCNTs are observed for the $TbCl_3$-filled SWCNTs, and the smallest differences are observed for the TbI_3-filled nanotubes. In **Figure 4(b)**, the shift of the G-band peaks and the relative area intensity of the G^--peak of the pristine and filled SWCNTs are presented. It is visible that the largest changes as compared to the pristine SWCNTs are again observed for the $TbCl_3$-filled SWCNTs and the smallest changes are observed

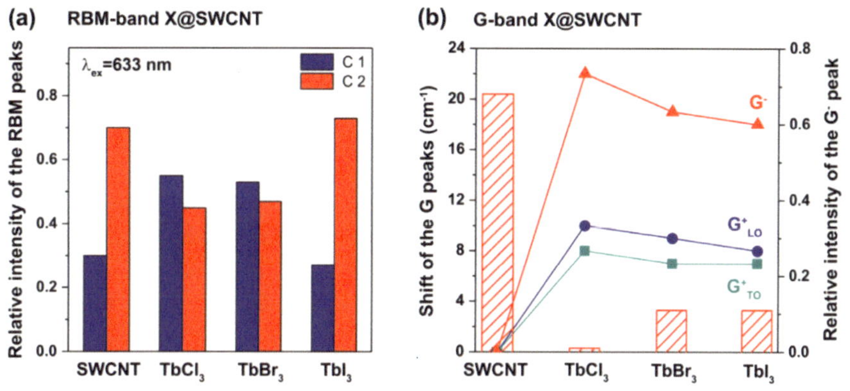

Figure 4.
The relative intensity of the RBM peaks, (a) shift of the G-band peaks and relative intensity of the G^- peak (b) in Raman spectra of the pristine SWCNTs and nanotubes filled with $TbCl_3$, $TbBr_3$, and TbI_3. The data are replotted from ref. [66].

for TbI_3-filled nanotubes. For example, the shift of the G^--peak decreases from 22 to 19 to 16 cm^{-1} in the line $TbCl_3$-$TbBr_3$-TbI_3. The relative intensity of the G^--peak decreases from 0.70 for the pristine SWCNTs to 0.01 for $TbCl_3$, 0.11 for $TbBr_3$ and 0.11 for TbI_3. On the basis of these data, the authors of Ref. [66] concluded that $TbCl_3$ causes the largest doping of SWCNTs, whereas TbI_3 results in the smallest doping.

Authors also investigated the influence of metal cation of metal halogenides on the changes of the electronic properties of SWCNTs. The above-described analysis of RBM and G-bands of Raman spectra was conducted for SWCNTs filled with $FeBr_2$, $CoBr_2$ and $NiBr_2$ [50] and $TmCl_3$, $TbCl_3$, and $PrCl_3$ [65]. It was revealed that the doping level of SWCNTs increases in the lines with $NiBr_2$-$CoBr_2$-$FeBr_2$ and $PrCl_3$-$TbCl_3$-$TmCl_3$.

Thus, Raman spectroscopy allowed revealing p-doping of SWCNTs and elucidating the differences in the doping efficiency of different metal halogenides on nanotubes. However, similar to OAS, Raman spectroscopy does not give quantitative information on the doping level of SWCNTs.

XPS spectroscopy allows quantifying the doping level of nanotubes. **Figure 5(a)** presents the typical C 1s XPS spectrum of the metal halogenide-filled SWCNTs by an example of $MnCl_2$-filled nanotubes in comparison with the spectrum of the pristine SWCNTs [46]. The spectrum of the pristine SWCNTs is a single peak

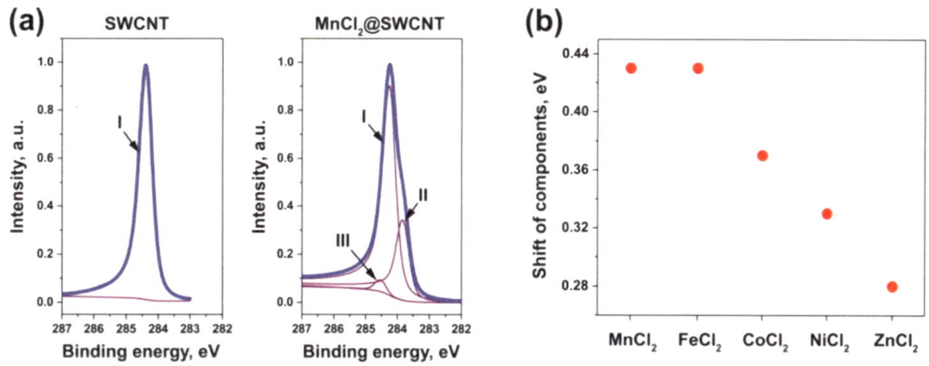

Figure 5.
(a) The C 1s XPS spectra of the pristine and $MnCl_2$-filled SWCNTs fitted with individual components. The components are denoted. The data are replotted from ref. [46]. (b) The shift of the component II relatively to the position of the component I in the C 1s XPS spectra of SWCNTs filled with $MnCl_2$, $FeCl_2$, $CoCl_2$, $NiCl_2$, and $ZnCl_2$. The data from refs. [46, 47, 49, 55] are taken into consideration for the preparation of the figure.

positioned at a binding energy of 284.38 eV. The spectrum of the filled nanotubes can be fitted with three components. The component I resembles the position and full width at half maximum of the pristine SWCNTs and it is assigned to the unfilled SWCNTs. The component II is downshifted by 0.43 eV relative to component I. This is attributed to an increase in the work function of the filled SWCNTs, which is caused by the charge transfer-induced downshift of the Fermi level of nanotubes. The origin of the component III is possibly caused by local interactions of carbon atoms of SWCNTs with atoms of the incorporated salt. Similar modifications were observed in the C 1s XPS spectra of SWCNTs filled with $MnBr_2$ [45, 46], $FeCl_2$, $FeBr_2$, FeI_2 [47], $CoBr_2$ [48], $NiCl_2$, $NiBr_2$ [49], $ZnCl_2$ [55, 56], $ZnBr_2$, ZnI_2 [55], $CdCl_2$ [56, 62], $CdBr_2$, CdI_2 [62], AgCl, AgBr, AgI [58], CuCl, CuBr, CuI [51], $RbAg_4I_5$ [57], $TbCl_3$ [56], $TmCl_3$ [69], and $PrCl_3$ [67]. The measured shifts of the component II relative to the component I amounted to 0.3–0.4 eV, and they were attributed to p-doping of SWCNTs by the encapsulated compounds.

Figure 5(b) compares the shifts of the component II relative to component I for SWCNTs filled with $MnCl_2$, $FeCl_2$, $CoCl_2$, $NiCl_2$, and $ZnCl_2$ [46, 47, 49, 55]. It is visible that the shift decreases in the line with $MnCl_2$-$FeCl_2$-$CoCl_2$-$NiCl_2$-$ZnCl_2$. This may testify that the doping level of SWCNTs decreases in this line. This conclusion is in agreement with the above-discussed data of Raman spectroscopy, which showed that among bromides of Fe, Co, and Ni, the doping efficiency decreased in the line with $FeBr_2$-$CoBr_2$-$NiBr_2$.

Thus, a comprehensive characterization of the electronic properties of SWCNTs filled with halogenides of 3d-, 4d-, and 4f-metals by OAS, Raman spectroscopy, and XPS showed that they lead to p-doping of SWCNTs accompanied by the downshift of their Fermi level. The differences in the doping level for different metal halogenides depending on metal cation and halogen anion were revealed.

3. SWCNTs filled with metallocenes

3.1 Synthesis of metallocene-filled SWCNTs

Metallocene molecules are not stable at high temperatures. The powders of substances decompose and could not be melted. At the same time, they sublime in vacuum at low temperatures. Taking into consideration this fact, researchers introduced metallocenes inside the nanotubes by the gas phase method. In a typical experiment, the SWCNTs were mixed with the powder of substances, sealed in an ampoule under vacuum, and heated at low temperatures (50–200°C) during several days. This method allowed filling SWCNTs with ferrocene [44, 76–86], cobaltocene [87, 88], nickelocene [89–92], and cerocene [93, 94]. It should be noted that ferrocene was also incorporated inside SWCNTs by the liquid phase method using its solution in acetone [95].

3.2 Temperature-dependent inner tube growth inside metallocene-filled SWCNTs

The temperature-dependent inner tube growth inside furnace- or laser-annealed ferrocene- [76, 77, 79], cobaltocene- [87], and nickelocene-filled SWCNTs [89, 90] was investigated by Raman spectroscopy. Authors traced modifications of Raman spectra of the filled SWCNTs that occurred at increasing annealing temperature or laser power at fixed annealing time. **Figure 6** presents the example of the investigation of the inner tube growth inside nickelocene-filled SWCNTs [89]. **Figure 6(a)** shows the RBM-bands of Raman spectra of the pristine, filled SWCNTs, and the samples annealed at temperatures between 400 and 1200°C for 2 h acquired at laser wavelength of 633 nm. The RBM-band of the pristine SWCNTs includes the peak at frequencies ranging from 125 to 160 cm^{-1}. This peak is shifted by 4 cm^{-1} after filling with nickelocene, which is usually observed for molecule-filled SWCNTs. Annealing at 400°C and higher temperatures results in an appearance of new peaks centered at 212, 216, and 253 cm^{-1}. These peaks belong to inner nanotubes with chiralities of (12,3), (13,1) and (11,1) and diameters of 1.081, 1.064, and 0.909 nm, respectively. The gradual increase in annealing temperature leads to an increase of the intensity of the peaks. **Figure 6(b)** demonstrates the dependence of the relative intensity of the peak of the (12,3) and (13,1) inner tubes on annealing temperature. It is visible that the intensity increases at temperatures ranging from 400 to 700°C. It saturates at 700°C and stays almost unchanged at higher annealing temperatures. Thus, inner tubes grow in the temperature range between 400 and 700°C.

Authors of Ref. [90] compared the growth temperatures of eight different inner tubes with chiralities of (7,5), (8,4), (7,6), (10,3), (12,3), (11,5), (14,2), (12,6) inside nickelocene-filled SWCNTs. They evaluated the growth temperature as the temperature at which the intensity of the RBM peak of the inner tube reaches the half of its maximum. **Figure 7(a)** and **(b)** shows the dependences of the growth temperature of the inner tubes on their diameter and chiral angle. It is visible that the growth temperature increases with increasing tube diameter, but it does not depend on their chiral angle. The same trend was observed for inner tubes grown inside ferrocene- [77] and cobaltocene-filled SWCNTs [87].

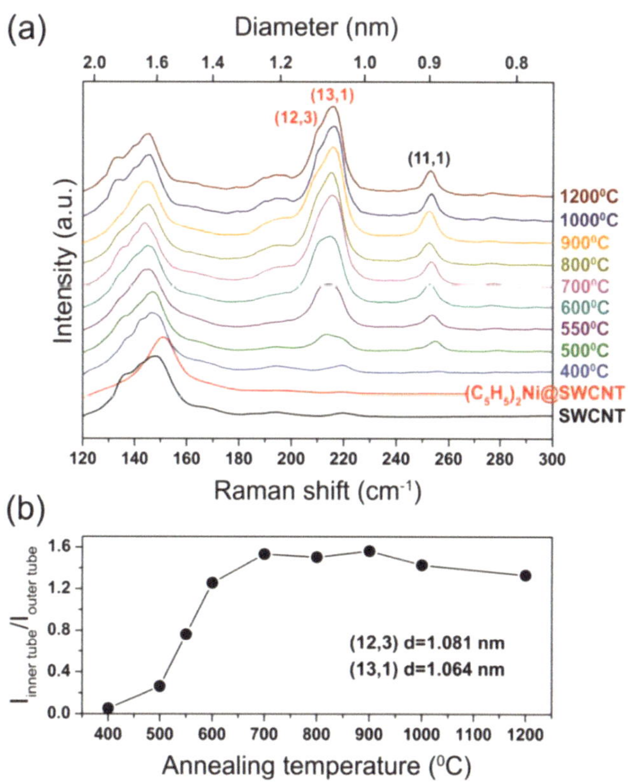

Figure 6.
(a) The RBM-band of Raman spectra of the pristine, nickelocene-filled SWCNTs, and the samples annealed at temperatures between 400 and 1200°C for 2 h acquired at laser wavelength of 633 nm. The peaks corresponding to the inner nanotubes with chiralities of (12,3), (13,1), and (11,1) are denoted. (b) The relative area intensity of the RBM peak of the inner tubes with chiralities of (12,3) and (13,1) plotted versus annealing temperature [89]—Published by The Royal Society of Chemistry.

3.3 Growth kinetics of inner tubes inside metallocene-filled SWCNTs

A detailed investigation of the growth kinetics of inner tubes inside in situ annealed nickelocene- [91], cobaltocene- [88], and ferrocene-filled SWCNTs [78] was conducted by Raman spectroscopy. Authors traced modifications of Raman spectra of the filled SWCNTs at increasing annealing time at a fixed temperature. **Figure 8** presents the example of the investigation of the growth kinetics of inner tubes inside nickelocene-filled SWCNTs [91]. **Figure 8a** shows the RBM-bands of Raman spectra of the pristine, filled SWCNTs, and the samples annealed at 540°C for the time periods between 2 and 4094 min acquired at laser wavelength of 568 nm. The spectrum of the pristine SWCNTs includes two peaks positioned at frequencies ranging from 125 to 185 cm^{-1}. They are shifted by 10 cm^{-1} for the filled SWCNTs. The spectra of the annealed samples include new peaks of inner tubes at frequencies between 205 and 295 cm^{-1}. The intensity of the peaks increases with increasing annealing time. **Figure 8(b)** demonstrates the dependence of the normalized area intensity of the RBM peaks on annealing time (growth curves) for nine different inner tubes with chiralities of (8,8), (12,3), (13,1), (9,6), (10,4), (11,2), (11,1), (9,3), and (9,2). It is visible that after an increase in the first minutes of annealing the intensity saturates and stays uncharged at further annealing. The growth curves differ for different inner tubes. The time period required for the saturation of the intensity gradually decreases with decreasing tube diameter.

The observed growth curves of inner tubes do not follow a self-exhausting growth model that was reported for the SWCNT growth in the chemical vapor deposition method. Authors of Ref. [91] modeled the growth of inner tubes by a new mathematical model including two growth rates α and β. The dependence of the amount of carbon in the form of grown inner tubes on time is expressed by the formula:

$$C(t) = A_0\left(1 - \chi e^{-\alpha t} - (1 - \chi)\, e^{-\beta t}\right),$$

where A_0 is the initial amount of carbon that can be transformed to inner tubes (at $t = 0$), α is the rate that determines the fast growth of inner tubes at the beginning, β is the rate that determines the slow growth over longer annealing hours, and χ describes which parts of carbon processed with rates α and β.

The fitting of the experimental growth curves with this model (**Figure 8(b)**) allowed calculating two rates α and β of the growth of inner tubes. **Figure 9** summarizes the calculated rates of the growth of inner tubes with chiralities of (8,8), (12,3), (13,1), (9,6), (10,4), (11,2), (11,1), (9,3), and (9,2) inside

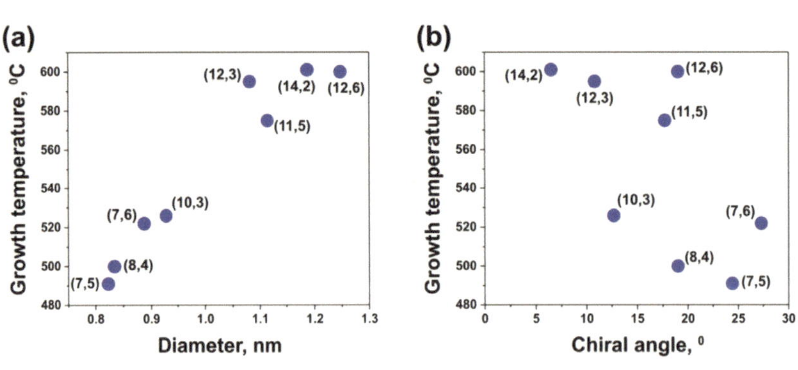

Figure 7.
The dependence of the growth temperature of inner tubes inside nickelocene-filled SWCNTs on their diameter (a) and chiral angle (b). The chirality of nanotube is indicated near the corresponding circle. The data are replotted from ref. [90].

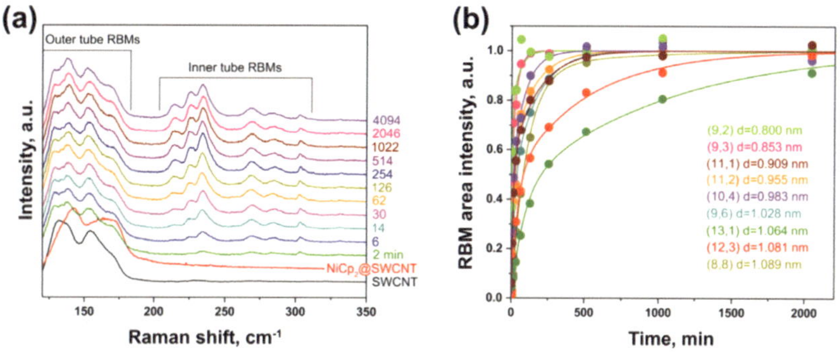

Figure 8.
(a) The RBM-band of Raman spectra of the pristine, nickelocene-filled SWCNTs, and the samples annealed at 540°C for time periods between 2 and 4094 min acquired at laser wavelength of 568 nm. The inner and outer tube RBMs are denoted. (b) The normalized RBM area intensity of inner tube peaks plotted versus annealing time. The data are replotted from Ref. [91].

nickelocene-filled SWCNTs at different annealing temperatures [91]. Two trends are observed in these plots. Firstly, the growth rates α and β increase with increasing annealing temperature from 480 to 600°C, which is caused by the fact that the inner tube growth is a thermally activated process. Secondly, they increase with decreasing the inner tube diameter from 1.1 to 0.8 nm. This can be explained by the increased catalytic activity of smaller-diameter nanoparticles [91]. It should be noted that the growth rates α and β do not depend on the chiral angle of inner tubes. The same trends were observed for the inner tube growth inside cobaltocene-filled SWCNTs [88].

Using logarithmic plots of the growth rates, two activation energies E_α and E_β of the growth of inner tubes inside nickelocene- [91] and cobaltocene-filled SWCNTs [88] were calculated. The values of E_α and E_β for the inner tubes with chiralities of (8,8), (12,3), (13,1), (9,6), (10,4), (11,2), (11,1), (9,3), and (9,2) amounted to 2.02–2.57 and 1.23–1.84 eV in the case of nickelocene, and 1.72–2.71 and 0.46–1.59 eV in the case of cobaltocene, respectively. Two activation energies were attributed to the energy barriers for solid-state diffusion of carbon through carbidic and purely metallic catalytic nanoparticles. **Figure 10(a)** and **(b)** presents the dependences of the activation energies of the inner tube growth inside nickelocene-filled SWCNTs on the tube diameter and chiral angle [91]. It is visible that the activation energy E_α gradually decreases with decreasing the tube diameter, which was explained by the size effect, whereas E_β does not show a clear dependence. Both activation energies do not seem to depend on chiral angle of inner tubes.

The identical growth mechanism of inner tubes inside nickelocene- and cobaltocene-filled SWCNTs allowed authors of Ref. [88] to compare the rates and activation energies of the growth on Ni and Co catalysts. The activation energies E_α of the inner tube growth on two catalysts were in line with each other, whereas E_β values were larger for Ni catalyst. This was in agreement with the slightly different activation energies reported for solid-state carbon diffusion through face-centered cubic nickel and cobalt with hexagonal close packed lattice. Major differences were observed for the growth rates of inner tubes on Ni and Co catalysts at a given temperature. The temperature at which inner tubes started to grow differed by 60°C. It amounted to 480°C for Ni and 540°C for Co catalyst. As a result, at a given temperature, the growth rates of inner tubes on Ni catalyst were significantly larger than those on Co catalyst. This was explained by different thermal stabilities of nickel and cobalt carbides, and different diffusion coefficients of carbon in the two metals.

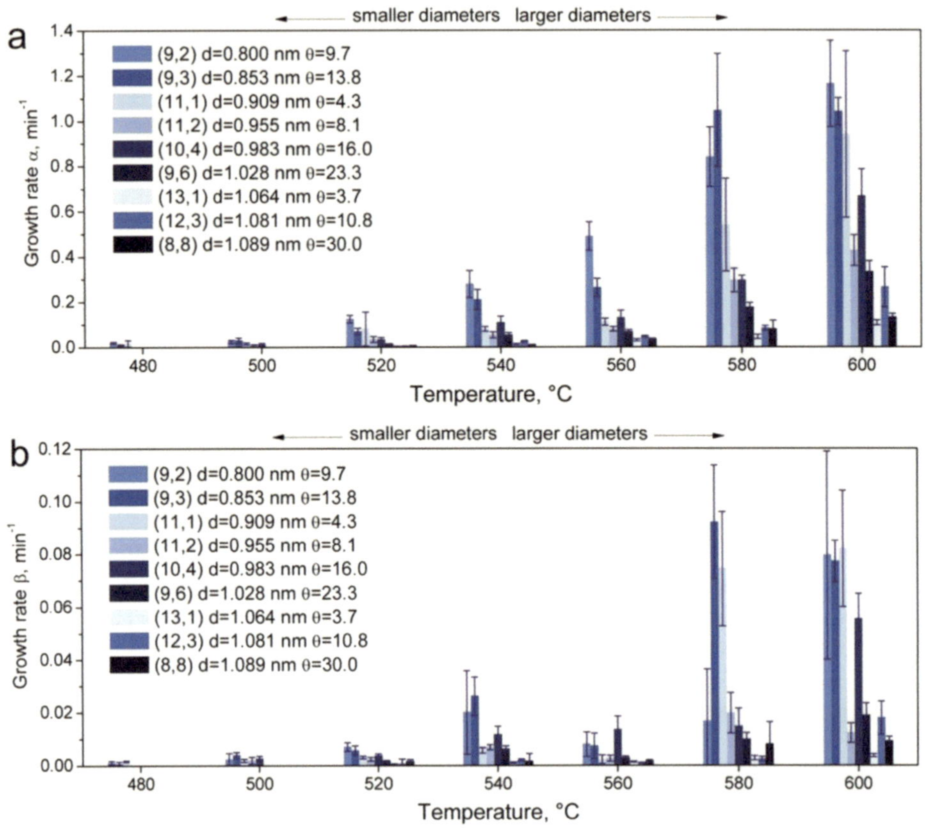

Figure 9.
The column bar diagrams showing the rates α (a) and β (b) of the growth of inner nanotubes inside nickelocene-filled SWCNTs at different annealing temperatures. At every annealing temperature, the column bar of the smallest diameter tube is shown at the leftmost side, and the largest diameter tube—at the rightmost side. The color of column bars reflects the chiral angle of inner tubes: the lightest color shade of blue corresponds to the smallest chiral angle and the darkest—to the largest chiral angle [91]—Published by The Royal Society of Chemistry.

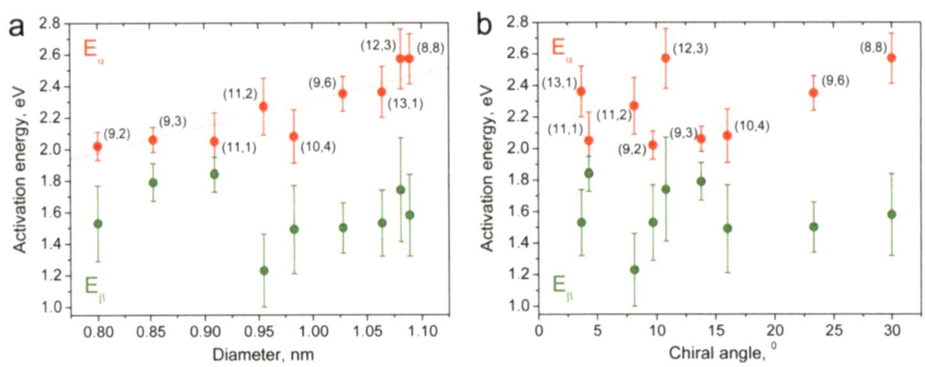

Figure 10.
The activation energies E_α and E_β of the growth of inner nanotubes inside nickelocene-filled SWCNTs plotted versus the tube diameter (a) and chiral angle (b). The chirality of nanotubes is indicated near the corresponding circle [91]—Published by The Royal Society of Chemistry.

3.4 Electronic properties of metallocene-filled SWCNTs

The electronic properties of SWCNTs filled with ferrocene [82], nickelocene [89, 90, 92], and cerocene [93, 94] were studied by XPS and UPS. It was shown that the encapsulated metallocenes cause n-doping of SWCNTs. Authors of Refs. [82, 89, 90, 92, 93] investigated the modification of the electronic properties of the filled SWCNTs upon annealing. **Figure 11(a)** shows the UPS spectra of the pristine, nickelocene-filled SWCNTs, and the samples annealed at temperatures between 250 and 1200°C for 2 h [89]. The spectrum of the pristine SWCNTs includes π- and σ-peaks positioned at binding energies of 3.18 and 8.0 eV, respectively. The spectrum of nickelocene-filled SWCNTs demonstrates the shift of the π-peak by 0.07 eV toward higher binding energies. The annealing of the filled SWCNTs at 250°C leads to a further upshift of the π-peak by 0.18 eV. At increasing annealing temperature, the π-peak gradually shifts toward lower binding energies and reaches the position of the pristine SWCNTs at 600°C. At further increase in annealing temperature, the π-peak downshifts and reaches the maximal shift of 0.18 eV at 1200°C. The change in the position of the π-peak testifies about the change in the doping level of SWCNTs upon annealing. Authors of Ref. [89] suggested that this change is caused by three processes: (i) the chemical modification of the filler of SWCNTs, (ii) the inner tube growth, and (iii) the evaporation of the filler. The annealing of the nickelocene-filled SWCNTs leads to the formation of nickel carbide that causes the largest n-doping level of SWCNTs. As it was discussed above, at 400°C, the inner

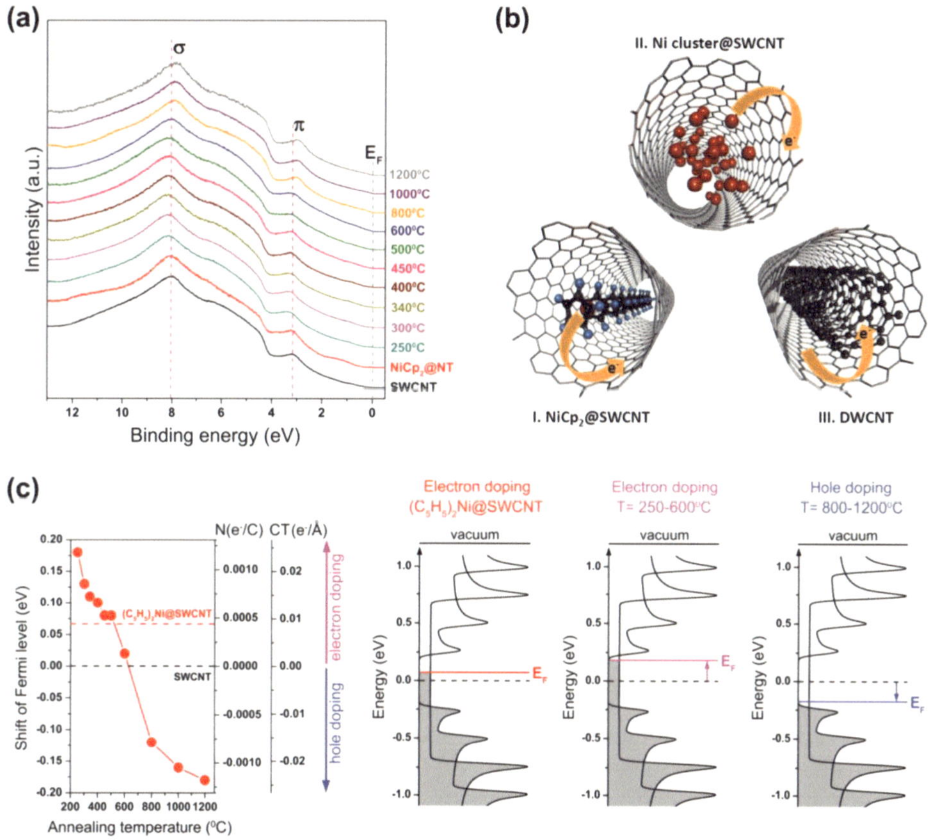

Figure 11.
(a) The UPS spectra of the pristine, nickelocene-filled SWCNTs, and the samples annealed at temperatures between 250 and 1200°C for 2 h. The π- and σ-peaks and the Fermi level (E_F) are denoted. (b) The schematics showing the charge transfer in nickelocene-filled SWCNTs (I), Ni cluster-filled SWCNTs (II) and DWCNTs (III). (c) The shift of the Fermi level, number of transferred electrons per carbon atom of SWCNTs ($N(e^-/C)$) and charge transfer density per nanotube length ($CT(e^-/Å)$) plotted versus annealing temperature as well as the schematics of electron and hole doping in the filled and annealed samples [89]—Published by The Royal Society of Chemistry.

tube growth starts and nickel carbide decomposes to pure nickel. This leads to the formation of nickel-filled DWCNTs. In DWCNTs, there is p-doping of outer tube by inner tube, which leads to decreasing the n-type doping. At high temperatures, nickel evaporates, and it leaves empty DWCNTs. It explains the largest p-doping of nanotubes at 1200°C. **Figure 11(b)** presents the schematics showing the charge transfer in the filled SWCNTs and annealed samples. **Figure 11(c)** demonstrates the shift of the Fermi level, calculated number of transferred electrons per carbon atom of SWCNTs ($N(e^{-}/C)$) and charge transfer density per nanotube length ($CT(e^{-}/Å)$) plotted versus annealing temperature as well as the schematics of n- and p-doping in the filled and annealed samples.

Similar thermally induced modifications of the electronic properties were reported for metallicity-mixed SWCNTs filled with ferrocene [82] and cerocene [93] as well as for metallicity-sorted semiconducting SWCNTs filled with nickelocene [92].

4. Conclusions and outlook

The literature survey conducted in this chapter shows that filling with metal halogenides leads to p-doping of SWCNTs. The doping level strongly depends on the metal cation and halogen anion. In contrast, filling with metallocenes leads to n-doping of SWCNTs. The high vacuum annealing of metallocene-filled SWCNTs further results in the growth of inner tubes with altered electronic properties. The growth kinetics of inner SWCNTs is characterized with two growth rates and activation energies. They show the dependence on the tube diameter and metal catalyst type.

The trends revealed in this chapter based on filling-induced modifications of the electronic properties of SWCNTs, and the growth kinetics of SWCNTs will provide the foundation for the dedicated preparation of SWCNTs with defined properties that are required for advanced applications.

However, despite the remarkable progress in the filling of SWCNTs and the controllable modification of their properties, there remain challenges that currently limit the applicability of filled SWCNTs in devices. The first issue is the scale of synthesis. Indeed, the filling routes of SWCNTs on a laboratory scale are well developed. However, for the implementation of filled SWCNTs in devices, it is crucial to up-scale the filling methods. The second issue is the filling yield. The filling is often not uniform throughout the entire sample batch. The optimization of reliable and reproducible strategies of filling of SWCNTs is necessary for application testing and the fabrication of real devices based on filled SWCNTs. The third issue involves the crystallinity of encapsulated substances. Although in many cases, well-ordered one-dimensional nanocrystals inside SWCNTs have been achieved by the melt method, there remains a lack of uniformity regarding the crystallization degree and phase composition of the filling compound. Therefore, a better understanding of the filling mechanism of SWCNTs is key to address these issues. It is important to further investigate the correlation between the synthesis parameters and the filling material. This should enable a considerable improvement on the filling ratio, crystallinity and uniformity, and thus will open new avenues for large-scale synthesis of filled SWCNTs.

Acknowledgements

This work was funded by the Deutsche Forschungsgemeinschaft (DFG ED 221/3-1).

Author details

Marianna V. Kharlamova* and Dominik Eder
Institute of Materials Chemistry, Vienna University of Technology, Vienna, Austria

*Address all correspondence to: mv.kharlamova@gmail.com

IntechOpen

© 2019 The Author(s). Licensee IntechOpen. This chapter is distributed under the terms of the Creative Commons Attribution License (http://creativecommons.org/licenses/by/3.0), which permits unrestricted use, distribution, and reproduction in any medium, provided the original work is properly cited. cc BY

References

[1] Endo M, Strano MS, Ajayan PM. Potential applications of carbon nanotubes. In: Jorio A, Dresselhaus G, Dresselhaus MS, editors. Carbon Nanotubes: Topics in Applied Physics. Berlin: Springer-Verlag; 2008. pp. 13-61. DOI: 10.1007/978-3-540-72865-8_2

[2] Chiang WH, Sankaran RM. Linking catalyst composition to chirality distributions of as-grown single-walled carbon nanotubes by tuning Ni_xFe_{1-x} nanoparticles. Nature Materials. 2009;**8**(11):882-886. DOI: 10.1038/nmat2531

[3] Ghorannevis Z, Kato T, Kaneko T, Hatakeyama R. Narrow-chirality distributed single-walled carbon nanotube growth from nonmagnetic catalyst. Journal of the American Chemical Society. 2010;**132**(28): 9570-9572. DOI: 10.1021/ja103362j

[4] He MS, Jiang H, Liu BL, Fedotov PV, Chernov AI, Obraztsova ED, et al. Chiral-selective growth of single-walled carbon nanotubes on lattice-mismatched epitaxial cobalt nanoparticles. Scientific Reports. 2013;**3**:1460. DOI: 10.1038/srep01460

[5] Yang F, Wang X, Zhang DQ, Yang J, Luo D, Xu ZW, et al. Chirality-specific growth of single-walled carbon nanotubes on solid alloy catalysts. Nature. 2014;**510**(7506):522-524. DOI: 10.1038/nature13434

[6] Zhao QC, Xu ZW, Hu Y, Ding F, Zhang J. Chemical vapor deposition synthesis of near-zigzag single-walled carbon nanotubes with stable tube-catalyst interface. Science Advances. 2016;**2**(5):e1501729. DOI: 10.1126/sciadv.1501729

[7] Fagan JA, Haroz EH, Ihly R, Gui H, Blackburn JL, Simpson JR, et al. Isolation of >1 nm diameter single-wall carbon nanotube species using aqueous two-phase extraction. ACS Nano. 2015;**9**(5):5377-5390. DOI: 10.1021/acsnano.5b01123

[8] Green AA, Hersam MC. Nearly single-chirality single-walled carbon nanotubes produced via orthogonal iterative density gradient ultracentrifugation. Advanced Materials. 2011;**23**(19):2185-2190. DOI: 10.1002/adma.201100034

[9] Liu HP, Nishide D, Tanaka T, Kataura H. Large-scale single-chirality separation of single-wall carbon nanotubes by simple gel chromatography. Nature Communications. 2011;**2**:309. DOI: 10.1038/ncomms1313

[10] Tu XM, Manohar S, Jagota A, Zheng M. DNA sequence motifs for structure-specific recognition and separation of carbon nanotubes. Nature. 2009;**460**(7252):250-253. DOI: 10.1038/nature08116

[11] Yomogida Y, Tanaka T, Zhang MF, Yudasaka M, Wei XJ, Kataura H. Industrial-scale separation of high-purity single-chirality single-wall carbon nanotubes for biological imaging. Nature Communications. 2016;**7**:12056. DOI: 10.1038/ncomms12056

[12] Kharlamova MV. Electronic properties of pristine and modified single-walled carbon nanotubes. Physics-Uspekhi. 2013;**56**(11):1047-1073. DOI: 10.3367/UFNe.0183.201311a.1145

[13] Kharlamova MV. Advances in tailoring the electronic properties of single-walled carbon nanotubes. Progress in Materials Science. 2016;**77**:125-211. DOI: 10.1016/j.pmatsci.2015.09.001

[14] Smith BW, Monthioux M, Luzzi DE. Encapsulated C_{60} in carbon

nanotubes. Nature. 1998;**396**(6709): 323-324. DOI: 10.1038/24521

[15] Sloan J, Hammer J, Zwiefka-Sibley M, Green MLH. The opening and filling of single walled carbon nanotubes (SWTs). Chemical Communications. 1998;(3):347-348. DOI: 10.1039/A707632K

[16] Govindaraj A, Satishkumar BC, Nath M, Rao CNR. Metal nanowires and intercalated metal layers in single-walled carbon nanotube bundles. Chemistry of Materials. 2000;**12**(1): 202-205. DOI: 10.1021/cm990546o

[17] Tonkikh AA, Tsebro VI, Obraztsova EA, Suenaga K, Kataura H, Nasibulin AG, et al. Metallization of single-wall carbon nanotube thin films induced by gas phase iodination. Carbon. 2015;**94**:768-774. DOI: 10.1016/j.carbon.2015.07.062

[18] Borowiak-Palen E, Ruemmeli MH, Gemming T, Pichler T, Kalenczuk RJ, Silva SRP. Silver filled single-wall carbon nanotubes-synthesis, structural and electronic properties. Nanotechnology. 2006;**17**(9):2415-2419. DOI: 10.1088/0957-4484/17/9/058

[19] Corio P, Santos AP, Santos PS, Temperini MLA, Brar VW, Pimenta MA, et al. Characterization of single wall carbon nanotubes filled with silver and with chromium compounds. Chemical Physics Letters. 2004;**383**(5-6):475-480. DOI: 10.1016/j.cplett.2003.11.061

[20] Kharlamova MV, Niu JJ. Comparison of metallic silver and copper doping effects on single-walled carbon nanotubes. Applied Physics A. 2012;**109**(1):25-29. DOI: 10.1007/s00339-012-7091-3

[21] Kharlamova MV, Niu JJ. Donor doping of single-walled carbon nanotubes by filling of channels with silver. Journal of Experimental and Theoretical Physics. 2012;**115**(3):485-491. DOI: 10.1134/S1063776112080092

[22] Kharlamova MV, Niu JJ. New method of the directional modification of the electronic structure of single-walled carbon nanotubes by filling channels with metallic copper from a liquid phase. JETP Letters. 2012;**95**(6):314-319. DOI: 10.1134/S0021364012060057

[23] Fujimori T, Morelos-Gomez A, Zhu Z, Muramatsu H, Futamura R, Urita K, et al. Conducting linear chains of sulphur inside carbon nanotubes. Nature Communications. 2013;**4**:2162. DOI: 10.1038/ncomms3162

[24] Philp E, Sloan J, Kirkland AI, Meyer RR, Friedrichs S, Hutchison JL, et al. An encapsulated helical one-dimensional cobalt iodide nanostructure. Nature Materials. 2003;**2**(12):788-791. DOI: 10.1038/nmat1020

[25] Sloan J, Novotny MC, Bailey SR, Brown G, Xu C, Williams VC, et al. Two layer 4: 4 co-ordinated KI crystals grown within single walled carbon nanotubes. Chemical Physics Letters. 2000;**329**(1-2):61-65. DOI: 10.1016/S0009-2614(00)00998-2

[26] Sloan J, Kirkland AI, Hutchison JL, Green MLH. Integral atomic layer architectures of 1D crystals inserted into single walled carbon nanotubes. Chemical Communications. 2002;(13):1319-1332. DOI: 10.1039/B200537A

[27] Sloan J, Friedrichs S, Meyer RR, Kirkland AI, Hutchison JL, Green MLH. Structural changes induced in nanocrystals of binary compounds confined within single walled carbon nanotubes: A brief review. Inorganica Chimica Acta. 2002;**330**:1-12. DOI: 10.1016/S0020-1693(01)00774-5

[28] Sloan J, Grosvenor SJ, Friedrichs S, Kirkland AI, Hutchison JL, Green MLH.

A one-dimensional BaI_2 chain with five- and six-coordination, formed within a single-walled carbon nanotube. Angewandte Chemie International Edition. 2002;**41**(7):1156-1159. DOI: 10.1002/1521-3773(20020402)41:7 <1156:AID-ANIE1156>3.0.CO;2-N

[29] Sloan J, Kirkland AI, Hutchison JL, Green MLH. Aspects of crystal growth within carbon nanotubes. Comptes Rendus Physique. 2003;**4**(9):1063-1074. DOI: 10.1016/S1631-0705(03)00102-6

[30] Carter R, Sloan J, Kirkland AI, Meyer RR, Lindan PJD, Lin G, et al. Correlation of structural and electronic properties in a new low-dimensional form of mercury telluride. Physical Review Letters. 2006;**96**(21):215501. DOI: 10.1103/PhysRevLett.96.215501

[31] Carter R, Suyetin M, Lister S, Dyson MA, Trewhitt H, Goel S, et al. Band gap expansion, shear inversion phase change behaviour and low-voltage induced crystal oscillation in low-dimensional tin selenide crystals. Dalton Transactions. 2014;**43**(20):7391-7399. DOI: 10.1039/C4DT00185K

[32] Kharlamova MV. Novel approach to tailoring the electronic properties of single-walled carbon nanotubes by the encapsulation of high-melting gallium selenide using a single-step process. JETP Letters. 2013;**98**(5):272-277. DOI: 10.1134/S0021364013180069

[33] Kharlamova MV. Comparative analysis of electronic properties of tin, gallium, and bismuth chalcogenide-filled single-walled carbon nanotubes. Journal of Materials Science. 2014;**49**(24):8402-8411. DOI: 10.1007/s10853-014-8550-3

[34] Yashina LV, Eliseev AA, Kharlamova MV, Volykhov AA, Egorov AV, Savilov SV, et al. Growth and characterization of one-dimensional SnTe crystals within the single-walled carbon nanotube channels. Journal of Physical Chemistry C. 2011;**115**(9):3578-3586. DOI: 10.1021/jp1107087

[35] Costa PMFJ, Sloan J, Rutherford T, Green MLH. Encapsulation of Re_xO_y clusters within single-walled carbon nanotubes and their in tubulo reduction and sintering to Re metal. Chemistry of Materials. 2005;**17**(26):6579-6582. DOI: 10.1021/cm0510209

[36] Hulman M, Kuzmany H, Costa PMFJ, Friedrichs S, Green MLH. Light-induced instability of PbO-filled single-wall carbon nanotubes. Applied Physics Letters. 2004;**85**(11):2068-2070. DOI: 10.1063/1.1790603

[37] Burteaux B, Claye A, Smith BW, Monthioux M, Luzzi DE, Fischer JE. Abundance of encapsulated C_{60} in single-wall carbon nanotubes. Chemical Physics Letters. 1999;**310**(1-2):21-24. DOI: 10.1016/S0009-2614(99)00720-4

[38] Chamberlain TW, Camenisch A, Champness NR, Briggs GAD, Benjamin SC, Ardavan A, et al. Toward controlled spacing in one-dimensional molecular chains: Alkyl-chain-functionalized fullerenes in carbon nanotubes. Journal of the American Chemical Society. 2007;**129**(27):8609-8614. DOI: 10.1021/ja071803q

[39] Gimenez-Lopez MD, Chuvilin A, Kaiser U, Khlobystov AN. Functionalised endohedral fullerenes in single-walled carbon nanotubes. Chemical Communications. 2011;**47**(7):2116-2118. DOI: 10.1039/C0CC02929G

[40] Kataura H, Maniwa Y, Kodama T, Kikuchi K, Hirahara K, Suenaga K, et al. High-yield fullerene encapsulation in single-wall carbon nanotubes. Synthetic Metals. 2001;**121**(1-3):1195-1196. DOI: 10.1016/S0379-6779(00)00707-4

[41] Suenaga K, Okazaki T, Wang CR, Bandow S, Shinohara H, Iijima S. Direct imaging of $Sc_2@C_{84}$ molecules

encapsulated inside single-wall carbon nanotubes by high resolution electron microscopy with atomic sensitivity. Physical Review Letters. 2003;**90**(5):055506. DOI: 10.1103/PhysRevLett.90.055506

[42] Guan LH, Shi ZJ, Li MX, Gu ZN. Ferrocene-filled single-walled carbon nanotubes. Carbon. 2005;**43**(13):2780-2785. DOI: 10.1016/j.carbon.2005.05.025

[43] Li LJ, Khlobystov AN, Wiltshire JG, Briggs GAD, Nicholas RJ. Diameter-selective encapsulation of metallocenes in single-walled carbon nanotubes. Nature Materials. 2005;**4**(6):481-485. DOI: 10.1038/nmat1396

[44] Shiozawa H, Pichler T, Gruneis A, Pfeiffer R, Kuzmany H, Liu Z, et al. A catalytic reaction inside a single-walled carbon nanotube. Advanced Materials. 2008;**20**(8):1443-1449. DOI: 10.1002/adma.200701466

[45] Kharlamova MV, Eliseev AA, Yashina LV, Lukashin AV, Tretyakov YD. Synthesis of nanocomposites on basis of single-walled carbon nanotubes intercalated by manganese halogenides. Journal of Physics Conference Series. 2012;**345**:012034. DOI: 10.1088/1742-6596/345/1/012034

[46] Kharlamova MV. Electronic properties of single-walled carbon nanotubes filled with manganese halogenides. Applied Physics A. 2016;**122**(9):791. DOI: 10.1007/s00339-016-0335-x

[47] Kharlamova MV, Brzhezinskaya MM, Vinogradov AS, Suzdalev IP, Maksimov YV, Imshennik VK, et al. The formation and properties of one-dimensional $FeHal_2$ (HaI=Cl, Br, I) nanocrystals in channels of single-walled carbon nanotubes. Nanotechnologies in Russia. 2009;**4**(9-10):634-646. DOI: 10.1134/S1995078009090080

[48] Kharlamova MV, Eliseev AA, Yashina LV, Petukhov DI, Liu CP, Wang CY, et al. Study of the electronic structure of single-walled carbon nanotubes filled with cobalt bromide. JETP Letters. 2010;**91**(4):196-200. DOI: 10.1134/S0021364010040089

[49] Kharlamova MV, Yashina LV, Eliseev AA, Volykhov AA, Neudachina VS, Brzhezinskaya MM, et al. Single-walled carbon nanotubes filled with nickel halogenides: Atomic structure and doping effect. Physica Status Solidi B. 2012;**249**(12):2328-2332. DOI: 10.1002/pssb.201200060

[50] Kharlamova MV. Raman spectroscopy study of the doping effect of the encapsulated iron, cobalt, and nickel bromides on single-walled carbon nanotubes. Journal of Spectroscopy. 2015;**2015**:653848. DOI: 10.1155/2015/653848

[51] Eliseev AA, Yashina LV, Verbitskiy NI, Brzhezinskaya MM, Kharlamova MV, Chernysheva MV, et al. Interaction between single walled carbon nanotube and 1D crystal in CuX@SWCNT (X = Cl, Br, I) nanostructures. Carbon. 2012;**50**(11):4021-4039. DOI: 10.1016/j.carbon.2012.04.046

[52] Fedotov PV, Tonkikh AA, Obraztsova EA, Nasibulin AG, Kauppinen EI, Chuvilin AL, et al. Optical properties of single-walled carbon nanotubes filled with CuCl by gas-phase technique. Physica Status Solidi B. 2014;**251**(12):2466-2470. DOI: 10.1002/pssb.201451240

[53] Chernysheva MV, Eliseev AA, Lukashin AV, Tretyakov YD, Savilov SV, Kiselev NA, et al. Filling of single-walled carbon nanotubes by CuI nanocrystals via capillary technique. Physica E: Low-dimensional Systems and Nanostructures. 2007;**37**(1-2):62-65. DOI: 10.1016/j.physe.2006.10.014

[54] Kumskov AS, Zhigalina VG, Chuvilin AL, Verbitskiy NI, Ryabenko AG, Zaytsev DD, et al. The structure of 1D and 3D CuI nanocrystals grown within 1.5-2.5 nm single wall carbon nanotubes obtained by catalyzed chemical vapor deposition. Carbon. 2012;**50**(12):4696-4704. DOI: 10.1016/j.carbon.2012.05.061

[55] Kharlamova MV, Yashina LV, Volykhov AA, Niu JJ, Neudachina VS, Brzhezinskaya MM, et al. Acceptor doping of single-walled carbon nanotubes by encapsulation of zinc halogenides. European Physical Journal B. 2012;**85**(1):34. DOI: 10.1140/epjb/e2011-20457-6

[56] Kharlamova MV. Comparison of influence of incorporated 3d-, 4d-and 4f-metal chlorides on electronic properties of single-walled carbon nanotubes. Applied Physics A. 2013;**111**(3):725-731. DOI: 10.1007/s00339-013-7639-x

[57] Falaleev NS, Kumskov AS, Zhigalina VG, Verbitskiy II, Vasiliev AL, Makarova AA, et al. Capsulate structure effect on SWNTs doping in $Rb_xAg_{1-x}I$@SWNT composites. CrystEngComm. 2017;**19**(22):3063-3070. DOI: 10.1039/C7CE00155J

[58] Eliseev AA, Yashina LV, Brzhezinskaya MM, Chernysheva MV, Kharlamova MV, Verbitsky NI, et al. Structure and electronic properties of AgX (X = Cl, Br, I)-intercalated single-walled carbon nanotubes. Carbon. 2010;**48**(10):2708-2721. DOI: 10.1016/j.carbon.2010.02.037

[59] Kharlamova MV, Kramberger C, Mittelberger A, Yanagi K, Pichler T, Eder D. Silver chloride encapsulation-induced modifications of Raman modes of metallicity-sorted semiconducting single-walled carbon nanotubes. Journal of Spectroscopy. 2018;**2018**:5987428. DOI: 10.1155/2018/5987428

[60] Kharlamova MV, Kramberger C, Domanov O, Mittelberger A, Yanagi K, Pichler T, et al. Fermi level engineering of metallicity-sorted metallic single-walled carbon nanotubes by encapsulation of few-atom-thick crystals of silver chloride. Journal of Materials Science. 2018;**53**(18):13018-13029. DOI: 10.1007/s10853-018-2575-y

[61] Kharlamova MV, Kramberger C, Domanov O, Mittelberger A, Saito T, Yanagi K, et al. Comparison of doping levels of single-walled carbon nanotubes synthesized by arc-discharge and chemical vapor deposition methods by encapsulated silver chloride. Physica Status Solidi B. 2018;**255**(12):1800178. DOI: 10.1002/pssb.201800178

[62] Kharlamova MV, Yashina LV, Lukashin AV. Charge transfer in single-walled carbon nanotubes filled with cadmium halogenides. Journal of Materials Science. 2013;**48**(24):8412-8419. DOI: 10.1007/s10853-013-7653-6

[63] Kharlamova MV, Kramberger C, Pichler T. Semiconducting response in single-walled carbon nanotubes filled with cadmium chloride. Physica Status Solidi B. 2016;**253**(12):2433-2439. DOI: 10.1002/pssb.201600300

[64] Zakalyukin RM, Mavrin BN, Dem'yanets LN, Kiselev NA. Synthesis and characterization of single-walled carbon nanotubes filled with the superionic material SnF_2. Carbon. 2008;**46**(12):1574-1578. DOI: 10.1016/j.carbon.2008.06.055

[65] Kharlamova MV. Rare-earth metal halogenide encapsulation-induced modifications in Raman spectra of single-walled carbon nanotubes. Applied Physics A. 2015;**118**(1):27-35. DOI: 10.1007/s00339-014-8880-7

[66] Kharlamova MV, Kramberger C, Mittelberger A. Raman spectroscopy study of the doping effect of the encapsulated terbium halogenides

on single-walled carbon nanotubes. Applied Physics A. 2017;**123**(4):239. DOI: 10.1007/s00339-017-0873-x

[67] Kharlamova MV, Volykhov AA, Yashina LV, Egorov AV, Lukashin AV. Experimental and theoretical studies on the electronic properties of praseodymium chloride-filled single-walled carbon nanotubes. Journal of Materials Science. 2015;**50**(16):5419-5430. DOI: 10.1007/s10853-015-9086-x

[68] Ayala P, Kitaura R, Nakanishi R, Shiozawa H, Ogawa D, Hoffmann P, et al. Templating rare-earth hybridization via ultrahigh vacuum annealing of $ErCl_3$ nanowires inside carbon nanotubes. Physical Review B. 2011;**83**(8):085407. DOI: 10.1103/PhysRevB.83.085407

[69] Kharlamova MV, Yashina LV, Lukashin AV. Comparison of modification of electronic properties of single-walled carbon nanotubes filled with metal halogenide, chalcogenide, and pure metal. Applied Physics A. 2013;**112**(2):297-304. DOI: 10.1007/s00339-013-7808-y

[70] Fedoseeva YV, Orekhov AS, Chekhova GN, Koroteev VO, Kanygin MA, Seovskiy BV, et al. Single-walled carbon nanotube reactor for redox transformation of mercury dichloride. ACS Nano. 2017;**11**(9):8643-8649. DOI: 10.1021/acsnano.7b04361

[71] Dresselhaus MS, Dresselhaus G, Jorio A, Souza AG, Saito R. Raman spectroscopy on isolated single wall carbon nanotubes. Carbon. 2002;**40**(12):2043-2061. DOI: 10.1016/S0008-6223(02)00066-0

[72] Araujo PT, Maciel IO, Pesce PBC, Pimenta MA, Doorn SK, Qian H, et al. Nature of the constant factor in the relation between radial breathing mode frequency and tube diameter for single-wall carbon nanotubes. Physical Review B. 2008;**77**(24):241403R. DOI: 10.1103/PhysRevB.77.241403

[73] Fouquet M, Telg H, Maultzsch J, Wu Y, Chandra B, Hone J, et al. Longitudinal optical phonons in metallic and semiconducting carbon nanotubes. Physical Review Letters. 2009;**102**(7):075501. DOI: 10.1103/PhysRevLett.102.075501

[74] Brown SDM, Corio P, Marucci A, Dresselhaus MS, Pimenta MA, Kneipp K. Anti-stokes Raman spectra of single-walled carbon nanotubes. Physical Review B. 2000;**61**(8):R5137-R5140. DOI: 10.1103/PhysRevB.61.R5137

[75] Jorio A, Souza AG, Dresselhaus G, Dresselhaus MS, Swan AK, Unlu MS, et al. G-band resonant Raman study of 62 isolated single-wall carbon nanotubes. Physical Review B. 2002;**65**(15):155412. DOI: 10.1103/PhysRevB.65.155412

[76] Briones A, Liu XJ, Kramberger C, Saito T, Pichler T. Nanochemical reactions by laser annealing of ferrocene filled single-walled carbon nanotubes. Physica Status Solidi B. 2011;**248**(11):2488-2491. DOI: 10.1002/pssb.201100114

[77] Kharlamova MV, Sauer M, Saito T, Krause S, Liu X, Yanagi K, et al. Inner tube growth properties and electronic structure of ferrocene-filled large diameter single-walled carbon nanotubes. Physica Status Solidi B. 2013;**250**(12):2575-2580. DOI: 10.1002/pssb.201300089

[78] Kharlamova MV, Kramberger C, Saito T, Shiozawa H, Pichler T. In situ Raman spectroscopy studies on time-dependent inner tube growth in ferrocene-filled large diameter single-walled carbon nanotubes. Physica Status Solidi B. 2014;**251**(12):2394-2400. DOI: 10.1002/pssb.201451166

[79] Liu XJ, Kuzmany H, Saito T, Pichler T. Temperature dependence of inner tube growth from ferrocene-filled single-walled carbon nanotubes. Physica

Status Solidi B. 2011;**248**(11):2492-2495. DOI: 10.1002/pssb.201100116

[80] Sauer M, Shiozawa H, Ayala P, Ruiz-Soria G, Kataura H, Yanagi K, et al. In situ filling of metallic single-walled carbon nanotubes with ferrocene molecules. Physica Status Solidi B. 2012;**249**(12):2408-2411. DOI: 10.1002/pssb.201200127

[81] Shiozawa H, Pichler T, Pfeiffer R, Kuzmany H, Kataura H. Ferrocene encapsulated in single-wall carbon nanotubes: A precursor to secondary tubes. Physica Status Solidi B. 2007;**244**(11):4102-4105. DOI: 10.1002/pssb.200776137

[82] Shiozawa H, Pichler T, Kramberger C, Gruneis A, Knupfer M, Buchner B, et al. Fine tuning the charge transfer in carbon nanotubes via the interconversion of encapsulated molecules. Physical Review B. 2008;**77**(15):153402. DOI: 10.1103/PhysRevB.77.153402

[83] Kocsis D, Kaptas D, Botos A, Pekker A, Kamaras K. Ferrocene encapsulation in carbon nanotubes: Various methods of filling and investigation. Physica Status Solidi B. 2011;**248**(11):2512-2515. DOI: 10.1002/pssb.201100160

[84] Liu XJ, Kuzmany H, Ayala P, Calvaresi M, Zerbetto F, Pichler T. Selective enhancement of photoluminescence in filled single-walled carbon nanotubes. Advanced Functional Materials. 2012;**22**(15): 3202-3208. DOI: 10.1002/adfm.201200224

[85] Plank W, Pfeiffer R, Schaman C, Kuzmany H, Calvaresi M, Zerbetto F, et al. Electronic structure of carbon nanotubes with ultrahigh curvature. ACS Nano. 2010;**4**(8):4515-4522. DOI: 10.1021/nn100615d

[86] Sauer M, Shiozawa H, Ayala P, Ruiz-Soria G, Liu XJ, Chernov A, et al. Internal charge transfer in metallicity sorted ferrocene filled carbon nanotube hybrids. Carbon. 2013;**59**:237-245. DOI: 10.1016/j.carbon.2013.03.014

[87] Kharlamova MV, Kramberger C, Saito T, Shiozawa H, Pichler T. Growth dynamics of inner tubes inside cobaltocene-filled single-walled carbon nanotubes. Applied Physics A. 2016;**122**(8):749. DOI: 10.1007/s00339-016-0282-6

[88] Kharlamova MV, Kramberger C, Sato Y, Saito T, Suenaga K, Pichler T, et al. Chiral vector and metal catalyst-dependent growth kinetics of single-wall carbon nanotubes. Carbon. 2018;**133**:283-292. DOI: 10.1016/j.carbon.2018.03.046

[89] Kharlamova MV, Sauer M, Saito T, Sato Y, Suenaga K, Pichler T, et al. Doping of single-walled carbon nanotubes controlled via chemical transformation of encapsulated nickelocene. Nanoscale. 2015;**7**(4): 1383-1391. DOI: 10.1039/C4NR05586A

[90] Kharlamova MV, Sauer M, Egorov A, Kramberger C, Saito T, Pichler T, et al. Temperature-dependent inner tube growth and electronic structure of nickelocene-filled single-walled carbon nanotubes. Physica Status Solidi B. 2015;**252**(11):2485-2490. DOI: 10.1002/pssb.201552206

[91] Kharlamova MV, Kramberger C, Saito T, Sato Y, Suenaga K, Pichler T, et al. Chirality-dependent growth of single-wall carbon nanotubes as revealed inside nano-test tubes. Nanoscale. 2017;**9**(23):7998-8006. DOI: 10.1039/C7NR01846K

[92] Kharlamova MV, Kramberger C, Sauer M, Yanagi K, Saito T, Pichler T. Inner tube growth and electronic properties of metallicity-sorted nickelocene-filled semiconducting single-walled carbon nanotubes. Applied Physics A. 2018;**124**(3):247. DOI: 10.1007/s00339-018-1679-1

[93] Shiozawa H, Pichler T, Kramberger C, Rummeli M, Batchelor D, Liu Z, et al. Screening the missing electron: Nanochemistry in action. Physical Review Letters. 2009;**102**(4):046804. DOI: 10.1103/PhysRevLett.102.046804

[94] Shiozawa H, Kramberger C, Rummeli M, Batchelor D, Kataura H, Pichler T, et al. Electronic properties of single-walled carbon nanotubes encapsulating a cerium organometallic compound. Physica Status Solidi B. 2009;**246**(11-12):2626-2630. DOI: 10.1002/pssb.200982344

[95] Shiozawa H, Giusca CE, Silva SRP, Kataura H, Pichler T. Capillary filling of single-walled carbon nanotubes with ferrocene in an organic solvent. Physica Status Solidi B. 2008;**245**(10):1983-1985. DOI: 10.1002/pssb.200879626

Chapter 4

Carbon Nanotube from Unconventional Precursor-Optimization of Synthesis Parameters

Karthikeyan Srinivasan,
Angulakshmi Sathyamoorthi Venkatasamy
and Mageswari Subramanian

Abstract

Carbon is a versatile element of distinctive properties and has been described as the key element of living substance. Carbon nanostructures have attracted lots of interest, due their prominent properties. Spray pyrolysis method is adopted for synthesis of carbon nanotubes (CNTs). Contrast to any petroleum product, there is no fear of its ultimate shortage as it is a renewable source and can be obtained easily by cultivating as much quantity as required. Synthesize well crystalline multiwalled carbon nanotubes (MWNTs) from unconventional precursor of methyl ester of *Helianthus annuus* oil by optimize the parameters such as reaction temperature, catalyst composition and feed rate of carbon precursor in order to obtain good yield with desirable morphology.

Keywords: carbon nanotube, spray pyrolysis, optimization

1. Introduction

Carbon nanotube (CNTs) can be considered to be a potential candidate of the forthcoming century due to its extraordinary properties [1–5]. In the year 1985, an important breakthrough in carbon research was realized by the work of Kroto et al. [6], which resulted in the discovery of a large family of all carbon molecules, called 'fullerenes'. Discovery of CNTs is attributed to Iijima [7] as the first scientist who was looking for new carbon structures, in the deposit formed on graphite cathode surfaces during the electric-arc evaporation (or discharge) that is commonly employed to produce fullerene soot. The most important methods of synthesis of CNTs are electric arc discharge, laser evaporation and chemical vapor deposition methods (CVD) [8, 9]. Arc-discharge method is the easiest and most common method of producing CNTs. Ando has carried out the arc-discharge evaporation of pure graphite rods in various kinds of ambient gases (He, Ar and CH_4) since CNTs were first discovered [10]. Laser technique is not economically advantageous because the process involves high purity graphite rods, high power lasers and low yield of CNTs. The CVD is another popular method for producing CNTs in which

IntechOpen

hydrocarbon vapor is thermally decomposed in the presence of a catalyst [11]. Several researchers describe the method for synthesizing CNTs in large scale from petroleum-based precursors such as benzene, xylene and hexane. In current, CNTs synthesized from unconventional precursors such as *camphor*, *eucalyptus* oil, *Pine* oil, *Citrus limonum* oil and *Brassica juncea* oil [12–16]. The main objective is to explore whether pyrolysis of natural carbon precursor of methyl ester of *Helianthus annuus* oil over Fe, Co and Mo catalysts supported on silica can be usefully employed for synthesis of good quality carbon nanotubes at low temperature conditions using spray pyrolysis method. Response surface methodology-based Box-Behnken design was employed in order to optimize the process parameters for synthesis of MWNTs.

2. Experimental methods

The spray pyrolysis method is similar to CVD method and the only difference with CVD is the vaporization and pyrolysis of carbon source occurs simultaneously in spray pyrolysis whereas these processes occur independently in two steps in CVD method. The Fe-Co-Mo catalysts supported on silica (0.5 g) was placed in the quartz boat and then inserted into the center of a quartz tube placed in the electrical heating furnace. The carrier gas nitrogen was flushed out before switch on the reaction furnace to remove air and create nitrogen atmosphere. The temperature was raised from room temperature up to the desired CNTs growing temperature. Subsequently, the carbon precursor methyl ester of *Helianthus annuus* oil was introduced into the quartz tube through spray nozzle at the rate of 20 mL/h. The deposition time lasted for 60 min at the chosen temperature. Nitrogen flow was maintained until the furnace was cooled to room temperature. The product collected was then weighed and stored in air tight container for further characterizations.

3. Results and discussion

The morphology of MWNTs synthesized at 550, 650 and 750°C using methyl ester of *Helianthus annuus* oil as a precursor is studied. The MWNTs grown at 550°C are mostly tangled with diameter in the range of 22–48 nm as shown in scanning electron microscope (SEM) image in **Figure 1a**. A nice growth of web like structure of MWNTs with diameter of 20–45 nm is observed at 650°C (**Figure 1b**). The SEM image of MWNTs synthesized at 750°C is shown in **Figure 1c**. The MWNTs with diameter of 75–100 nm is observed at 750°C (**Figure 1c**). The results are in good agreement with Li et al. [17].

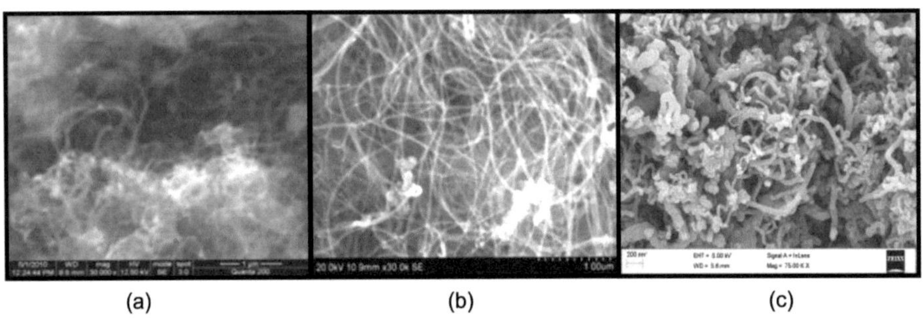

Figure 1.
(a–c) SEM images of MWNTS grown at 550, 650 and 750°C.

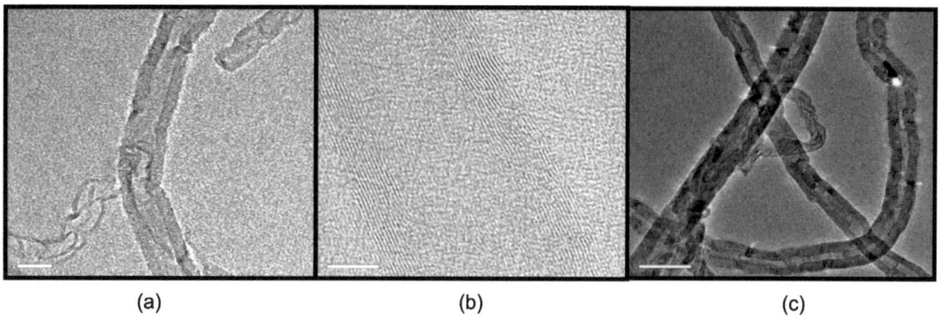

Figure 2.
(a–c) HRTEM images of MWNTS grown at 550, 650 and 750°C.

High resolution transmission electron microscope (HRTEM) recorded for the MWNTs synthesized at 550°C is shown in **Figure 2a**. HRTEM analysis shows a rope like tubular structure of MWNTs grown on the surface of chosen catalyst clusters. The HRTEM image (**Figure 2b**) clearly shows well-graphitized layers of MWNTs with inner and outer diameter in the range of 8–13 and 16–24 nm respectively, grown from catalytic decomposition of methyl ester of *Helianthus annuus* oil at 650° C. The selectivity towards MWNTs formation is observed in the best optimized condition of the present study at 650°C. The carbon deposit obtained at 750°C is found to be consisting of MWNTs of diameter 65–75 nm (**Figure 2c**).

The crystalline nature of the sample synthesized is studied using Raman spectrum and it is depicted in **Figure 3a**. In this study, the D and G peaks are observed at about 1370 and 1563 cm^{-1} for the samples synthesized at 550°C. The intensity ratio value of the G and D band, i.e., I_G/I_D value 0.71 provides important information relative to the purity and structural quality of the nanotubes that the MWNTs are made up of defective layers. D and G peaks are observed at about 1335 and 1545 cm^{-1} for the samples prepared at 650°C as shown in **Figure 3b**. The I_G/I_D ratio calculated from the peak area is 1.8. A further increase in temperature from 650 to 750° C results in a rapid drop in the I_G/I_D ratio to one (**Figure 3c**). Among the chosen experimental temperatures, the highest I_G/I_D ratio is observed for 650°C. This indicates the highest quality and purity of samples formed at 650°C. The absence of peaks below 300 cm^{-1} in Raman spectrum of the carbon deposits obtained in this study reveals the absence of single walled carbon nanotubes (SWNTs) [18]. The catalytic vapor deposition of methyl ester of *Helianthus annuus* oil over Fe catalyst supported on silica at 650°C results in a considerable quality of carbon nanotubes deposit (**Figure 4a**). However, as-grown CNTs are diameter in the range of 60–90 nm with poor structure and contains abundant amount of amorphous carbon. In **Figure 4b** MWNTs grown is in the range of diameter of 28–70 nm using Fe-Co catalyst.

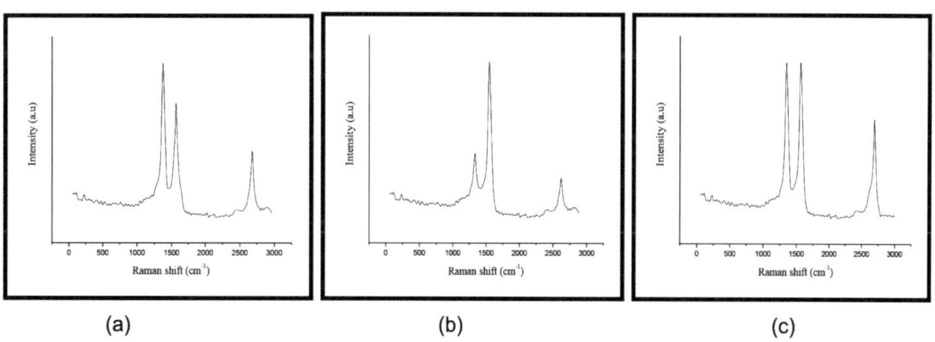

Figure 3.
(a–c) Raman spectrum of MWNTS grown at 550, 650 and 750°C.

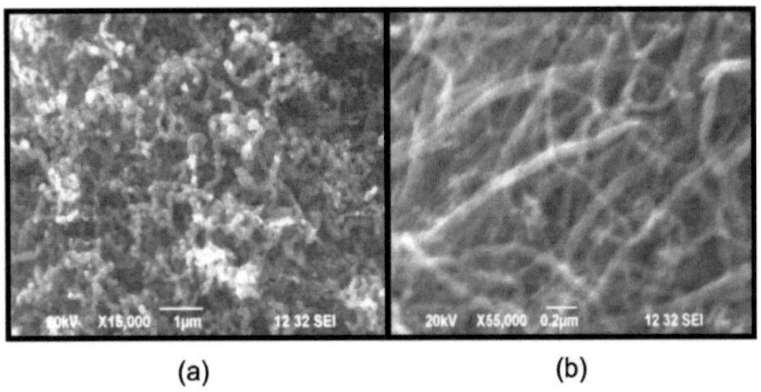

Figure 4.
(a and b) SEM images of MWNTS grown using Fe, Fe-Co catalyst supported on silica.

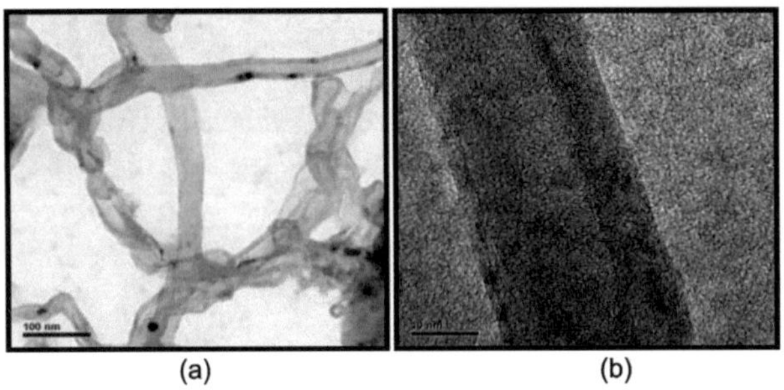

Figure 5.
(a and b) HRTEM images of MWNTS grown using Fe, Fe-Co catalyst supported on silica.

The HRTEM image (**Figure 5a**) of MWNTs synthesized over Fe catalyst supported on silica shows the poor crystallization of walls and a layer of amorphous carbon on outer surface of the tube. The outer diameter of MWNTs are around 40–60 nm respectively. The tubular structure of CNTs (**Figure 5b**) grown over Fe-Co catalysts supported on silica are MWNTs with thick in size and covered with layer of amorphous carbon. The inner and the outer diameter of the MWNTs are 15 and 20–30 nm, respectively.

The Raman spectrum recorded for the MWNTs obtained over Fe catalysts supported on silica where shown in **Figure 6a**. The spectrum shows G-band at 1554 cm^{-1} and a peak at 1339 cm^{-1} corresponds to D-band. The I_G/I_D value of 0.65

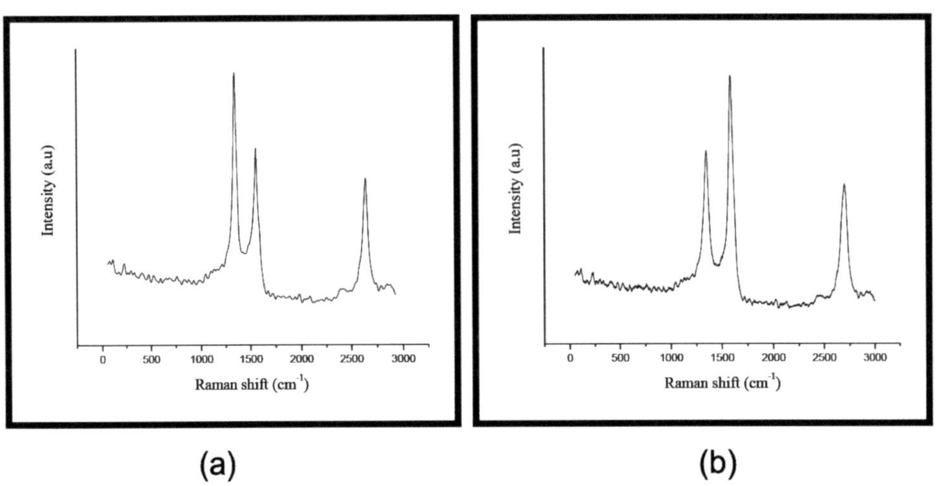

Figure 6.
(a and b) Raman spectrum of MWNTS grown using Fe, Fe-Co catalyst supported on silica.

evidences the imperfection in graphitization of MWNTs layers. Using Fe-Co catalyst spectrum show G-band at 1584 cm^{-1} and a peak at 1349 cm^{-1} corresponds to D-band. The value of I_G/I_D for the MWNTs grown on Fe-Co catalysts supported on silica is 1.1.

SEM image of MWNTs formed at the flow rate of 10 mL/h are of 50–80 nm in diameter with spaghetti like structure (**Figure 7a**). The increase in precursor flow rate to 30 mL per hour has resulted MWNTs are thick in size with diameter in the range of 40–70 nm as shown in **Figure 7b**. This may be due to higher rate of decomposition of precursor [19].

Methyl ester of *Helianthus annuus* oil at a flow rate of 10 mL/h produced MWNTs of poor quality in graphitization is evident from HRTEM image shown in **Figure 8a**. **Figure 8b** shows the defective structure of MWNTs synthesized using precursor flow rate of 30 mL/h. The tube diameter is relatively thicker in the range of 40–60 nm with amorphous carbon at the outer wall of the tube [20]. The G-band is observed at 1576 cm^{-1} and D-band is observed at 1351 cm^{-1} for the precursor flow rate of 10 mL is shown in **Figure 9a**. The I_G/I_D ratio calculated is 1.1. This indicated that the MWNTs formed are of moderate graphitization. In the flow rate

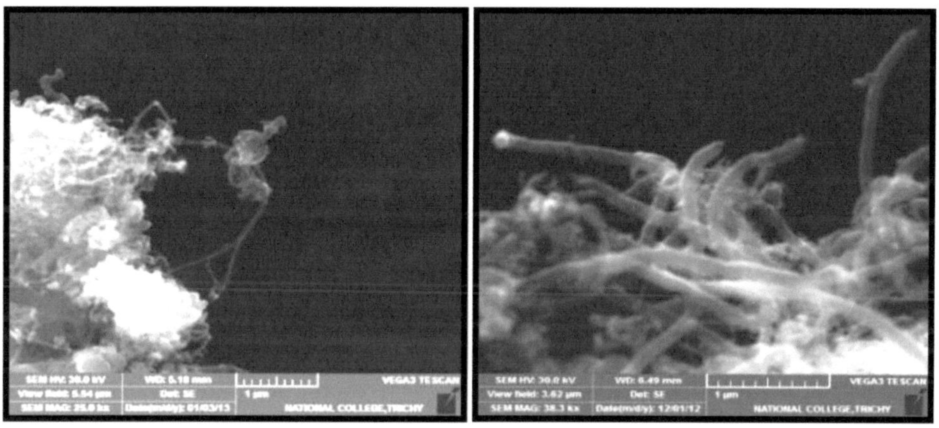

Figure 7.
(a and b) SEM images of MWNTS grown at the flow rate of 10 and 30 mL.

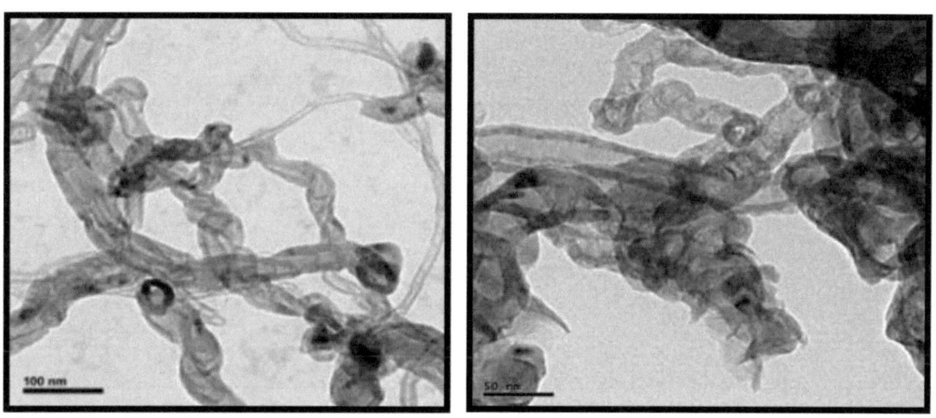

Figure 8.
(a and b) HRTEM images of MWNTS grown at the flow rate of 10 and 30 mL.

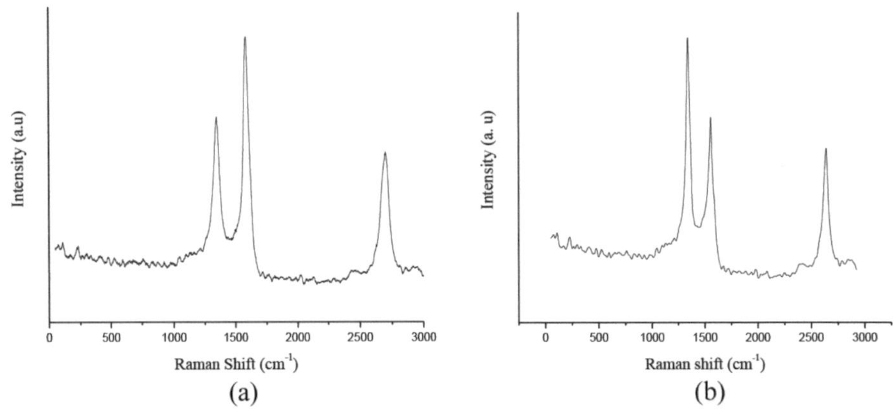

Figure 9.
(a and b) Raman spectrum of MWNTS grown at the flow rate of 10 and 30 mL.

of 30 mL/h appearance of D and G peaks at 1341 and 1554 cm^{-1}, respectively, with I_G/I_D value of 0.65 indicates the formation of MWNTs with defective graphitic layers (**Figure 9b**).

4. Box-Behnken design and data analysis for the yield percentage of MWNTs

Reaction temperature (°C), composition of catalyst (g) and feed rate of precursor (mL) were considered as independent process variables, and their individual and interactive effects on the yield percentage (as a response) of MWNTs were investigated using the Box-Behnken design approach. Syntheses of MWNTs experiments are conducted according to the design matrix and the corresponding results are tabulated in **Table 1**. The quadratic equation for predicting the optimum point was obtained according to the Box-Behnken design and input variables and then the empirical relationship between the response and the independent variables in the coded units for the yield percentage of MWNTs from the chosen precursor methyl

Run	Factor 1	Factor 2	Factor 3	Response 1
	A: reaction temperature (°C)	B: catalyst composition (g)	C: feed rate of precursor (mL)	Yield (%)
1	550	0.5	10	15
2	650	0.75	30	60
3	750	0.75	20	55
4	550	0.5	30	35
5	650	0.5	20	78
6	650	0.75	10	40
7	650	0.5	20	75
8	650	0.5	20	76
9	750	0.5	10	42
10	650	0.5	20	72

Run	Factor 1	Factor 2	Factor 3	Response 1
	A: reaction temperature (°C)	B: catalyst composition (g)	C: feed rate of precursor (mL)	Yield (%)
11	650	0.5	20	71
12	550	0.75	20	20
13	550	0.25	20	10
14	750	0.25	20	50
15	650	0.25	10	37
16	650	0.25	30	50
17	750	0.5	30	45

Table 1.
Box-Behnken design matrix and corresponding response for methyl ester of Helianthus annuus oil.

ester of *Helianthus annuus* oil were presented on the basis of the experimental results as follows:

$$Y = 74.4 + 14A + 3.5B + 7C - 1.25AB - 4.25AC + 1.75BC - 26.575A2 - 14.075B2 - 13.575C2 \quad (1)$$

5. Analysis of variance (ANOVA)

The statistical significance of the quadratic model was evaluated by the ANOVA. The ANOVA results for the quadratic equation summarized in **Table 2** for the yield percentage of MWNTs from the chosen precursor of methyl ester of *Helianthus annuus* oil.

Source	Sum of squares	df	Mean square	F-value	p-value Prob > F
Model	7195.064706	9	799.4516	24.68532	<0.0001
A—temperature	1568	1	1568	48.41641	0.0002
B—catalyst composition	98	1	98	3.026026	0.1255
C—feed rate of carbon precursor	392	1	392	12.1041	0.0103
AB	6.25	1	6.25	0.192986	0.6737
AC	72.25	1	72.25	2.230922	0.1789
BC	12.25	1	12.25	0.378253	0.5580
A^2	2973.602632	1	2973.603	91.81834	<0.0001
B^2	834.1289474	1	834.1289	25.75608	0.0014
C^2	775.9184211	1	775.9184	23.95866	0.0018
Residual	226.7	7	32.38571		
Lack of fit	193.5	3	64.5	7.771084	0.0382
Pure error	33.2	4	8.3		
Cor total	7421.764706	16			

Table 2.
ANOVA for RSM parameters fitted to a polynomial equation for methyl ester of Helianthus annuus oil.

ANOVA indicates that the actual relationship between the response and significant variables represented by the above quadratic equations are accurate. The significance of the coefficient term is determined by the values of F and p and the larger the F-value and smaller the value of p, the more significant is the co-efficient term. The p is lower than 0.05, suggesting the model to be statistically significant. For the present synthesis process, the ANOVA results indicated the Model F value was 24.68 for methyl ester of *Helianthus annuus* oil, suggesting only 0.01% chance of a "Model F value" so large could occur due to noise and most of the variation in the response could be explained by the regression equation and the model was significant.

6. Three dimensional response surface plots

To study the interaction between all three variables, three dimensional surfaces and two dimensional contours were plotted by keeping one variable constant at central level and the other two varying within the experimental ranges.

In **Figure 10**, the response surface and contour plots were developed for methyl ester of *Helianthus annuus* oil as a function of temperature and catalyst composition while feed rate of precursor was kept constant as 20 mL. In this experiment, **Figure 10** indicates that the response is sensitive to the reaction temperature. The yield percentage of MWNTs increases with increase in temperature and attains peak at optimum temperature (650°C) for methyl ester of *Helianthus annuus* oil. Low yield obtained at 550 and 750°C is possibly due to the fact that the catalyst could not be activated and high rate of pyrolysis followed by encapsulation of catalyst respectively. High yield obtained at 650°C in this study is attributed to almost equal rate of pyrolysis of precursor and CNTs growth. The highest yield obtained at high temperature (750°C) was probably due to condition may be attributed to thermal energy of precursor vapors which favors high rate of cracking on the catalyst [21].

The response surface and contour plots were developed for methyl ester of *Helianthus annuus* oil using the variables temperature and feed rate of carbon precursor by keeping catalyst composition (0.5 g) as constant. As can be understood from **Figure 11** increase of precursor feed rate from 10 to 20 mL increases the yield percentage of MWNTs. Further increase of flow rate to 30 mL leads to reduction in the yield of MWNTs.

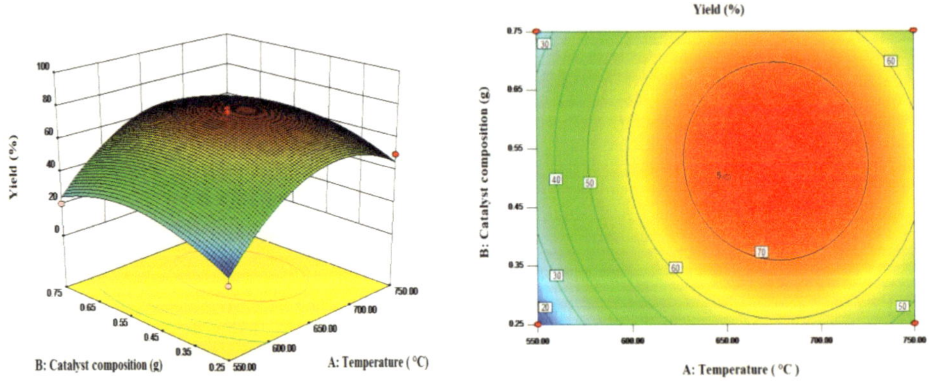

Figure 10.
Response surface and contour plots for the yield of MWNTs as the function of temperature and catalyst composition.

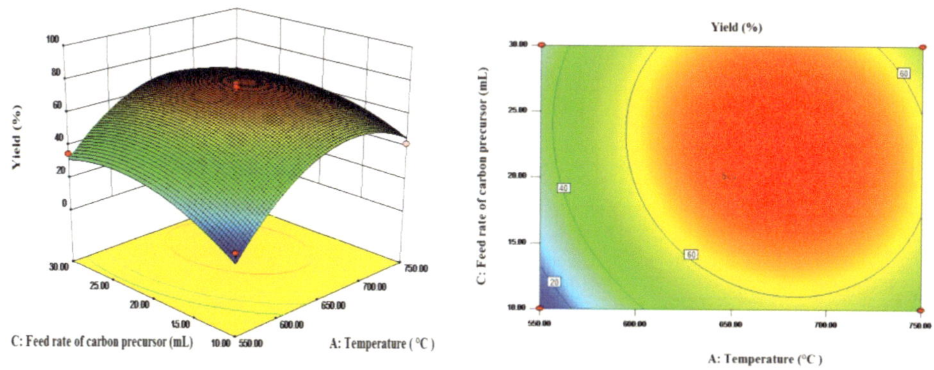

Figure 11.
Response surface and contour plots for the yield of MWNTs as the function of temperature and feed rate of precursor.

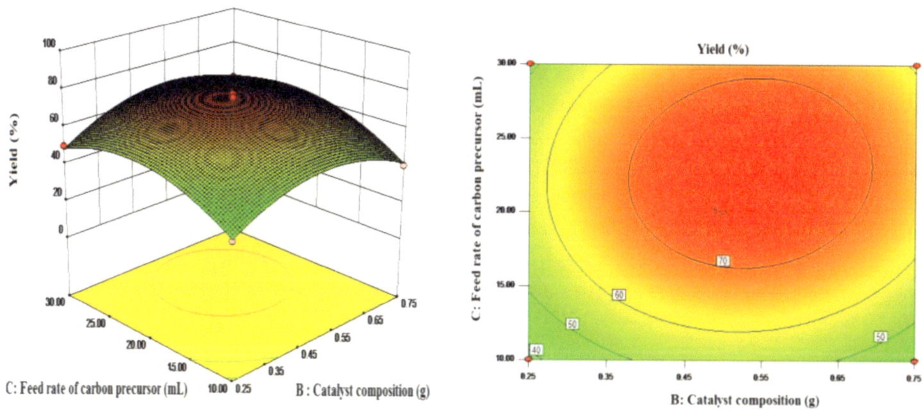

Figure 12.
Response surface and contour plots for the yield of MWNTs as the function of catalyst composition and feed rate of precursor.

The interactive effect of the feed rate of carbon precursor and catalyst composition on the percentage of yield of MWNTs at constant temperature of 650°C is illustrated in **Figure 12**. Piedigrosso et al. reported that amount of nanotubes formed over silica supported Co catalyst depends on content of Co [22]. A strong relationship between the catalyst and yield of MWNTs deposit was observed in this study. It is seen from **Figure 12** that the yield percentage of MWNTs increases with increasing catalyst composition attains maximum and starts decreases. Increase in yield at optimum condition may be due to synergistic advantages of high catalytic decomposition, effectiveness in growing CNTs and promotional character of Fe, Co and Mo respectively.

7. Optimization of process variables

The numerical optimization was used to optimize the yield percentage of MWNTs from methyl ester of *Helianthus annuus* oil, and optimum values are presented in **Table 3**. A desirability value of 1.0 was obtained after optimizing the process parameters. The experimental values of yield percentage of MWNTs grown from methyl ester of *Helianthus annuus* oil under the optimal conditions closely

Variables	Optimum values
Reaction temperature (°C)	674.29
Catalyst composition (g)	0.53
Feed rate of precursor (mL)	22.28
Yield percentage (predicted)	77.11
Yield percentage (actual)	78

Table 3.
Obtained optimum values of the process variables and responses.

agree with the predicted values (**Table 3**) obtained from the model and validate the findings of response surface optimization.

8. Conclusion

The present work reveals the natural precursor of methyl ester of *Helianthus annuus* oil as carbon precursor for synthesis of carbon nanotubes. Utilization of Fe, Co and Mo catalysts supported on silica for synthesis of well-graphitized carbon nanotubes with high yield at low temperature conditions using spray pyrolysis method is successfully reported. Optimization studies by applying Response Surface Methodology, a Box-Behnken design for optimizing the process parameters such as reaction temperature, catalyst composition and feed rate of precursor clearly indicated the efficiency of yield percentage of MWNTs.

Author details

Karthikeyan Srinivasan[1]*, Angulakshmi Sathyamoorthi Venkatasamy[2] and Mageswari Subramanian[3]

1 Department of Chemistry, Chikkanna Government Arts College, Tirupur, India

2 Department of S&H, Kathir College of Engineering, Coimbatore, India

3 Department of Chemistry, Vivekanandha College of Engineering for Women, Namakkal, India

*Address all correspondence to: environkarthi@gmail.com

IntechOpen

© 2019 The Author(s). Licensee IntechOpen. This chapter is distributed under the terms of the Creative Commons Attribution License (http://creativecommons.org/licenses/by/3.0), which permits unrestricted use, distribution, and reproduction in any medium, provided the original work is properly cited.

References

[1] Wang QH, Corrigan JD, Dai JY, Chang RPH, Krauss AR. Field emission from nanotube bundle emitters at low fields. Applied Physics Letters. 1997;**70**: 3308-3310

[2] Awasthi K, Rajeshkumar, Tiwari RS, Srivastava ON. Large scale synthesis of bundles of aligned carbon nanotubes using a natural precursor: Turpentine oil. Journal of Experimental Nanoscience. 2010;**5**(6):498-508

[3] Angulakshmi VS, Rajasekar K, Sathishkumar C, Karthikeyan S. Growth of vertically aligned carbon nanotubes on a silicon substrate by a spray pyrolysis method. New Carbon Materials. 2013;**28**(4):284-288

[4] Karthikeyan S, Mageswari S, Kalaiselvan S, Syed Shabudeen PS, Sivakumar N. Optimization of growth temperature of multi-walled carbon nanotubes fabricated by chemical vapour deposition and their application for arsenic removal. International Journal of Material Science Poland. 2014;**32**(4):709-718

[5] Venckatesh R, Kalaiselvan S, Balachandran K, Karthikeyan S. Botanical hydrocarbon sources based MWCNTs synthesized by spray pyrolysis method for DSSC applications. Silicon. 2016;**10**(2):211-217

[6] Kroto HW, Heath JR, O'Brien SC, Curl RF, Smalley RE. C60: Buckministerfullerene. Nature. 1985; **318**(6042):162-163

[7] Iijima S. Helical microtubules of graphitic carbon. Nature. 1991; **354**(6348):56-58

[8] Ebbesen TW, Ajayan PM. Large-scale synthesis of carbon nanotubes. Nature. 1992;**358**(6383):220-222

[9] Kong J, Cassell AM, Dai H. Chemical vapor deposition of methane for single-walled carbon nanotubes. Chemical Physics Letters. 1998;**292**(4–6):567-574

[10] Ando Y. The preparation of carbon nanotubes. Fullerene Science and Technology. 1994;**2**(2):173-180

[11] Kumar M, Ando Y. Chemical vapor deposition of carbon nanotubes: A review on growth mechanism and mass production. Journal of Nanoscience and Nanotechnology. 2010;**10**(6):3739-3758

[12] Kumar M, Ando Y. A simple method of producing aligned carbon nanotubes from an unconventional precursor – *camphor*. Chemical Physics Letters. 2003;**374**(5–6):521-526

[13] Ghosh P, Afre RA, Soga T, Jimbo T. A simple method of producing single walled carbon nanotubes from a natural precursor: *Eucalyptus* oil. Materials Letters. 2007;**61**(17):3768-3770

[14] Karthikeyan S, Mahalingam P. Studies of yield and nature of multi-walled carbon nanotubes synthesized by spray pyrolysis of pine oil at different temperatures. International Journal of Nanotechnology and Applications. 2010;**4**(3):189-197

[15] Mageswari S, Syed Shabudeen PS, Karthikeyan S. Entangled multi-walled carbon nanotubes from Citrus limonum oil. Rasayan Journal of Chemistry. 2014; **7**(3):281-286

[16] Kalaiselvan S, Karthik M, Vladimir R, Karthikeyan S. Growth of bamboo like carbon nanotubes from brassica Juncea as natural precursor. Journal of Environmental Nanotechnology. 2014; **3**(2):92-100

[17] Li WZ, Wen JG, Ren ZF. Effect of temperature on growth and structure of

carbon nanotubes by chemical vapor deposition. Applied Physics A. 2002; **74**(3):397-402

[18] Afre RA, Soga T, Jimbo T, Kumar M, Ando Y, Sharon M, et al. Carbon nanotubes by spray pyrolysis of turpentine oil at different temperatures and their studies. Microporous and Mesoporous Materials. 2006;**96**(1–3): 184-190

[19] Shaijumon MM, Ramaprabhu S. Studies of yield and nature of carbon nanostructures synthesized by pyrolysis of ferrocene and hydrogen adsorption studies of carbon nanotubes. International Journal of Hydrogen Energy. 2005;**30**(3):311-317

[20] Fonseca A, Hernadi K, Piedigrosso P, Colomer JF, Mukhopadhyay K, Doome R, et al. Synthesis of single and multi-wall carbon nanotubes over supported catalysts. Applied Physics A. 1998;**67**(1):11-22

[21] Kumar M, Ando Y. Controlling the diameter distribution of carbon nanotubes grown from *camphor* on a zeolite support. Carbon. 2005;**43**(3): 533-540

[22] Piedigrosso P, Konya Z, Colomer JF, Fonseca A, Tendeloo GV, Nagy JB. Production of differently shaped multi-wall carbon nanotubes using various cobalt supported catalysts. Physical Chemistry Chemical Physics. 2000;**2**(1): 163-170

Chapter 5

Vertically Aligned Carbon Nanotubes Production by PECVD

Oleg I. Il'in, Marina V. Il'ina, Nikolay N. Rudyk,
Alexandr A. Fedotov and Oleg A. Ageev

Abstract

This chapter presents the results of experimental studies of the PECVD technological mode parameters' influence on the formation of catalytic centers and carbon nanotubes' (CNTs') growth processes. This chapter also presents the ability to regulate the growth parameter for the controlled production of CNTs with the required geometric parameters, properties, and growth mechanisms. The results of experimental studies of the heating temperature and activation time effects on the catalytic center formation will be presented. This chapter also shows the effects of growth temperature, heating rate, and the activation time on the geometric and structural parameters of the carbon nanotubes. Experimental studies were carried out with the use of AFM, SEM, TEM, and EXAFS techniques. The results can be used in the development of technological processes for creating ultrafast energy-efficient electronic component base with carbon nanostructures, particularly nanoelectromechanical switches, flexo- and piezoelectric generators, gas sensors, and high-performance emitters.

Keywords: carbon nanotubes, PECVD, nanotechnology, technological modes, growth mechanisms

1. Introduction

Unique properties of carbon nanotubes (CNTs) [1–5] open wide possibilities for their application as functional elements of carbon micro- and nanoelectronics devices [6–13]. Thus, for integration of the CNT growing methods in the mass production technology, most of the interests are ordered arrays of single-walled CNT, and based on them, they are located at the place of use in accordance with the design of the developed device [14–18]. In this regard, there is necessary to provide the requirements for the structure, properties, and geometrical parameters of CNT [19].

For make requirements demanded to CNT, the plasma-enhanced chemical vapor deposition method (PECVD) is most promising [20, 21]. The initial stage of the CNTs' growing process is based on the thin layer formation of the transition metal, which are formed catalytic centers (CC) during subsequent destruction. On the surface of CC takes place dissociation of condensed carbon-containing gas molecules and transportation of free carbon atoms to the base of growing CNTs. The parameters of the catalytic centers (size, dispersion, chemical composition, etc.) [22, 23] determine the parameters of carbon nanotubes (diameter, height, chirality, electrical properties, growth kinetics, etc.) that underlie their instrument application [20, 24]. In practice, CNT arrays are most often grown on nanoscale CC obtained

IntechOpen

by coagulation of a nanoscale thickness metal film during annealing [25, 26]. For exclusion of the interaction between the CC and the substrate, it is necessary to form buffer layers based on metal or dielectric films.

In the PECVD method, there are a lot number of interrelated parameters (temperature, pressure, gas flows, gas types, types of catalyst film material, etc.) that influence the processes of CC formation and CNT growth [27, 28].

This chapter describes the results of experimental research of the PECVD technological modes that influence on the formation of CC and vertically aligned carbon nanotubes' (VA CNTs') growth for the controlled production of nanotubes with the required geometric parameters and properties.

2. Features of the PECVD process parameters' influence on the formation of catalytic centers and carbon nanotubes' growth

2.1 Experimental samples and equipment

Experimental research on the formation of catalytic centers and the growth of VA CNT was performed using the PECVD module of cluster nanotechnology complex NANOFAB NTK-100 (NT-MDT, Russia). A silicon wafer with a deposited chromium film and a nickel catalytic film with thicknesses of 20 and 10 nm, respectively, was used as the initial substrate. Cr and Ni films were deposited by magnetron sputtering on an AUTO 500 (BOC Edwards, UK). The catalytic centers formation was performed in an argon and ammonia atmosphere. The CNT growth was performed in ammonia and acetylene atmosphere. In all experimental researches, the pressure in the chamber was 4.5 Torr. The structural analysis of the VA CNT arrays was conducted by the transmission electron microscopy (TEM) using the Tecnai Osiris (FEI, Netherlands) and the Raman spectrometer Renishaw InVia Reflex (Renishaw plc, UK). Analysis of TEM images and Raman spectra showed that the experimental samples were multiwalled carbon nanotubes [29]. Surface investigations of the obtained VA CNT arrays were carried out using a scanning electron microscope (SEM) Nova Nanolab 600 (FEI Company, Netherlands). Investigation of structure and analysis of processes in catalyst films and sublayer was performed by extended X-ray absorption fine structure spectroscopy (EXAFS) with special source of synchrotron radiation (Kurchatov Institute, Russia). Analysis of geometric parameters of CC was conducted by atomic force microscopy using the Ntegra Probe Nanolaboratory (PNL) (NT-MDT, Russia). To process the experimental data, the ImageAnalysis application package was used.

2.2 The effect of heating temperature on the catalytic center formation

As a catalyst for growing CNT, most often used transition metals are Fe [30–34], Ni [35, 36], Co [37–39], and binary compounds based on them such as Fe-Mo [40], Co-Fe [41], Fe-Cu [42]. At the same time, an important factor to create devices with CNTs is the growth of nanotubes on the sublayer ohmic contact. This factor creates additional difficulties in technological process by increasing kinetic interactions at high growth temperatures, such as doping and formation of intermetallic compounds between the catalyst and the sublayer material.

Experimental research was performed on the samples with Ni/Cr/Si structure. The samples were heated in the temperature range 700–800 with 50°C interval in PECVD module in the atmosphere of inert gas (Q_{Ar} = 40 sccm) and ammonia (Q_{NH3} = 15 sccm). Argon is used to purge the chamber and remove residual air.

Ammonia creates a reducing atmosphere in the reactor that prevents oxidation of the catalytic film. During the annealing fragmentation and corrugating, Ni film occurs with catalytic centers formation.

On the initial samples, according to the SEM data, the surface of the structure was smooth. After heating the samples to a given temperature, the metal film was fragmented with CC formation (**Figure 1**).

Analysis and statistical processing of the obtained AFM images allowed to obtain the dependence of the catalytic centers height and diameter by substrate heating temperature (**Figure 2**).

From the obtained dependences (**Figure 2**), it can be seen that the heating temperature has a significant effect on the geometric dimensions of the CC. The difference between the temperature expansion coefficients of Si substrate (10.9×10^{-6}°C^{-1}) and Cr sublayer film (3.1×10^{-6}°C^{-1}) promotes the appearance of significant mechanical stresses during annealing in the film/substrate contact, which causes fragmentation and rupture of the metal film into separate islands. CC formed at a temperature of 700°C is characterized by a larger scatter of geometric parameters, which is associated with the initiation of the film breakdown process and insufficiently intense surface diffusion of Ni atoms. As the temperature rises to 750°C, diffusion exchange of atoms is activated through the contact plane between the film and the substrate. As the plasticity of the metal increases, the consolidation of small CC into larger ones with a decrease in the scatter of their height is observed.

When the temperature rises above 750°C, a simultaneous process of sublimation and surface diffusion is observed, which leads to a decrease in the diameter and height of the CC. At the same time, there is a decrease in the number of small CCs, which were observed at a temperature of 700°C.

To analyze the processes occurring in the films of the catalyst and the sublayer during heating, the sample with 750°C heated temperature was investigated by EXAFS. The obtained X-ray absorption spectra (**Figure 3**) were compared with spectra of the reference substrates.

The results of the research show that in the Ni/Cr/Si structure, Ni is mainly in the oxidized state (the volume of pure Ni is ~30%). The Cr absorption spectrum (**Figure 3b**) coincides with the reference spectrum of Cr_2O_3.

For the subsequent growth of CNTs, it is necessary to form QC from pure Ni; therefore, it is necessary to carry out the process of "activation" (i.e., reduction) of Ni from NiO before synthesizing CNTs.

This research was supported by the Russian Science Foundation under grant no. 18-79-00176.

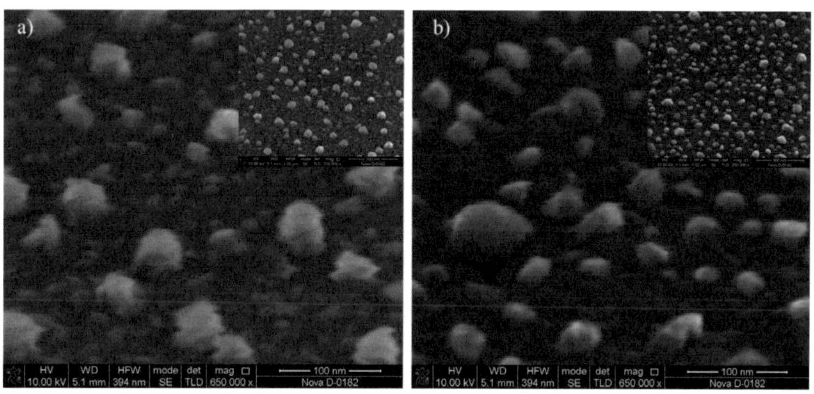

Figure 1.
SEM images of CC formed at different heating temperatures: (a) 700°C, (b) 800°C.

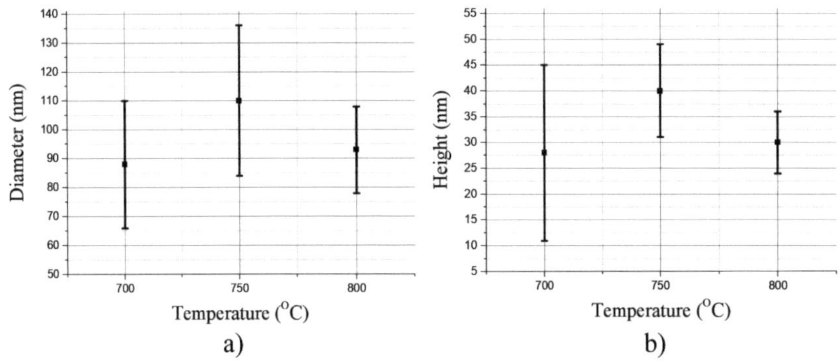

Figure 2.
Dependencies of diameter (a) and height (b) from the substrate heating temperature.

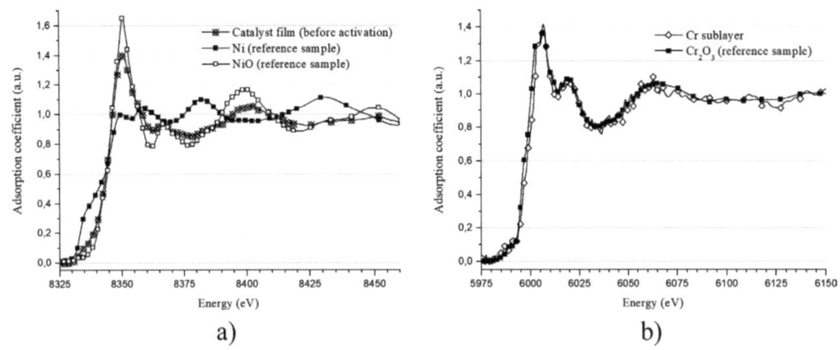

Figure 3.
EXAFS spectra of Ni/Cr/Si structures after baking at the heating stage, for the absorption energy range of Ni (a) and Cr (b).

2.3 The effect of activation time on the catalytic center formation

The ammonia gas line in the PECVD module allows for the "activation" process. "Activation" is the exposure of samples in an NH_3 atmosphere (210 sccm) obtained at the heating stage. The use of a DC high-voltage source allows to initiate the plasma at the "activation" stage. To analyze the effect of "activation" stage on the catalytic center parameters, a series of experiments were carried out to determine the dependence of the effect of exposure time in ammonia with and without plasma initiation.

In the beginning, experimental samples were created at the heating stage (heating to 750°C for 20 min, in an atmosphere of Ar (40 sccm) and NH_3 (15 sccm)). Upon completion of the heating stage, the "activation" stage began. Only NH_3 (210 sccm) was fed into the chamber, while the heating temperature did not change. The exposure time of the samples was 1, 3, and 5 min. Sample nos. 1–3 were exposed without plasma initiation, nos. 4–6 with initiation ammonia plasma.

Research of CC parameters was provided by AFM and SEM. The results of numerical processing AFM images are present on **Figure 4**.

Based on the data obtained, it can be concluded that the "activation" process has a significant impact on the evolution of catalytic centers. Moreover, the greatest contribution is made by the impact time and not by the presence of plasma in the process.

Comparison of dependencies of **Figures 2** and **4** shows that exposure in ammonium atmosphere leads to additional etching of Ni CC and reducing their diameter and height. At the activation stage are formed more ordered and uniform CC arrays in comparison with arrays obtained at the heating stage. It is shown that an increase in the processing time of a CC at the "activation" stage allows not only to reduce the CC sizes from 122 ± 19 to 65 ± 9 nm (in the 1–5 min exposure time range), but also to increase their homogeneity.

Investigation of the activation process effect on the reduction of NiO to Ni was not carried out due to possible oxidation of CC during the chamber vacuum is released.

This study was supported by the Russian Science Foundation under grant no. 18-79-00176.

2.4 The effect of growth temperature on carbon nanotubes geometric dimensions

The growth stage is a technological step which begun after the "activation" stage. Into the chamber of the PECVD module a mixture of NH3 (210 sccm) and C2H2 (70 sccm) gases is fed. Plasma is initiated for the vertical alignment of CNT relative to the substrate in the growth process.

SEM images of the grown CNTs are present in **Figure 5**. Due to the features of the CNTs growing process in PECVD module, set temperature in the heating step also serves "activation" and growth temperatures.

The SEM images show the integral temperature effect, which is associated both with the formation of CC at the stages of heating and activation, and with the CNT growth processes at set temperature. For samples obtained at a temperature of 650°C, characterized by the height of the CNT array 65 ± 5 nm, with a diameter of 25 ± 3 nm and the presence of disoriented CNTs. CNTs are grown by the "tip" mechanism. Increasing the temperature to 700°C leads to the elimination of the formation of disoriented CNTs and CNT geometric dimensions are in the same range (diameter 25 ± 4 nm, height 66 ± 5 nm). This effect may be associated with better hydrogen desorption in the decomposition of acetylene on CC surface. This leads to the formation of a less "defective" carbon layer on the CC surface, which does not give branching and disorientation of CNT. A raise temperature to 750°C leads to an increase in the mobility of the CC and association smaller CC into larger. The parameters of CNT were: diameter 44 ± 3 nm and height 80 ± 9 nm. Also observed growth of individual CNTs with a diameter of 70 ± 3 nm and height 350 ± 10 nm. At 800°C diameter and height of CNT arrays were 51 ± 6 nm and 100 ± 12, respectively. There is also observed of individual CNT with a height of 600 ± 24 nm and diameter of 52 ± 6 nm. The observed with increasing temperature, raising CNT diameter indicating continued association small CC into larger in heating and activation stages. The almost complete absence of CNTs with a diameter of less than 25 nm indicates

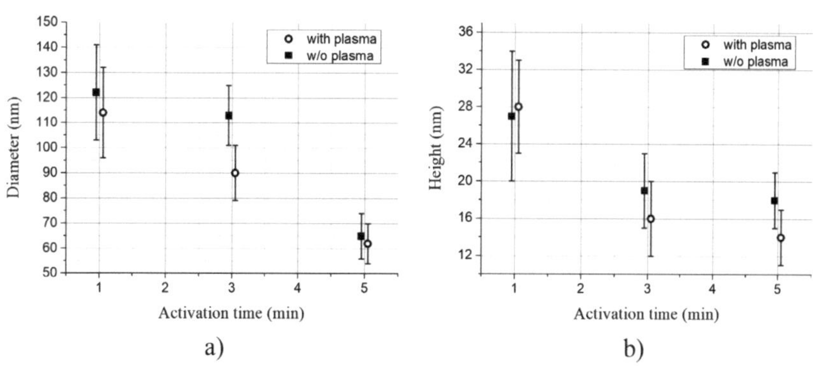

Figure 4.
Dependencies of diameter (a) and height (b) on time of activate.

the ongoing process of absorption of small CC by larger ones or possible sublimation of small CC at a given temperature. However, with increasing temperature the process of acetylene desorption from the sample surface is accelerated. Thus, not having time to react with the CC carbon-containing gas is pumped out by vacuum system.

Structural analysis of the VA CNT arrays grown at temperatures of 700, 750, and 800°C was carried out using the Raman spectroscopy (**Figure 6**).

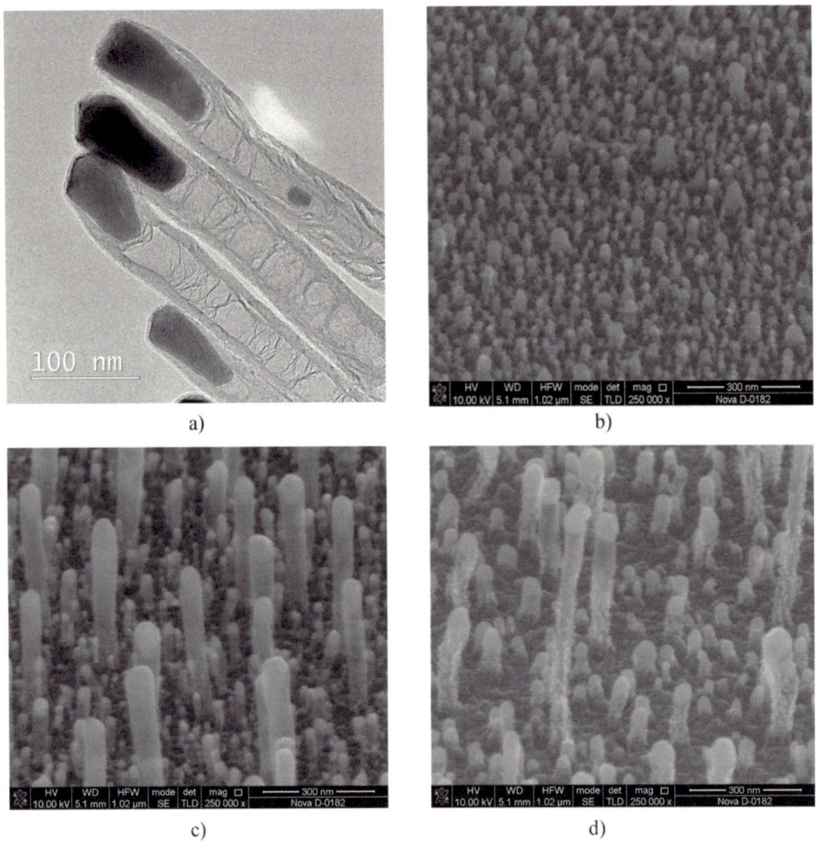

Figure 5.
Images of VA CNT arrays obtained at different growth temperatures: (a) TEM image, 750°C; and SEM images: (b) 700°C; (c) 750°C; (d) 800°C.

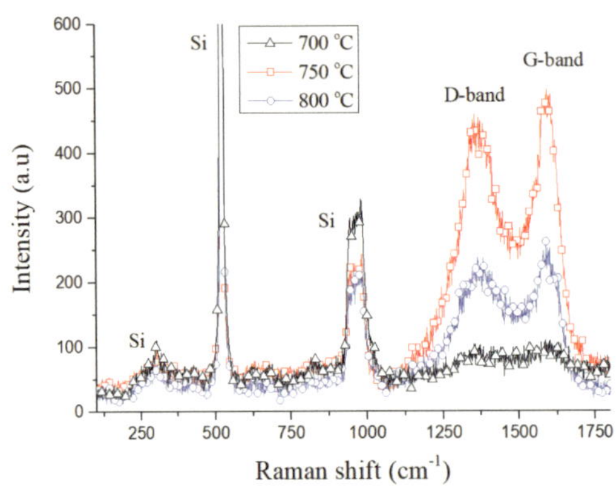

Figure 6.
Raman spectra of the VA CNT arrays grown at different temperatures.

The absence of the RBM mode in the range of 0–200 cm^{-1} and the presence of D- and G-modes on the spectrum shows that the grown CNTs are multiwalled. The ratio of the peaks I_D/I_G shows that the defectiveness of the tubes is 0.91, 0.88 and 0.90 for growth temperatures of 700, 750 and 800°C, respectively. High "defects" of CNT is related with the features of samples structure and technique for obtaining the spectra at which the laser beam falls along the normal to the substrate surface. The presence of Ni CC in the top of the CNTs can lead to an increase in the amplitude of the D-mode and, as a result, cause symmetry breaking of the graphite layer with sp2 hybridization of carbon atoms, which is recorded in the spectrum. However, a comparative analysis found that CNT obtained at a temperature of 750°C characterized by the lowest "defectiveness". The ratio of the amplitude of the main mode of Si to the G-mode of a CNT qualitatively shows that the concentration of CNTs grown at a temperature of 750°C is 3.1% and 5.2% higher than that of samples grown at 800 and 700°C, respectively.

CNTs grown at a temperature of 750°C were investigated using EXAFS. X-ray absorption spectra are shown in **Figure 7**.

The EXAFS results show that Ni on CNT samples have almost completely recovered, with a metal percentage of 90–95%. A decrease in the spectrum intensity of the chromium absorption energy range of a sample with CNT (**Figure 7b**) compared to a sample with CC (**Figure 3b**) suggests the presence of a phase transformation of Cr_2O_3 to CrO, or the formation of a solid solution of $Cr_2O_3 + Cr_3C_2$. The reduction of the oxidized chromium sublayer to metallic Cr does not occur at the all stages of PECVD process. A comparative analysis of the diameter of CC formed at the end of the "activation" stage and the diameter of grown CNTs allows to conclude that CC with a diameter of less than 70–90 nm produce growth of single CNTs.

This study was supported by the Russian Science Foundation under grant no. 18-79-00176.

2.5 Effect of heating rate on the carbon nanotube parameters

To investigate the effect of heating rate on the geometric dimensions of CNTs, a series of experiments on heating samples to temperatures of 650, 750, and 800°C in 20 and 45 min were carried out (**Figurc 8**).

Based on the data obtained from SEM images (**Figure 9**), it can be concluded that at low heating temperatures (650°C), the heating rate does not have a significant impact on the parameters of the grown CNTs.

In sample nos. 1 and 2, the CNTs diameter was about 30 nm and height about 45 nm. When the temperature rises to 750°C in sample no. 3, compared with nos. 1 and 2, there is an increase of CNTs height (300 ± 35 nm) and diameter (54 ± 8 nm). An increase in heating time to 45 min (sample no. 4) leads to a decrease in the dispersion of the CCs diameter at the heating stage. The CNT arrays on sample no. 4 have a greater homogeneity of nanotubes (as compared with samples No 3), and diameter was 45 ± 4 nm and height 280 ± 30 nm. These geometrical dimensions indicate the absorption of small CCs by large CCs at the heating and activation stages, with the subsequent sublimation of CCs. Heating the sample no. 5 to 800°C in 20 min allowed to form CNT arrays with diameter of 60 ± 7 nm and height of 120 ± 10 nm. Separate CNTs with heights up to 600 nm were observed. A decrease in the sample no. 6 heating rate (800°C for 45 min) leads to the formation of CNTs with a diameter of 65 ± 6 nm and a height of 85 ± 10 nm. It can be assumed that at a temperature of 800°C, an increase in CC formation time leads to raising in the number of silicon atoms "jumps" and their exit through the Cr_xSi_y/Cr layer to nickel CCs. In this case, Ni interacts well with Si with formation of nickel silicides, which causes a decrease in the catalytic activity of Ni. An increase in the heating time to

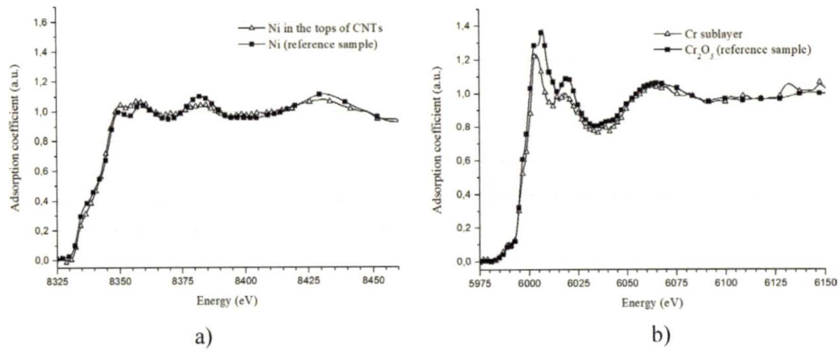

Figure 7.
EXAFS spectra of Ni/Cr/Si structure after growth process for the absorption energy range of Ni (a) and Cr (b).

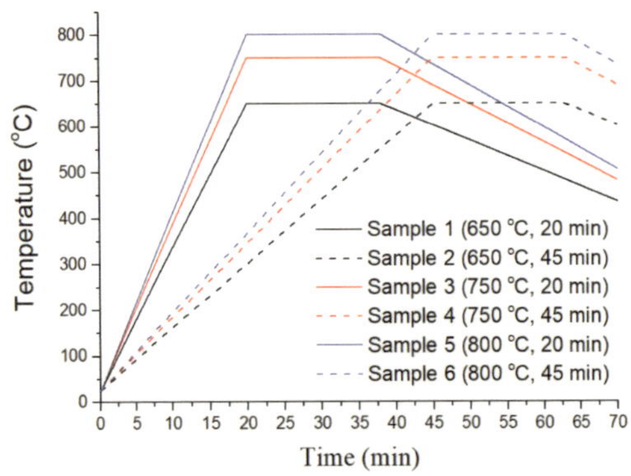

Figure 8.
Schematic representation of the CNT growing process on samples with different heating rates.

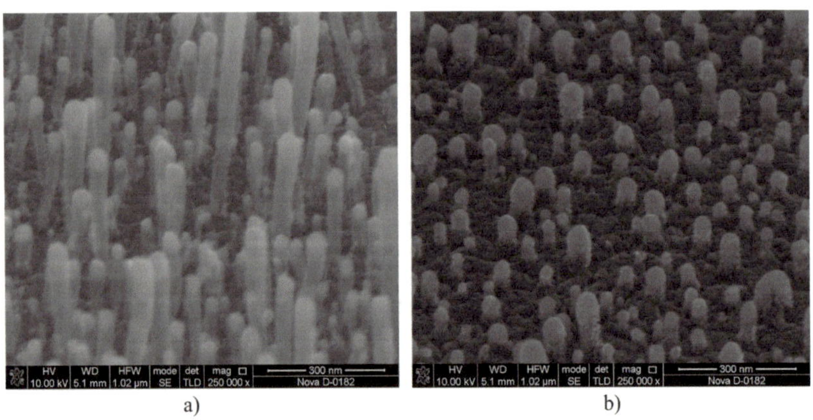

Figure 9.
SEM images of substrates with CNT heated for 45 min to temperatures: (a) 750°C, (b) 800°C.

45 min becomes sufficient to saturate nickel CCs with silicon, and a temperature of 800°C leads to a better dissolution of carbon in the poisoned CC. As a result, CC is oversaturated by carbon, with subsequent carbidization of the catalyst and stopping the growth process.

It should be noted that heating above 650°C does not have a significant effect on the diameter of the grown CNTs. However, control of the heating rate and growth temperature allows stabilizing the height of the CNT at 280 ± 30 nm and reducing the variation of nanotube diameters.

2.6 The effect of activation time on the carbon nanotube parameters

To investigate the effect of the activation process on the CNTs growth without plasma, the values of diameter, height, and density of CNT for all samples were determined in [27] (**Table 1**).

An increase in the diameter and a decrease density of CNTs at intervals of 1–3 and 6–10 min is associated with the predominance of the absorption processes of small centers with large centers. But at the time intervals of 3–6 and 10–15 min, the etching processes of CC in the ammonia atmosphere prevail. At the same time, the intervals of 3–6 and 10–15 min are dominated by the processes of CC etching in the ammonia atmosphere. All obtained CNTs are grown by the "top" mechanism. It should be noted that when the activation time is more than 1 min, the CNTs in the growth process begin to unite into bundles, which is not observed when CC is activated in the plasma. With an increase in the activation time, a decrease in the deviation of the growing tubes from the normal to the substrate is observed.

To assess the effect of the growth process on the parameters of CNT, an experiment without the "activation" stage was conducted. SEM images of the obtained CNTs are shown in **Figure 10**.

From **Figure 10**, it can be seen that the CNTs obtained without "activation" stage are grown by the "root" mechanism, and the open CNTs peaks begin to branch with graphite flakes formation. Thus, control of the activation process allows to control CNT growth mechanism.

2.7 The effect of growth time on the carbon nanotube parameters

To research the growth time effect on the geometrical dimensions of CNTs, samples with CNTs, whose growth time was 5, 10, 15, and 30 min, were investigated by [28].

As a result of processing the obtained SEM images (**Figure 11**), the dependences of the diameter and height of the CNT on the growth time (**Figure 12**) are shown.

From the obtained dependences (**Figure 12**), it is seen that with an increase in the growth time to 15 min, the diameter and height of the CNT vary linearly with time. In the interval of growth from 15 to 30 min CNTs diameter does not change significantly, and the height dependence has a tendency to saturate. Thus, it can be assumed that in the initial stages simultaneously with the increase in length, the formation of new walls of CNTs occurs, which leads to an increase in the external diameter of the nanotubes. When in these technological modes the critical size of the CNT is reached, the disorientation of the CNTs begins (**Figure 11c**).

The investigation of CNTs by the Raman spectroscopy (**Figure 13**) showed that the spectra of CNTs grown within 5 min are almost the same as the spectra of CNTs grown within 10 min. CNTs grown for 15 min show a noticeable

Activation time, min	CNTs' diameter, nm	CNTs' height, nm	CNTs' density, mkm^{-1}
1	26 ± 4	77 ± 8	410 ± 38
3	35 ± 8	250 ± 20	250 ± 25
6	24 ± 4	200 ± 9	171 ± 18
10	51 ± 11	260 ± 22	136 ± 16
15	50 ± 5	310 ± 25	228 ± 23

Table 1.
Parameters of CNT grown at different activation time.

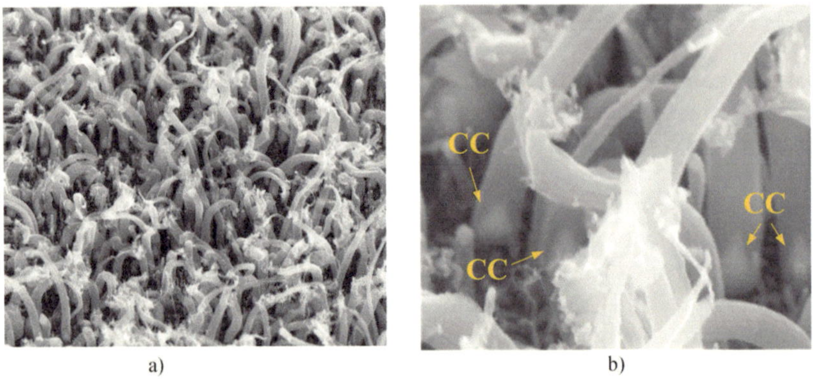

Figure 10.
CNT arrays grown without activation at different magnification: (a) 40000×, (b) 200000×.

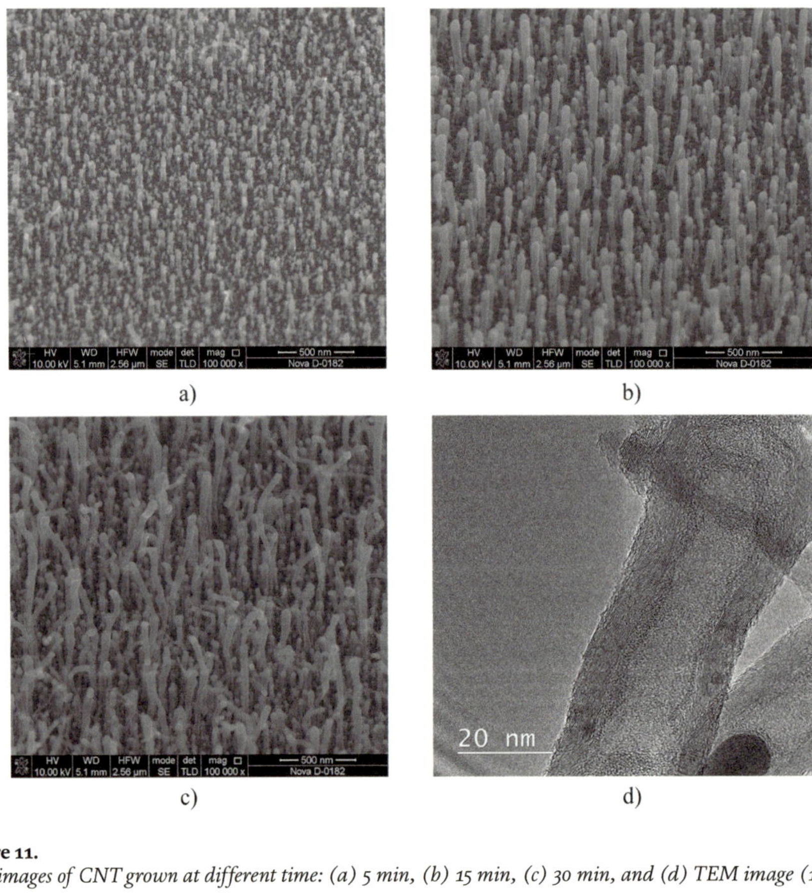

Figure 11.
SEM images of CNT grown at different time: (a) 5 min, (b) 15 min, (c) 30 min, and (d) TEM image (15 min).

increase in the intensity of the spectrum. That is, an increase in the height of the CNT leads to an increase in Raman intensity. The ratio of I_D/I_G peaks was found to be 0.95, 0.86, and 0.91 for growth times of 5, 10, and 15 min, respectively. The high "defectiveness" of CNTs grown within 5 min is explained by the small height and diameter of CNTs, where most of the volume of CNTs is the catalytic center, which makes a large contribution to the intensity of the D-mode.

With a raised growth time of 10 min, an increase in the aspect ratio and a decrease in "defects" of CNTs are obtained. However, a further increase in growth time causes the process of undercutting of the base of a CNT by plasma

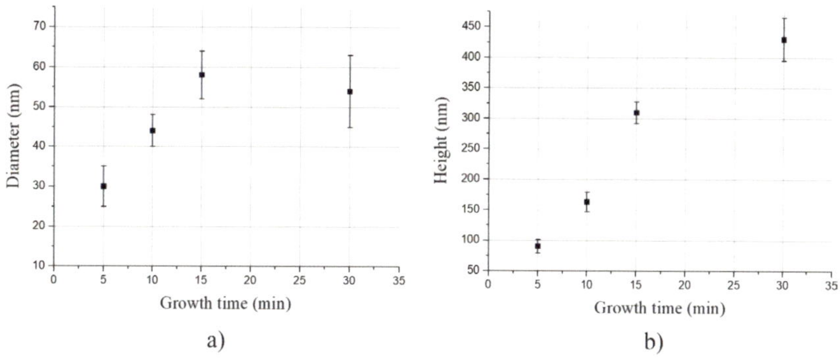

Figure 12.
Dependence of diameter (a) and height (b) of CNT on growth time.

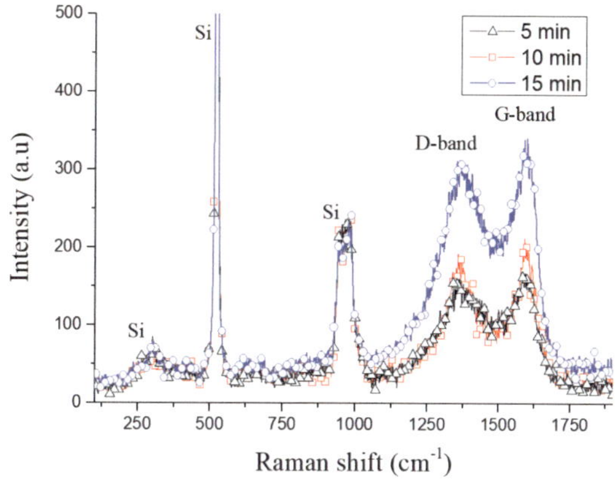

Figure 13.
Raman spectra of the VA CNT arrays with 5, 10, and 15 min grown time.

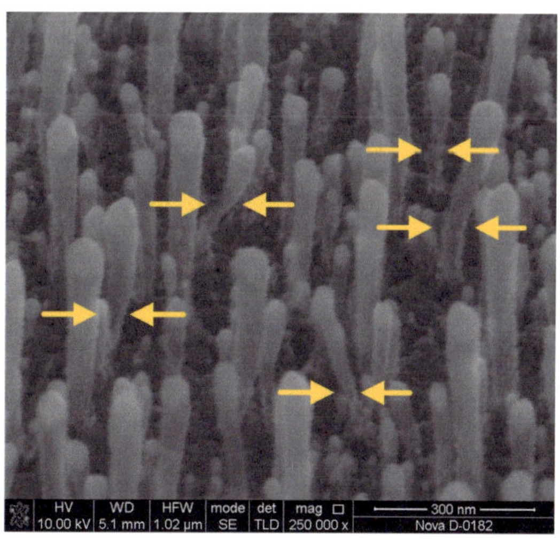

Figure 14.
CNTs grown for 15 min.

(**Figure 14**). The processing time in plasma increases with increasing time of growth itself, which leads to disruption of carbon bonds and the growth of the D-mode in Raman spectra.

Thus, the growth time has a significant impact on both the height of the CNTs and their "defectiveness." Therefore, to obtain CNTs of greater height with less "defectiveness," it is necessary to select modes with lower plasma power or growth time, as well as increase the pressure of carbon-containing gas to accelerate the transport of acetylene to the CCs.

This study was supported by the Russian Science Foundation under grant no. 18-79-00176.

3. Conclusion

As a result of the performed research, the PECVD modes of production CC with controlled parameters were determined. It is shown that heating at a temperature of 700–800°C, with simultaneous inlet of argon and ammonia into the PECVD chamber, makes it possible to obtain CCs with a diameter of 110 ± 11 nm and a height of 40 ± 7 nm. It was established that ammonia at the "activation" stage allows reducing the diameter (up to 90 ± 18 nm) and height (up to 18 ± 3 nm), as well as increasing the homogeneity of the array of the obtained catalytic centers. It was confirmed that on the heating stage, Ni CCs are in an oxidized state with an Ni content of ~30%, but Ni is reduced to almost pure metal (95%) at the "activation" stage.

It is established that the growth temperature of 750°C shows the smallest "defectiveness" of CNTs with a diameter of 44 ± 3 nm and a height of 80 ± 9 nm. It is shown that at low temperatures (650°C), the heating rate does not have a significant impact on the parameters of the grown CNTs, herewith diameter and height of the CNT was about 30 and 45 nm, respectively. Heating above 650°C does not affect the diameter of the obtained CNT; however, control of the heating rate and growth temperature allows stabilizing the height of the CNT at 280 ± 30 nm with a diameter of 45 ± 4 nm.

The PECVD modes of growth CNTs with controlled "tip" and "root" mechanisms are determined. It was established that the control of technological parameters at the "activation" stage allows to control the diameter (24 ± 4 – 51 ± 11 nm) and density (136 ± 16 – 410 ± 38 μm^{-2}) of CNTs.

The obtained results can be used in the development of technological processes for the creation of ultrafast energy-efficient electronic components based on carbon nanostructures, particularly nanoelectromechanical switches [43, 44], flexo- and piezogenerators [45, 46], gas sensors [47, 48], and highly efficient autoelectronic emitters [49, 50].

Acknowledgements

Chapter paragraphs 2.2–2.4 and 2.7 were supported by the Russian Science Foundation under grant no. 18-79-00176.

Chapter paragraphs 2.5 and 2.6 were financially supported by the Russian Foundation for Basic Research (project no. 16-29-14023 ofi_m) and Internal grant of the Southern Federal University (project no. VnGr-07/2017-26).

Conflict of interest

The authors declare no conflict of interest.

Thanks

The authors are grateful to Zubavichus Ya.V. (NRC "Kurchatov Institute") for research and interpretation of the obtained EXAFS spectra and to Levshov D.I. (Southern Federal University) for assistance in research by Raman spectroscopy.

Author details

Oleg I. Il'in[1*], Marina V. Il'ina[1], Nikolay N. Rudyk[1], Alexandr A. Fedotov[1] and Oleg A. Ageev[2]

1 Institute of Nanotechnologies, Electronics and Electronic Equipment Engineering, Southern Federal University, Taganrog, Russian Federation

2 Research and Education Center "Nanotechnologies", Southern Federal University, Taganrog, Russian Federation

*Address all correspondence to: oiilin@sfedu.ru

IntechOpen

© 2019 The Author(s). Licensee IntechOpen. This chapter is distributed under the terms of the Creative Commons Attribution License (http://creativecommons.org/licenses/by/3.0), which permits unrestricted use, distribution, and reproduction in any medium, provided the original work is properly cited.

References

[1] Lim YD, Grapov D, Hu L, et al. Enhanced field emission properties of carbon nanotube bundles confined in SiO_2 pits. Nanotechnology. 2018;**29**(7):075205. DOI: 10.1088/1361-6528/aaa1bb

[2] Wang J, Jin X, Liu Z, et al. Growing highly pure semiconducting carbon nanotubes by electrotwisting the helicity. Nature Catalysis. 2018;**1**(5): 326-331. DOI: 10.1038/s41929-018-0057-x

[3] Lutz C, Ma Z, Thelen R, et al. Analysis of carbon nanotube arrays for their potential use as adhesives under harsh conditions as in space technology. Tribology Letters. 2019;**67**(1):10. DOI: 10.1007/s11249-018-1121-z

[4] Syurik J, Ageev OA, Cherednichenko DI, Konoplev BG, Alexeev A. Non-linear conductivity dependence on temperature in graphene-based polymer nanocomposite. Carbon. 2013;**63**:317-323. DOI: 10.1016/j.carbon.2013.06.084

[5] Il'ina M, Il'in O, Blinov Y, Konshin A, Konoplev B, Ageev O. Piezoelectric response of multi-walled carbon nanotubes. Materials (Basel). 2018;**11**(4):638. DOI: 10.3390/ma11040638

[6] Bokobza L. Multiwall carbon nanotube elastomeric composites: A review. Polymer (Guildf). 2007;**48**(17):4907-4920. DOI: 10.1016/j.polymer.2007.06.046

[7] De Volder MFL, Tawfick SH, Baughman RH, Hart AJ. Carbon nanotubes: Present and future commercial applications. Science (80-). 2013;**339**(6119):535-539. DOI: 10.1126/science.1222453

[8] Franklin AD, Luisier M, Han S-J, et al. Sub-10 nm carbon nanotube transistor. Nano Letters. 2012;**12**(2): 758-762. DOI: 10.1021/nl203701g

[9] Lim YD, Hu L, Avramchuck AV, et al. Temperature-dependent selective growth of carbon nanotubes in Si/SiO_2 structures for field emitter array applications. Materials Research Bulletin. 2017;**95**:129-137. DOI: 10.1016/j.materresbull.2017.07.022

[10] Peng H, Li Q, Chen T, editors. Industrial Applications of Carbon Nanotubes. A volume in: Micro and Nano Technologies. Amsterdam: Elsevier; 2017. DOI: 10.1016/C2015-0-00493-2. Available from: https://www.sciencedirect.com/book/9780323414814/industrial-applications-of-carbon-nanotubes

[11] Alekseev A, Chen D, Tkalya EE, et al. Local organization of graphene network inside graphene/polymer composites. Advanced Functional Materials. 2012;**22**(6):1311-1318. DOI: 10.1002/adfm.201101796

[12] Syurik YV, Ghislandi MG, Tkalya EE, et al. Graphene network organisation in conductive polymer composites. Macromolecular Chemistry and Physics. 2012;**213**(12):1251-1258. DOI: 10.1002/macp.201200116

[13] Sanchez Esqueda I, Yan X, Rutherglen C, et al. Aligned carbon nanotube synaptic transistors for large-scale neuromorphic computing. ACS Nano. 2018;**12**(7):7352-7361. DOI: 10.1021/acsnano.8b03831

[14] Bonard J-M, Croci M, Klinke C, Kurt R, Noury O, Weiss N. Carbon nanotube films as electron field emitters. Carbon. 2002;**40**(10):1715-1728. DOI: 10.1016/S0008-6223(02)00011-8

[15] Zou J, Zhang K, Li J, et al. Carbon nanotube driver circuit for 6 × 6 organic

light emitting diode display. Scientific Reports. 2015;**5**(1):11755. DOI: 10.1038/srep11755

[16] Tolt ZL, Mckenzie C, Espinosa R, Snyder S, Munson M. Carbon nanotube cold cathodes for application in low current x-ray tubes. Journal of Vacuum Science & Technology B: Microelectronics and Nanometer Structures. 2008;**26**(2):706. DOI: 10.1116/1.2802092

[17] Heo S, Kim H, Ha J, Cho S. A vacuum-sealed miniature X-ray tube based on carbon nanotube field emitters. Nanoscale Research Letters. 2012;**7**(1):258. DOI: 10.1186/1556-276X-7-258

[18] Park S, Vosguerichian M, Bao Z. A review of fabrication and applications of carbon nanotube film-based flexible electronics. Nanoscale. 2013;**5**(5):1727. DOI: 10.1039/c3nr33560g

[19] Levchenko I, Han Z-J, Kumar S, Yick S, Fang J, Ostrikov K. Large arrays and networks of carbon nanotubes: Morphology control by process parameters. In: Syntheses and Applications of Carbon Nanotubes and Their Composites. IntechOpen. London, UK; 2013. DOI: 10.5772/52674

[20] Tanaka K, Iijima S, editors. Carbon Nanotubes and Graphene. 2nd ed. Amsterdam: Elsevier; 2014. DOI: 10.1016/C2011-0-07380-5. Available from: https://www.sciencedirect.com/book/9780080982328/carbon-nanotubes-and-graphene#book-info

[21] Ageev OA, Balakirev SV, Bykov AV, et al. Development of new metamaterials for advanced element base of micro- and nanoelectronics, and microsystem devices. In: Advanced Materials. Springer Proceedings in Physics. Springer International Publishing Switzerland. 2016;563-580. DOI:10.1007/978-3-319-26324-3_40

[22] Bulyarskii SV, Basaev AS. Coalescence of clusters of catalysts for the growth of carbon nanotubes during their formation under homogeneous and heterogeneous conditions. Physics of the Solid State. 2015;**57**(6):1055-1059. DOI: 10.1134/S1063783415060074

[23] Fedotov AA. Study of growth processes of PECVD carbon nanotubes, used in micro-and nanosystem technologies. In: Piezoelectrics and Nanomaterials: Fundamentals, Developments and Applications. Nova Science Publishers, Inc. New York, USA; 2015. pp. 61-84

[24] Gao J, Zhong J, Bai L, Liu J, Zhao G, Sun X. Revealing the role of catalysts in carbon nanotubes and nanofibers by scanning transmission X-ray microscopy. Scientific Reports. 2015;**4**(1):3606. DOI: 10.1038/srep03606

[25] Neyts EC. The role of ions in plasma catalytic carbon nanotube growth: A review. Frontiers of Chemical Science and Engineering. 2015;**9**(2):154-162. DOI: 10.1007/s11705-015-1515-5

[26] Shariat M, Shokri B, Neyts EC. On the low-temperature growth mechanism of single walled carbon nanotubes in plasma enhanced chemical vapor deposition. Chemical Physics Letters. 2013;**590**:131-135. DOI: 10.1016/j.cplett.2013.10.061

[27] Il'in OI, Il'ina MV, Rudyk NN, Fedotov AA, Levshov DI, Ageev OA. The influence of activation and growth time on the geometry and structural perfection of multi-walled carbon nanotubes. Journal of Physics Conference Series. 2018;**1038**:012062. DOI: 10.1088/1742-6596/1038/1/012062

[28] Il'in OI, Il'ina MV, Rudyk NN, Fedotov AA, Ageev OA. The growth temperature effect on vertically aligned carbon nanotubes parameters.

Nanosystems: Physics, Chemistry, Maths. 2018;**9**(1):92-94. DOI: 10.17586/2220-8054-2018-9-1-92-94

[29] Il'ina MV, Il'in OI, Blinov YF, et al. Memristive switching mechanism of vertically aligned carbon nanotubes. Carbon. 2017;**123**:514-524. DOI: 10.1016/j.carbon.2017.07.090

[30] Amama PB, Pint CL, McJilton L, et al. Role of water in super growth of single-walled carbon nanotube carpets. Nano Letters. 2009;**9**(1):44-49. DOI: 10.1021/nl801876h

[31] de los Arcos T, Garnier MG, Seo JW, Oelhafen P, Thommen V, Mathys D. The influence of catalyst chemical state and morphology on carbon nanotube growth. The Journal of Physical Chemistry. B. 2004;**108**(23):7728-7734. DOI: 10.1021/jp049495v

[32] Moisala A, Nasibulin AG, Kauppinen EI. The role of metal nanoparticles in the catalytic production of single-walled carbon nanotubes—A review. Journal of Physics: Condensed Matter. 2004;**15**(42):1-26. DOI: 10.1088/0953-8984/15/42/003

[33] Park J-B, Choi G-S, Cho Y-S, et al. Characterization of Fe-catalyzed carbon nanotubes grown by thermal chemical vapor deposition. Journal of Crystal Growth. 2002;**244**(2):211-217. DOI: 10.1016/S0022-0248(02)01661-5

[34] Yamada T, Namai T, Hata K, et al. Size-selective growth of double-walled carbon nanotube forests from engineered iron catalysts. Nature Nanotechnology. 2006;**1**(2):131-136. DOI: 10.1038/nnano.2006.95

[35] Ho GW, Wee ATS, Lin J, Tjiu WC. Synthesis of well-aligned multiwalled carbon nanotubes on Ni catalyst using radio frequency plasma-enhanced chemical vapor deposition. Thin Solid Films. 2001;**388**(1-2):73-77. DOI: 10.1016/S0040-6090(01)00828-8

[36] Ren ZF, Huang ZP, Wang DZ, et al. Growth of a single freestanding multiwall carbon nanotube on each nanonickel dot. Applied Physics Letters. 1999;**75**(8):1086-1088. DOI: 10.1063/1.124605

[37] Bethune DS, Kiang CH, de Vries MS, et al. Cobalt-catalysed growth of carbon nanotubes with single-atomic-layer walls. Nature. 1993;**363**(6430):605-607. DOI: 10.1038/363605a0

[38] Teo KBK, Chhowalla M, Amaratunga GAJ, et al. Uniform patterned growth of carbon nanotubes without surface carbon. Applied Physics Letters. 2001;**79**(10):1534-1536. DOI: 10.1063/1.1400085

[39] Wen H-C, Yang K, Ou K-L, Wu W-F, Luo R-C, Chou C-P. Carbon nanotubes grown using cobalt silicide as catalyst and hydrogen pretreatment. Microelectronic Engineering. 2005;**82**(3-4):221-227. DOI: 10.1016/j.mee.2005.07.028

[40] Thomas Koilraj T, Kalaichelvan K. Synthesis of carbon nanotubes using Fe-Mo/AL_20_3 bimetallic catalyst by CVD method. IEEE-International Conference on Advanced Engineering Science and Management (ICAESM-2012). 2012:429-433

[41] Jeong SW, Son SY, Lee DH. Synthesis of multi-walled carbon nanotubes using Co–Fe–Mo/Al_2O_3 catalytic powders in a fluidized bed reactor. Advanced Powder Technology. 2010;**21**(2):93-99. DOI: 10.1016/j.apt.2009.10.008

[42] He M, Chernov AI, Obraztsova ED, Jiang H, Kauppinen EI, Lehtonen J. Synergistic effects in FeCu bimetallic

catalyst for low temperature growth of single-walled carbon nanotubes. Carbon. 2013;**52**:590-594. DOI: 10.1016/j.carbon.2012.10.029

[43] Pesetski AA, Baumgardner JE, Krishnaswamy SV, et al. A 500 MHz carbon nanotube transistor oscillator. Applied Physics Letters. 2008;**93**(12):123506. DOI: 10.1063/1.2988824

[44] Chen Q, Yuan X, Cole M, Zhang Y, Meng L, Yan Y. Theoretical study of a 0.22 THz backward wave oscillator based on a dual-gridded, carbon-nanotube cold cathode. Applied Sciences. 2018;**8**(12):2462. DOI: 10.3390/app8122462

[45] Il'ina MV, Il'in OI, Blinov YF, Smirnov VA, Ageev OA. Nonuniform elastic strain and memristive effect in aligned carbon nanotubes. Technical Physics. 2018;**63**(11):1672-1677. DOI: 10.1134/S1063784218110129

[46] Ilina MV, Blinov YF, Ilin OI, Rudyk NN, Ageev OA. Piezoelectric effect in non-uniform strained carbon nanotubes. IOP Conference Series: Materials Science and Engineering. 2017;**256**:012024. DOI: 10.1088/1757-899X/256/1/012024

[47] Rudyk NN, Il'in OI, Il'ina MV, Fedotov AA, Klimin VS, Ageev OA. Carbon nanotubes based vacuum gauge. Journal of Physics Conference Series. 2017;**917**:082008. DOI: 10.1088/1742-6596/917/8/082008

[48] Xiao Z, Kong LB, Ruan S, et al. Recent development in nanocarbon materials for gas sensor applications. Sensors and Actuators B: Chemical. 2018;**274**:235-267. DOI: 10.1016/j.snb.2018.07.040

[49] Zhbanov A, Pogorelov E, Chang Y-C. Carbon nanotube field emitters. In: Carbon Nanotubes. IntechOpen. London, UK; 2010. DOI: 10.5772/39432

[50] Yan X, Wu Y, Wang B, et al. Fabrication of carbon nanotube on nickel–chromium alloy wire for high-current field emission. Applied Surface Science. 2018;**450**:38-45. DOI: 10.1016/j.apsusc.2018.04.114

Section 3

Characterization of Carbon Nanotubes

Chapter 6

Functionalized Carbon Nanotubes for Detection of Volatile Organic Pollutant

Arti Dinkarrao Rushi, Kunal Prasanta Datta, Prasanta Sudarshan Ghosh, Ashok Mulchandani and Mahendra D. Shirsat

Abstract

Ever since carbon nanotubes were discovered, it has attracted researchers from all corners due to the fascinating characteristics they offer. Detection of environmental pollutants is one of the prominent sectors where carbon nanotubes have been extensively applied. Pristine carbon nanotubes attracted sincere attention at the initial phase; however, the inherent shortcomings as a chemical sensor were immediately recognized. Thereafter, improvement of selective and sensitive behavior of carbon nanotubes toward varied class of analytes has remained a core area of carbon nanotube-based research. Tailoring/covering sidewalls of carbon nanotubes with different functionalizing entities has shown remarkable outcomes in improving behavior of nanotube sensors in various domains. Organic conducting polymers, macrocyclic compounds, metal organic frameworks, metal nanoparticles, etc. have been extensively reported to functionalize carbon nanotube surface. This chapter discusses porphyrin-functionalized single-walled carbon nanotubes that are found to be selective toward volatile organic compounds, and most importantly, good selectivity and sensitivity have been observed for selected group of analytes.

Keywords: carbon nanotubes, porphyrin, volatile organic compounds, sensor, chemiresistor

1. Introduction

Environment encompasses all living and non-living things coexisting with invisible strings at balance. "Pollution" is a major component that disturbs this balance and a serious threat to environmental ecosystem. Human practices exercising untowardly comfort are adding up to the existing lot of "pollutants," and this is causing adverse effects on all living creatures. Some of the dominating pollutants that are human activity generated include oxides of sulfur and nitrogen, mono- and di-oxides of carbon, particulate matter, and volatile organic compounds (VOCs).

Out of the major air pollutants, VOCs cover a large area of chemical species that are present in nature. This class of pollutants is ubiquitous in nature and also copious in amount. Most of the VOCs are colorless, having a sweet smell. Although this odor is pleasant, inhalation of the same is very harmful to health. Even at room temperature, VOCs are present with high vapor pressure which is the cause of their volatile nature. High concentration of atmospheric VOCs degrades the air quality. When atmospheric VOCs get added to the rain water, the aqueous as well as terrestrial surroundings get polluted.

The classification of VOCs can be done according to their composition like (i) aromatic hydrocarbons (e.g., benzene, toluene, xylene), (ii) aliphatic hydrocarbons (e.g., hexane), and (iii) oxygenated compounds (e.g., ketones, acetone). According to their toxicity level [1], VOCs are categorized as follows: (i) extremely hazardous that shows adverse effect upon exposure even at low concentration of occurrence (e.g., benzene); (ii) class A compounds that may cause significant harm to the environment (e.g., acetaldehyde, aniline, benzyl chloride); and (iii) class B compounds that have lower post-exposure effects.

VOCs are dominant in industries as major solvents especially in the production of cleansers, paints, glues, carpets, rubbers, furniture, varnishes, wax, cosmetics, and air fresheners. Therefore, these products become the major sources of their emission. Also, fuel combustion and vehicle combustion have been found to be most prominent sources of VOC emission. Several types of foods and plants are also critical sources of VOCs.

Although VOCs are having many applications in our day-to-day life, their exposure is harmful. Small-scale exposure to VOCs causes ocular and throat irritation, vomiting, nausea, headache, loss of coordination, skin allergy [2–4], etc. Chronic effects that have been observed with VOCs include damage to the liver, lungs, kidney, reproductive system, respiratory system [5], and central nervous system in human beings. It is well known that some of the VOCs are carcinogenic [6, 7] too. Long-term exposures to VOCs are often deadly for all living organisms. Several VOCs (such as methane) are the main contributors of global warming. In addition, there are some VOCs that play a key role in destroying the stratospheric ozone. Because of the depletion of ozone layer, harmful ultraviolet radiations (280–315 nm) emitted by the Sun enter the atmosphere which is a potential cause of skin cancer [8] in most of the cases. When plants are continually exposed to the VOCs, it is generally observed that they suffer from slow stem growth and infertility in respective species. Also, flower and fruit growth is affected upon exposure to VOCs [9]. These data may not fully cover the scopes of damage caused by VOCs but may be sufficient enough to realize the need of some protective measurements against exposure to VOCs.

It is accepted that there should be all-out effort at war footing to curb the root causes of VOCs emission, yet, at the same time, reliable means are required to ensure a safe level of inhabitancy. As efficient control of VOC emission is a tough deal, exposure of the same to living systems is certain and obvious. In such cases, it is necessary to get real-time information of their concentration in the environment so that safety measures could be adapted in time to avoid potential hazard. To design some safety measure, reliable detection of VOCs at their permissible exposure limit (PEL) concentration is very necessary. There are three standard ways of reporting the concentrations of VOCs: (i) the concentration of individual VOCs; (ii) the sum of the concentrations of specific, individual VOCs in a sample; and (iii) as total organic carbon (TOC), which is the concentration of organic carbon in the gas stream. The concentrations are expressed in milligrams per cubic meter (mg/m^3). For TOC monitoring, among the most widely accepted techniques are flame ionization detectors (FIDs), catalytic oxidation, and photoionization detection (PID). For monitoring of individual VOCs concentration—(i) sorbent tube followed by gas chromatography (GC) separation [10–14], (ii) non-dispersive infrared (NDIR) detection [15–17], (ii) differential optical absorption spectrometry (DOAS) [18], (iv) Fourier-transform infrared (FTIR) spectroscopy, and

(v) mass spectroscopy [19] are among the mostly used standard techniques. The reliability, precision, and sensitivity of most of these techniques are unquestionable. However, limitations are there in terms of inherent delay between sample collection and measurement/analysis. The typical procedure of detection of VOCs with these techniques starts from sample collection from the affected sites which would be stored in canisters/Teflon bags, and then analysis of the samples is carried out in the laboratory environment. Thus, possibility of degradation of sample and incapability of accessing real-time data are inherent shortcomings, even with these most powerful analytical tools. Moreover, direct-reading analytical instruments such as FID and PID detectors do not only respond to VOCs but also to other organic compounds. As these instruments are calibrated with only one compound, the signal represents all compounds of the mixture as an equivalent of this compound. The output signal gives no information about the qualitative composition of the mixture. More profusely, all the abovementioned techniques are very costly and require continuous consumable and trained manpower. Gadgets that are capable of real-time monitoring come costly and are usually not very user-friendly.

Therefore, there is a need of some reliable, efficient, cheap, and user-friendly detection device for the real-time monitoring of the VOC level toxic air.

Sensors constitute the intelligent class that can be useful for the detection of VOCs as they are capable of transforming some change in a target physical/chemical quantity to any format of signals that are easy to be read or logged [20]. For an efficient operation, a sensor should be:

i. highly sensitive,

ii. highly selective,

iii. durable,

iv. of small size,

v. handy, and

vi. with operation ease.

Taking into consideration all of the above points and in order to obtain efficient sensing, it is very important to choose proper sensing platform, appropriate sensing materials [20–21], and optimized synthesis parameters.

To apply a sensor for

i. real-time monitoring of the VOCs,

ii. efficient power management, and

iii. enhancing the mobility of the sensing gadget,

it is essential to minimize the size of the sensor or, more comprehensively, the sensor backbone. The use of nanomaterials or, more precisely, carbon nanotubes (CNTs) is a perfect solution in order to control the physical footprint of the sensor.

In this chapter, CNT-based VOCs sensors, especially in chemiresistive modality, will be discussed. A detailed review of porphyrin-/metalloporphyrin-functionalized CNT-based VOC sensing has also been provided.

2. Carbon nanotubes

Carbon is the most versatile element in creation. The structural and characteristic specialties of carbon nanotubes are defined again, by the fascinating features of carbon atom. Carbon exhibits three allotropes in solid phase, viz. diamond, graphite, and Buckminsterfullerene [22] that show extreme contrast in their behavior.

CNT is made up of a seamless rolling of graphite sheet. During rolling up of a graphite sheet, σ-π rehybridization takes place due to the formation of a curvature resulting in π confinement. The high electrical conductivity of CNTs finds its root at this π-confinement. CNTs can be classified into three types on structural basis, viz. (i) single-walled carbon nanotubes (SWCNTs) that are hollow cylinders of individual graphene sheets, (ii) multi-walled carbon nanotubes (MWCNTs) that are group of coaxial CNTs, and (iii) double-walled carbon nanotubes that are group of two coaxial CNTs. The last type (i.e., DWCNTs) did not find much technological importance in contrast to the other types. Iijima, in 1991, observed MWCNTs for the first time [23] followed by independent and simultaneous observation of SWCNTs in 1993 by Iijima [24] and Bethune [25].

In order to maintain tubular structure during formation from a graphite sheet, SWCNTs should have at least 0.4 nm and at most about 3.0 nm diameter [26, 27] as suggested by structural calculations. A larger diameter SWCNT has a strong tendency to collapse, unless it is supported by other force or surrounded by neighboring force as in MWCNTs. Such low cross-section results in electron confinement and reduction of phase space for scattering that result in low scattering probability and high mobility for charge carriers. These facts indicate toward a faster charge transduction in CNTs, and while operating as a gas sensor, such characteristics reduce the response time for CNT-based sensors significantly. Excellent mechanical/thermal properties and chemical stability [28] also allow CNTs to withstand harsh operating conditions. However, the highest advantageous aspect of the 1-D nanostructure of CNTs lies in absolutely high aspect ratio with the diameter of few nanometers and lengths up to 100 microns that results in the presence of almost all atoms on the surface of the tube. Such molecular wires [29], with excellent quantum confinement, are extremely sensitive to local environment since any local charge can drastically modulate the carrier concentration along 1-D wire axis. Thus, CNTs are considered among the most potent nanostructured sensing elements.

The circular curvature in CNTs results in σ-π rehybridization. Three σ bonds, therefore, remain slightly out of plane, and automatically, for compensation, the π-orbital is asymmetrically distributed inside and outside the wall of nanotube (as depicted in **Figure 1**) [28].

A rich π-electron conjugation is thus formed due to the distorted electron cloud, outside the nanotube structure, rendering high electrochemically active surface. Any electron-donating or electron-withdrawing entity (e.g., NO_2, NH_3, O_2) finds it easy to either donate or withdraw electrons from CNT surface, thereby modulating

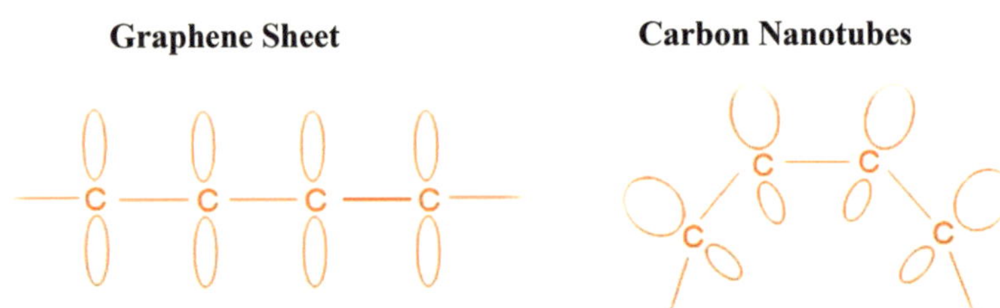

Figure 1.
π electron conjugation in graphene sheet and CNT.

overall electrical resistance [30–31]. Fundamental sensing mechanism during interaction of CNTs with electron-donating/electron-withdrawing analytes, therefore, can be given as below:

$$CNT + Gas \rightarrow CNT^{\delta e}\, Gas^{\delta +} \text{ or } CNT^{\delta +}\, Gas^{\delta e}$$

where δ denotes the number of charge(s) transferred during interaction.

The sensing mechanism for pristine SWCNTs, based on the above model, is quite simple, and since SWCNTs are p-type in nature, it is always easy to propose charge transfer-based sensing mechanism. Apart from attractive surface, gas molecules can be adsorbed at nanopores inside a tube, interstitials in tube bundles, and grooves above the gap between two neighboring tubes [28]. The adsorption capacity and mechanism of individual sites are unique, which may be affected by surface functionality and structural deformations [32]. It should be noted that such adsorption is succeeded by (i) charge transfer between a donor or acceptor type of molecule (or group) and individual SWCNTs, leading to the Fermi level modulation in semiconducting tubes, resulting in conductivity change [33], and/or (ii) creation of SWCNT molecule-SWCNT junction, leading to a hopping mechanism for intertube charge transfer between SWCNTs and an intertube modulation of the SWCNT network, such phenomenon being valid for both metallic and semiconducting SWCNTs [34]. Such capability, on the one hand, enhances lower detection limit for CNT-based sensors and, equally, results in strong adsorbance of the analyte, rendering poor recovery behavior. Li et al. [35] have reported almost 10 hours of recovery time for pristine nanotube-based sensors. Theoretical calculations further show that defect sites in CNTs result in stronger chemisorptions of analytes and large transfer interaction, in comparison to defect-free nanotubes.

A review of pristine CNT (SWCNTs as well as MWNTs)-based VOC sensors (this is applicable for other analytes too) shows that the limitation in sensing mechanism [36] and high surface-binding affinity to a large class of analytes possess toughest hurdles toward obtaining the best of CNTs as sensing agents. Also, Shirsat et al. [37] have suggested CNTs as not favorable platform for detection of VOCs because of lack of affinity.

In order to obtain better versatility in CNT-based sensors and to enable the same for selective sensing, a large spectrum of efforts have been carried out till date that includes (but not limited to) (i) creation of defect sites on CNT surface [38]; (ii) functionalization of CNT side wall with target-specific entities that contain conducting polymers [39, 40], metal nanoparticles [41], metal oxides [42], etc.; and (iii) formation of composites [43–45], etc. Functionalization of CNT side wall has the added advantage of improving sensor recovery characteristics since in such configuration, the analyte is not allowed to come in direct contact of CNT surface. The honeycomb structure of CNTs, along with high binding energy of gas analytes, results in slow desorption of analytes, thereby reflecting poor recovery behavior in pristine CNT-based sensors. Thus, functionalized structures of CNTs have attracted significant research interest toward alleviation of the problems of selectivity and recovery of pristine CNT-based sensors.

3. Porphyrins

A member of the family of macrocyclic compounds, porphyrin(s), has remained as the first preference for VOC sensing. As porphyrins show high impact on VOC sensing regime, a comprehensive study of its nature, properties, and sensing mechanism is given below.

Porphyrins and the related tetrapyrrolic macrocyclic compounds, chlorins, and corrins constitute an interesting class of macrocyclic structures that play crucial roles in nature and life [46]. Porphyrins, along with their metalloderivatives, exhibit an abundant series of physicochemical properties that enables them to cater as protagonists in flexible catalysis, especially, in the field of biology and chemistry [47]. Heme(the component of hemoglobin) and chlorophyll of chloroplasts are the examples of porphyrins that are present in nature. Nature, most beautifully, incorporates little-to-moderate changes in these macrocyclic structures to materialize the processes and versatility of life. Through the last two decades, the scientific community across the globe has witnessed abundance of examples that forward exploration of porphyrins/metalloporphyrins as building blocks for tailor-made applications [48]. The versatility in porphyrin molecular framework results in a wide range of mechanisms for analyte binding which include Van der Waals force, hydrogen bonding, π interactions, and coordination chemistry [49]. Detection of highly oxidizing/reducing gases [50], and especially odorous analytes, specific to the class of VOCs, has been reported, and these promising outcomes have triggered series of investigations toward sensitive and selective detection [51] of VOCs by porphyrins/metalloporphyrins.

The fundamental porphyrin structure (or porphine, as it is mostly called) is a cyclic tetrapyrrole ring with a highly delocalized planar network (**Figure 2**). The structure is analogous to other π—macrocycles having the characteristic central "core" which can be altered by the puckering of the porphyrin macrocycle (the range of variation of the core has been observed to lie between 2.098 and 1.929 Å).

This basic unit—a free base porphyrin (with no side substituent and no central metal ion)—has four modified pyrrole subunits interconnected at their α carbon atoms via methine bridges (=CH**—**). They are sharing a macrocycle of 20 carbon atoms and 4 nitrogen atoms, to form the conjugate system. The pyrrole rings build up closed aromatic plane forming the "porphyrin nucleus." With its conjugated 18-π electron system (conforming to Huckel 4n + 2 rule), it constitutes a macrocyclic system that is known to be highly stable, even against sulfuric acid and trifluoroacetic acid [52]. The 18-π electron system is further responsible for the aromaticity of porphyrins.

The "core" of a porphyrin structure is ideal for metal incorporation. The inside ring NH protons of porphyrin are acidic in nature and can get deprotonated to give porphyrinato ions [53]. The unshared pairs of electrons, thus created, are directed toward the center of the core. The ionic radii of many metals facilitate their "fit" to this cavity by coordination bond with the four porphyrin nitrogen atoms. Such metalation gives rise to innumerable metalloporphyrin (MP) structures.

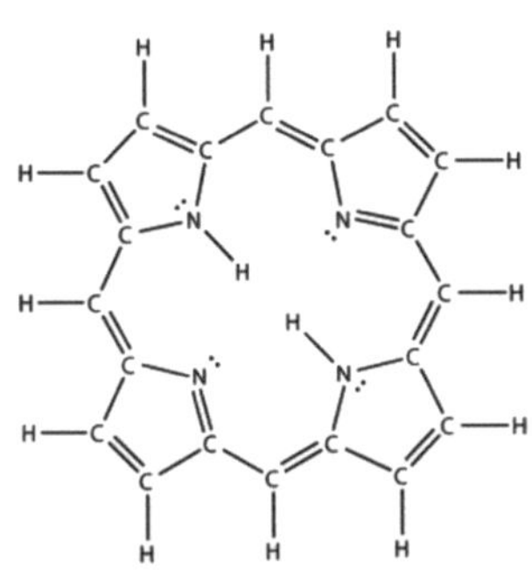

Figure 2.
Structure of porphine.

Sr. No.	Sensing material	Sensor type	VOC analytes	Concentration detected	Response recovery time	Ref.
1	Zinc-porphyrin and zinc-phthalocyanine	Optical	Pyridine, methanol	1 ppm	—	[56]
2	Tetraphenyl porphyrin	Optical	acetic acid, butyric acid hexanoic acid	855, 846, 822 ppm	10 min exposure time	[57]
3	ZnTriad porphyrin	Optical	Butylamine, hexylamine, propylamine	90 ppm	41–78 s resp.	[58]
4	5,10,15,20-Tetrakis(4-carboxyphenyl) porphyrin	Fluorescence	TNT vapor	10 ppb	—	[59]
5	Magnesium porphyrin	Optoelectronic	Methanol, propanol, ethanol, octanol, and methylbutanol	41 ppt	~300 s response time	[60]
6	Octaethylporphine-based transition-metal complexes (Ph(M),M = Co, Cu and Zn) and carbon black (CB)	Chemiresistive	Acetone, toluene, TNT	~30 ppt	~250 s response time	[61]
7	5,10,15,20-Tetrakis(4-carboxyphenyl)-21H,23H-porphyrin and its metal derivatives	Spectroscopic	Butylamine, hexylamine, hexanethiol, and dimethylformamide, acetone, diethylether	1000s of ppm	4–9 s	[62]

Table 1.
Review of some recent reports on porphyrin-/metalloporphyrin-based VOCs sensors.

The early efforts of porphyrin-/MP-based VOCs sensors employed the broad selectivity of this class to a wide range of analytes for successful implementation as electronic nose applications [54]. Through the 1980s and 1990s, most of the work on porphyrin-/metalloporphyrin-based VOC detection has been carried out through array-based detectors with the ultimate aim of mimicking the mammalian olfactory system followed by discovery of the fact that arrays of non-selective chemical sensors may show properties similar to that of natural olfaction [55].

Some reports on porphyrin/metalloporphyrin based VOCs sensing are given in (**Table 1**).

However, as a matter of utmost inherent deficiency, the electrical conductivity of porphyrins/metalloporphyrins is not adequate for electrical transduction, a fact that refrained the application of porphyrins/metalloporphyrins as chemiresistors. The alleviation routes include exercising these class of macrocycles as functionalizing entity for highly conducting backbones—most popularly, carbon nanotubes; thanks to the advancements in supramolecular chemistry.

4. Porphyrin-/MP-functionalized SWCNTs for VOC sensing

As discussed in the previous section, limited transduction mechanism available with the SWCNTs stands as a stiff constraint toward sensing applicability. At the same time, physical adsorption of analytes to the SWCNT/SWCNT bundles is another major difficulty that needs consideration for designing of real-time sensing elements. Surface functionalization can be a facile route to alleviate these problems and has been accepted with profound research interest worldwide. Specifically designed molecular recognition layer, when employed for surface functionalization/tailoring of SWCNTs, offers selective sensing of an analyte/a class of analytes and hinders the SWCNT surface from direct interaction with analytes.

SWCNTs, functionalized with porphyrins/MPs [63–68], have attracted significant research interest during the last decade, and encouraging research outcomes on VOC sensing based on such platform has been reported. Especially, chemiresistive modality [68] of sensing by SWCNTs functionalized with various porphyrins and MPs has shown pronounced sensing results. This particular approach has widened up possibilities of real-time VOC monitoring prospect employing electrical transduction modalities.

Surface functionalization of SWCNTs/MWCNTs is carried out either by covalent or non-covalent route. Highly stable interfaces with specific matrices are offered by covalent route of functionalization, and this pathway has been extensively adapted in the field of opto-electric/electronic study of the SWCNTs-porphyrin hybrids [69–71]. Diverse range of binding entities has been covalently linked to SWCNTs that widened the adaptation and popularity range of covalent functionalization technique(s). Harsh chemical treatments are, however, a requisite for creation of favorable moieties on the SWCNT surface that cause disruption surface conjugated π-system. Applications that ask for electrical and electrochemical activity of nanotube surface to be retained specifically [72, 73], thus, cannot accept covalent functionalization modality. It is also reported that covalent functionalization results in transformation of graphitic sp^2 surface to sp^3 orientation [66]. Therefore, even if charge transfer is possible through chemical linkages, covalently functionalized SWNT surfaces miserably fall short in charge transduction sort of operations, where extensive π-π delocalization at interfacial sites is a compulsion.

Non-covalent functionalization, on the other hand, has the specific advantage in that it can be carried out under relatively mild reaction conditions, maintaining the graphitic structure of CNTs [74]. The surface electronic properties of CNTs are, therefore, retained perfectly [75], and this ensures efficient sensing prospects. Non-covalent route of functionalization is a classic supramolecular approach where various adsorption forces, such as Van der Waals force, hydrogen bonds, electrostatic force, and π-stacking interactions [76–78], are responsible for surface tailoring of SWCNTs. Initial reports of formation of porphyrin-functionalized CNTs include SWCNT functionalization in solution phase by Nakashima et al. [79] where functionalization was affected by van der Waals forces. This particular report had truly opened a new era of functionalization of CNTs by porphyrin(s). Porphyrins constitute certain attractive choices for non-covalent functionalization of CNTs with highly polarisable aromatic porphyrinic cores that interact strongly with π—conjugated graphenic sidewalls [79, 80].

Functionalized structures exhibit multifold betterment in terms of sensitivity as well as selectivity as compared to pristine SWCNTs and/or porphyrin-based sensors [66, 68]. This has proved the synergistic impact of functionalization in sensing behavior and attracted high research interest. In porphyrin-functionalized SWCNT electrochemical sensor system, the transduction characteristics of the sensors are decided by SWCNTs, whereas porphyrins or metalloporphyrins define specific affinity toward the analyte(s) [81]. The enhanced sensitivity of the functionalized platforms, as compared to pristine SWCNT-based device, results from extended π-π conjugation formed at the interface of SWCNTs. Porphyrins, on the other hand, due to variants in metal core and side substituents, show selective characteristics toward VOCs [81].

Sr. No.	Materials	Modality	Analytes	Detection range	Ref.
1	SWCNTs functionalized by phthalocyanines and porphyrin	QCM, resistive	Benzene, toluene, xylene	QCM: 60–1200 ppm Resistive: 150–1200 ppm	[81]
2	CNTs modified with porphyrin	QMB	1-Butanol	46–230 ppb	[65]
3	SWCNTs functionalized by OEP, FeOEP, RuOEP, MnOEP, TPP, RuOEP, FeOEP	Chemiresistive, ChemFET	MEK, acetone, methanol, ethanol	Percentage saturated vapors of analyte	[66]
4	Single-walled carbon nanotube-poly(porphyrin) hybrid	Chemiresistive	Acetone	50–2000 ppm	[67]
6	SWCNT-MTPP (M = Co, Ni, Cu, Zn, Fe, Cr, Mn)	Chemiresistive	Pentane, hexane, cyclohexane, acetone, MEK, 3-pentanone, MeOH, EtOH, iPrOH, MePh, p-xylene	1000 ppm except for alkanes and for amines (2000 and 300 ppm, respectively)	[82]

Table 2.
Review of SWCNT porphyrin-based VOC sensor.

Some of the latest reports , comprising of porphyrin/metalloporphyrin functionalized SWCNTs based VOCs sensor , are given in (**Table 2**):

A typical approach of non-covalently functionalizing SWNCNTs with porphyrins/MPs and their application for VOCs detection is presented below.

5. Experimental details

Sensor substrate, Si/SiO_2 substrates, pre-patterned with gold (Au) micro fingertips, is a preferred platform for chemiresistive sensors. Such substrates are typically made up of four layers. The base substrate is highly doped p+ type Si with <100> orientation. 100 nm oxide layer is grown on the same by low pressure chemical vapor deposition (LPCVD). Chromium (Cr, 20 nm) and gold (Au, 180 nm) are successively deposited on the Si/SiO_2 substrate. Au is used as contact electrode material, while Cr serves the purpose of adhesive layer between Au and substrate. The microelectrode patterns are defined with ~3 μm gap (in this case) between nearest adjacent electrodes. Standard photolithography along with lift-off technique is employed for the purpose.

5.1 Sensor fabrication

According to the architecture of substrate and nature of the sensing material to be synthesized/deposited on substrate, the synthesis process has to be determined and optimized. For synthesis/deposition of the sensing materials, either a top-down or a bottom-up approach is adapted depending on the material and requisite level of spatial control. For the present discussion, a bottom-up approach has been adapted where the sensing material is deposited on the prefabricated microelectrode pattern on Si/SiO_2 substrate that itself offers contacts for integration with measurement electronics. SWCNTs, non-covalently functionalized with porphyrins/metalloporphyrins (MPs), were applied as sensing material(s) for the present course of investigations. The following sections will comprehensively discuss the steps undertaken for the fabrication of sensors employed in the present study.

SWCNTs were incorporated in the substrate in a manner so as to bridge the 3 μm gap between two adjacent Au fingertips with SWNTs. SWNTs were AC dielectrophoretically aligned between two Au fingertips to (i) achieve a device structure that is less random and (ii) attain significant control over the device base resistance.

To obtain aligned SWCNT configuration, first of all, suspension of SWNTs/N,N-dimethyl formamide (N,N-DMF) was prepared. Concentration of SWNTs in N,N-DMF was optimized through repeated experimentation. For synthesis of the devices to be used for sensing purpose, the concentration was optimized to be ~0.4 mg/20 ml. The optimized quantity of SWCNTs were dispersed by sonication for 90 min. to achieve homogeneity at medium ultrasonic power level (VWR 100C; variable power ultrasonic bath) followed by centrifugation (R-24; Remi centrifuge) at 15000 rpm for 90 min. The centrifugation was applied to get rid of SWCNTs agglomerates that were remaining in the suspension after ultrasonication. Decanted suspension obtained after centrifugation was stocked for further use. Prior to use, a requisite amount of the stock suspension was used to be sonicated for 30 min.

To align the SWCNTs, a 0.2 μL of SWNTs suspension was used to be placed on the 3 μm gap (dispersion through microsyringe) between two Au electrodes on a chip and subjected to an optimized AC signal. The dielectrophoretic alignment circuit was set up through electrical contacts to the microelectrode patterns established via probe station (ECOPIA; EPS 1000), an electronically gated timer (a modified RADIX T48LC microprocessor controlled timer), frequency generator (Aplab; 2219-U), and a PC-controlled I/V measurement unit (CHI 660C electrochemical workstation was used for this purpose). Optimized parameters for dielectrophoretic alignment were 4 MHz frequency with 2.03 Vp-p for regulated duration (through the timer; optimized during experimentation for achieving requisite device resistance) [26]. Following alignment, devices were washed with nanopure water and dried under gentle blow of nitrogen. After preliminary drying, aligned devices were annealed at 300 °C under reducing atmosphere (5% H_2 and 95% N_2) to ensure better contact between Au pads and SWNTs.

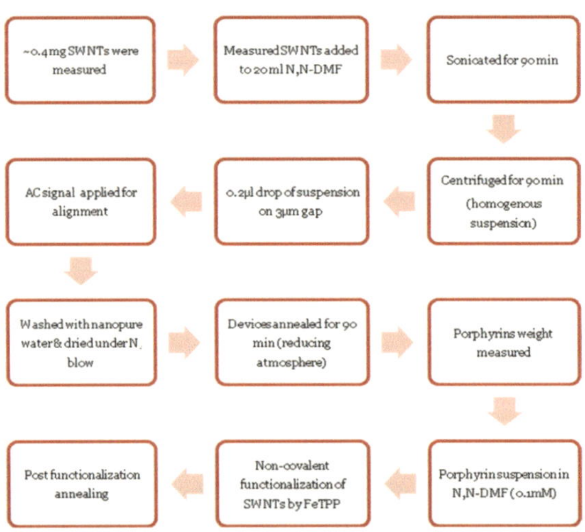

Figure 3.
Workflow for sensor fabrication process.

5.2 Non-covalent functionalization of SWCNTs by porphyrin/MPs

After obtaining aligned pristine SWCNT network-based device, non-covalent functionalization of the SWCNTs by iron tetraphenyl porphyrin (FeTPP) was carried out. For a typical VOC sensing application, directly procured FeTPP (from Sigma Aldrich; Switzerland) were used. Solutions of desired FeTPP were prepared in N,N-dimethyl formamide (0.1 mM) by vigorous mixing (REMI; Cyclomixer CM-101) for 30 min at room temperature.

For the functionalization of aligned SWCNTs, a 0.4 μL drop of the FeTPP solution was placed over the pristine aligned SWCNT network for 30 min (optimized casting duration), followed by wash under nanopure water, and, finally, annealing in N_2 atmosphere for 60 min at optimized temperature for each case. **Figure 3** gives a consolidated representation of the workflow for sensor fabrication process.

6. Results and discussion

6.1 Field emission scanning electron microscopy

Field emission scanning electron microscopy image of the fabricated devices was recorded with Nova NanoSEM 450. **Figure 4** shows FESEM image of typical SWNTs-FeTPP sensor. It could be clearly observed from the **Figure 4** that the SWCNTs functionalized with FeTPP could well bridge the gap between two Au microelectrodes. The average diameter of the functionalized structures could be estimated to be ~20 nm. Encouraging part of this observation might be in terms of a facile, repeatable, and batch fabrication pathway.

6.2 Current-voltage (I-V) characteristics

CHI 660C electrochemical workstation (CH Instruments; Texas, USA) was used for studying the current-voltage (I-V) characteristic by applying the linear sweep voltametry technique (−1 to +1 V; 10 mV/s). Before performing electrical

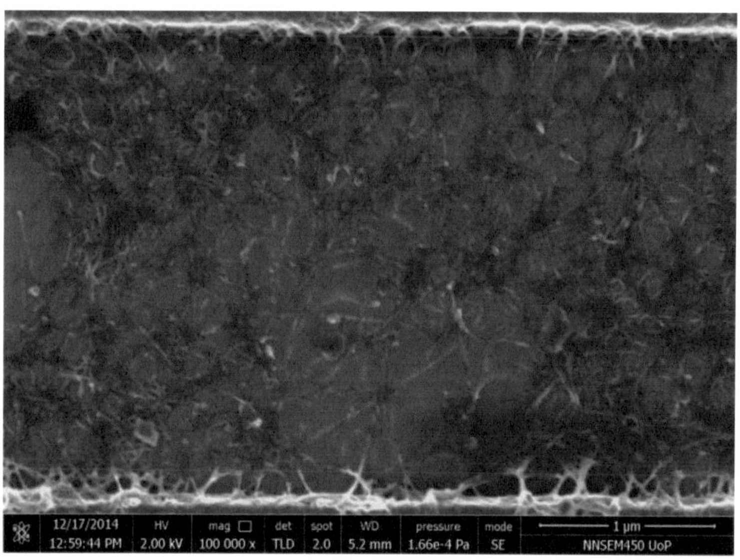

Figure 4.
FESEM image of the SWCNTs-FeTPP sensor.

measurements, the substrates were bonded to a custom PCB. Current-voltage characteristics of SWCNT-based devices before and after functionalization are shown in **Figure 5**. In case of SWCNT (pristine)-based sensor, current level is high as compared to functionalized sensors. **Figure 5** shows that after functionalization, decrease in current level was evident. Such decrease is attributed to the donation of electron from FeTPP to SWCNT backbone resulting in a decrease of majority charge carriers in p-type SWCNTs and creation of charge scattering sites at point of recombination [68].

6.3 Chemiresistive sensing characteristics

Chemiresistive sensing of the sensors at room temperature was carried out through a custom-built system where the sensor chip with electrical throughputs was atmospherically isolated in a flow cell (quartz c.a. 8 cc by volume). Dry air flow was employed initially through the flow cell (10–12 min) to achieve steady baseline for measurement. The fabricated sensors were exposed to various concentrations of VOCs by preparing calibrated dilution. Mass flow controllers (Alicat Scientific Inc., MC200) were employed to control the flow of both dry air and analytes. After each period of exposure, dry air was flown for sensor recovery. A constant current (5 μA; DC) was applied to the sensors, and consequent changes in potential, as a function of various concentrations of analytes, were recorded by a PC-controlled source-measure unit (Keithley 2400).

Figure 6 shows representation of normalized resistance ($\Delta R/R_0$) values for the sensors under analyte exposure of various concentrations, where ΔR representation shift of resistance value of the sensor from its baseline value and R_0 denoted the baseline resistance (the normalizing factor). Performance of the sensor(s) at four different concentrations (5, 15, 25, 35 ppm) has been reflected in terms of histogram. Performance of at least five devices has been observed for each analyte. Device-to-device variation was found to be encouragingly trivial. All analytes could be characterized with electron-donating nature [51].

The sensing performance clearly reflects the highest response for ethylbenzene, while toluene could also be detected at higher response values, in comparison to

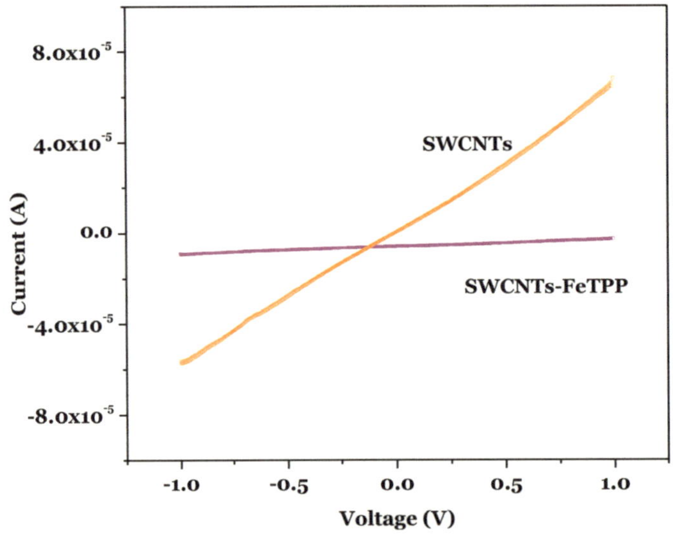

Figure 5.
I-V characteristics of the SWCNT and SWCNT-FeTPP sensor.

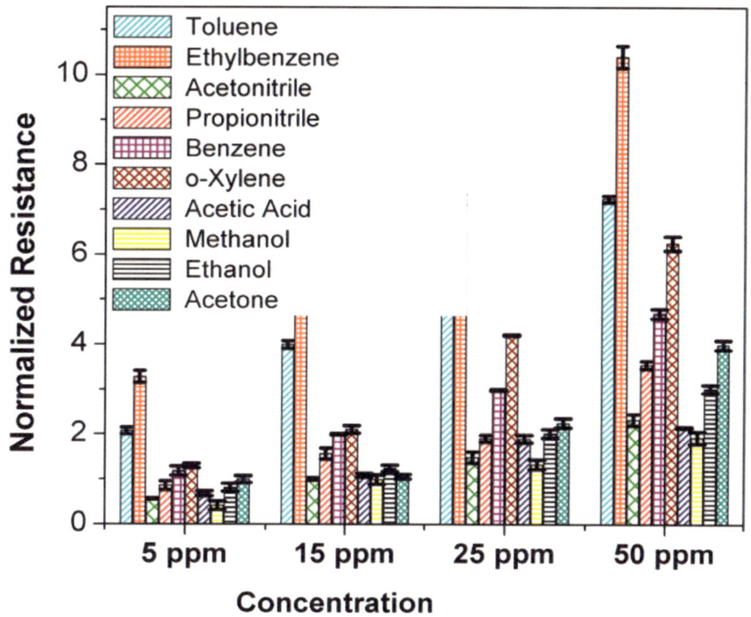

Figure 6.
Histogram of SWCNTs-FeTPP sensor toward various VOCs.

other analytes. As a common rationale, all the analytes undertaken for study have reflected an electron-donating behavior [68], and therefore, it is well envisaged that the structural signature of the analytes play important role in shaping the sensor response [83]. For example, the ethyl group present as side-substituent in benzene ring (of ethylbenzene) might have acted as a better electron donor than methyl group in toluene. Such behavior might be rationalized on the ground that methyl "carbon" of an ethyl group owns better electron density from adjacent hydrogens and is in a better position to donate. In o-xylene, the two alkyl substituents decrease effective electron density present on the central ring in comparison to toluene, resulting in lower sensing response than for toluene. The absence of any side substituent in benzene has resulted in inferior response for benzene, in comparison to its derivatives. Another issue that is pertinent from the observations is that the sensor has exhibited better performance for aromatic VOCs. Due to structural symmetry of aromatic side substituent in TPP and benzene ring, the π transitions might have been better favored. Therefore, from these observations, it is quite sure that if the structural geometry of target VOCs is properly analyzed, suitable porphyrin/metalloporphyrin can be employed as functionalizing entity for CNTs toward designing of explicitly selective VOC sensor backbones in chemiresistive modality.

7. Conclusions

Reliable detection of VOCs has remained a challenge because of the characteristics similarity among VOCs families. Porphyrin/metalloporphyrins, although, have shown prominent capacity in detection of VOCs; chemiresistor modality of sensing with this class of materials could be truly realized after CNTs were introduced in the realm. Porphyrin-/metalloporphyrin-functionalized SWCNTs have a significant potential in development of low footprint chemiresistors that enjoy simple functionality and can be embedded into standard electronics domain at less complexity, a clear advantage over porphyrin-/metalloporphyrin-based sensors that are not operated for direct electrical measurements, for example, mass transduction, spectroscopic measurements, etc. Also, in comparison to the broad selectivity of porphyrin-/metalloporphyrin-based VOCs sensors, functionalized CNT-based chemiresistors can offer better selectivity based on the structural geometry of the target VOCs and according to the design of the porphyrin/metalloporphyrin structure. Systematic study and an affluent database generated through the same, for porphyrin/metalloporphyrin, VOC correlation, can lead to development of future olfactory type solutions for VOC detection.

Conflict of interest

There is no "conflict of interest."

Author details

Arti Dinkarrao Rushi[1*], Kunal Prasanta Datta[2], Prasanta Sudarshan Ghosh[3], Ashok Mulchandani[4] and Mahendra D. Shirsat[5]

1 Department of Basic Sciences and Humanities, Maharashtra Institute of Technology, Aurangabad, MS, India

2 Deen Dayal Upadhayay KAUSHAL Kendra, Dr. Babasaheb Ambedkar Marathwada University, Aurangabad, MS, India

3 Department of Basic Sciences and Humanities, International Centre of Excellence in Engineering and Management, Aurangabad, MS, India

4 Department of Chemical and Environmental Engineering, University of California, Riverside, CA, USA

5 Department of Physics and RUSA Center for Advanced Sensor Technology, Dr. Babasaheb Ambedkar Marathwada University, Aurangabad, MS, India

*Address all correspondence to: artirushi33@gmail.com

IntechOpen

© 2019 The Author(s). Licensee IntechOpen. This chapter is distributed under the terms of the Creative Commons Attribution License (http://creativecommons.org/licenses/by/3.0), which permits unrestricted use, distribution, and reproduction in any medium, provided the original work is properly cited.

References

[1] Gianfranco M, Giovanni C, Rosanna F, Leonardo G, Antonio M, Davide P. Real-time monitoring of volatile organic compounds in hazardous sites, environmental monitoring. In: Ekundayo E, editor. Environmental Monitoring. Rijeka, Croatia: InTech; 2011. pp. 219-244

[2] Choi S, Kim N, H Cha H, Rhan HA. Micro sensor node for air pollutant monitoring: hardware and software issues. Sensors. 2009;**9**:7970-7987. DOI 10.3390/s91007970

[3] Sax N, Lewis R. Hawley's Condensed Chemical Directory. 11th ed. New York: VNR; 1987. pp. 177-178, 704-969

[4] Lim C, Yu J, Kin D, Byun H, Lee D. IEEE Sensors Journal. 2006

[5] Mizukoshi A, Kumagai K, Yamamoto N, Noguchi M, Yoshiuchi K, Kumano H, et al. A novel methodology to evaluate health impacts caused by voc exposures using real-time VOC and holter monitors. International Journal of Environmental Research and Public Health. 2010;**7**:4127-4138. DOI: 10.3390/ijerph7124127

[6] Xu Z, Mulchandani A, Chen W. Detection of Benzene, toluene, ethyl benzene, and xylenes (BTEX) using toluene dioxygenase-peroxidase coupling reactions. Biotechnology Progress. 2003;**19**:1812-1815

[7] Ghanem M, Guo Y, Hassard J, Osmond M, Richards M. Sensor Grids for Air Pollution Monitoring. https://www.researchgate.net/publication/2920029_Sensor_Grids_For_Air_Pollution_Monitoring/citations

[8] Henriksen T, Dahlback A, Larsen S, Moan J. Ultraviolet radiation and skin cancer: Effect of an ozone layer depletion. Photochemistry and photobiology. 1990;**51**:579-582. DOI: 10.1111/j.1751-1097.1990.tb01968.x

[9] Cape J. Effects of airborne volatile organic compounds on plants. Environmental Pollution. 2003;**122**:145-157. DOI: 10.1016/S0269-7491(02)00273-7

[10] Cape J, Leith I, Binnie J, Content J, Donkin M, Skewes M, et al. Effects of VOCs on herbaceous plants in an open-top chamber experiment. Environmental Pollution. 2003;**124**:341-353. DOI: 10.1016/S0269-7491(02)00464-5

[11] Buszewski B, Ligor T, Jezierski T. Identification of volatile lung cancer markers by gas chromatography-mass spectrometry: Comparison with discrimination by canines. Analytical and Bioanalytical Chemistry. 2012;**404**:141-146

[12] Reusser D, Field J. Determination of benzylsuccinic acid in gasoline-contaminated groundwater by solid-phase extraction coupled with gas chromatography–mass spectrometry. Journal of Chromatography A. 2002;**953**:215-225. DOI: 10.1016/S0021-9673(02)00107-3

[13] Potter D, Pawliszyn J. Detection of substituted benzenes in water at the pg/ml level using solid-phase microextraction and gas chromatography—ion trap mass spectrometry. Journal of Chromatography. 1992;**625**:247-255. DOI: 10.1016/0021-9673(92)85209-C

[14] Silva L, Ferreira F, Tresa A, Santos T, Duarte A. Carbon nanotube field-effect transistor detector associated to gas chromatography for speciation of benzene, toluene, ethylbenzene, (*o*-, *m*- and *p*-)xylene. Journal of Chromatography A. 2009;**1216**:6517-6521. DOI: https://doi.org/10.1016/j.chroma.2009.07.060

[15] Liu H, Chen T, Chen D, Chang D, Frost R. The development of a new method for the determination of concentrations of carbonaceous gases using a non-dispersive infrared sensor. Fuel. 2013;**104**:749-751. DOI: 10.1016/j.fuel.2012.07.063

[16] Sklorz A, Jansen S, Lang W. Detection limit improvement for NDIR ethylene gas detectors using passive approaches. Sensors and Actuators B. 2012;**175**:246-254. DOI: 10.1016/j.snb.2012.09.085

[17] Zhu Z, Xu Y, Jiang B. One ppm NDIR Methane gas sensor with single frequency filter denoising algorithm. Sensors. 2012;**12**:12729-12740. DOI: 10.3390/s120912729

[18] Kim K. Comparison of BTX measurements using a differential optical absorption spectroscopy and an on-line gas chromatography system. Environmental Engineering Science. 2004;**21**:181-194. DOI: 10.1089/109287504773087354

[19] Ketola R, Kiuru J, Tarkiainen V, Kokkonen J, Anen J, Kotiaho T. Detection of volatile organic compounds by temperature-programmed desorption combined with mass spectrometry and Fourier transform infrared spectroscopy. Analytica Chimica Acta. 2006;**562**:245-251. DOI: 10.1016/j.aca.2006.01.069

[20] Reemts J, Parisu J, Sclettwein. Electrochemical growth of gas-sensitive polyaniline thin films across an insulating gap. Thin Solid Films. 2004;**466**:320-325. DOI: 10.1016/j.tsf.2004.03.010

[21] McGovern S, Spinks G, Wallace G. Micro-humidity sensors based on a processable polyaniline blend. Sensors and Actuators B. 2005;**107**:657-665. DOI: 10.1016/j.snb.2004.11.046

[22] Helland A, Wick P, Koehler A, Schmid K, Som C. Reviewing the environmental and human health knowledge base of carbon nanotubes. Environmental Health Perspectives. 2007;**115**:1125-1131. DOI: 10.1289/ehp.9652

[23] Smalley R. Formation and Properties of C60 and the Fullerenes. National Institute of Standards and Technology; Dec 1990:6-7

[24] Iijima S. Helical microtubules of graphitic carbon. Nature. 1991;**354**: 56-68

[25] Iijima S, Ichihashi T. Single-shell carbon nanotubes of 1-nm diameter. Nature. 1993;**363**:603-605. DOI: 10.1038/363603a0

[26] Wei B, Vajtai R, Ajayan P. Reliability and current carrying capacity of carbon nanotubes. Applied Physics Letters. 2001;**79**:1172-1174. DOI: 10.1063/1.1396632

[27] Roberson D, Brenner D, Mintmire J. Energetics of nanoscale graphitic tubules. Physical Review B. 1992;**45**:12592. DOI: 10.1103/PhysRevB.45.12592

[28] Lucas A, Lambin P, Smalley R. On the energetics of tubular fullerenes. Journal of Physics and Chemistry of Solids. 1993;**54**:587-593. DOI: 10.1016/0022-3697(93)90237-L

[29] Meyappan M, editor. Carbon Nanotubes: Properties and Applications. Taylor and Francis: CRC Press; 2005

[30] Qi P, Vermesh O, Grecu M, Javey A, Wang Q, Dai H. Toward large arrays of multiplex functionalized carbon nanotube sensors for highly sensitive and selective molecular detection. Nano Letters. 2003;**3**:347-351. DOI: 10.1021/nl034010k

[31] Pradhan B. Materials Research Society Symposium Proceedings. 2001:633

[32] Sumanasekara G. Adu Fang C, Eklund P. Effects of gas adsorption and collisions on electrical transport in single-walled carbon nanotubes. Physical Review Letters. 2000;**85**:1096. DOI: 10.1103/PhysRevLett.85.1096

[33] Agnihotri S, Mota J, Rostam-Abadi M, Rood M. Temporal changes in nitrogen adsorption properties of single-walled carbon nanotubes. Carbon. 2004;**42**:2699-2710. DOI: 10.1016/j.carbon.2004.06.016

[34] Zhao J, Buldum A, Han J, Lu J. Gas molecule adsorption in carbon nanotubes and nanotube bundles. Nanotechnology. 2002;**13**:195. DOI: 10.1088/0957-4484/13/2/312

[35] Li J, Lu Y, Ye Q, Cinke M, Han J, Meyappan M. Carbon nanotube sensors for gas and organic vapor detection. Nanoletters. 2003;**3**:929-933. DOI: 10.1021/nl034220x

[36] Dai L. From conducting polymers to carbon nanotubes: A revolution of sensors based on architectural diversity of the π-conjugated structure. Energeia. 2005;**16**:1-6

[37] Shirsat M, Sarkar T, Kakoullis JR, Myung N, Konnanath B, Spanias A, et al. Porphyrins-functionalized single-walled carbon nanotubes chemiresistive sensor arrays for VOCs. The Journal of Physical Chemistry C. 2012;**116**:3845-3850. DOI: 10.1021/jp210582t

[38] Yang CM, Kanoh H, Kaneko K, Yudasaka M, Ijima S. Adsorption behaviors of HiPco single-walled carbon nanotube aggregates for alcohol vapors. Journal of Physical Chemistry B. 2002;**106**:8994-8999. DOI: 10.1021/jp025767n

[39] Star A, Han T, Joshi V, Gabriel J, Gruner G. Nanoelectronic carbon dioxide sensor. Advanced Materials. 2004;**16**:2049-2052. DOI:10.1002/adma.200400322

[40] Bekyarova E, Davis M, Burch T, Itkis ME, Zhao B, Sunshine S, et al. Chemically functionalized single-walled carbon nanotubes as ammonia sensors. Journal of Physical Chemistry B. 2004;**108**:19717-19720. DOI: 10.1021/jp0471857

[41] Star A, Joshi V, Skarupo S, Thomas D, Gabriel J. Gas sensor array based on metal-decorated carbon nanotubes. Journal of Physical Chemistry B. 2006;**110**:21014-21020. DOI: 10.1021/jp064371z

[42] Mubeen S, Zhang T, Chartuprayoon N, Rheem Y, Mulchandani A, Myung N, et al. Sensitive detection of H2S using gold nanoparticle decorated single-walled carbon nanotubes. Analytical Chemistry. 2010;**82**:250-257. DOI: 10.1021/ac901871d

[43] Hernandez S, Kakoullis J, Lim J, Mubeen S, Hangerter C, Mulchandani A, et al. Hybrid ZnO/SWNT nanostructures based gas sensor. Electroanalysis. 2012;**24**:1613-1620. DOI: 10.1002/elan.201200135

[44] Choi H, Lee J, Dong K, Ju B, Lee W. Gas sensing performance of composite materials using conducting polymer/single-walled carbon nanotubes. Macromolecular Research. 2012;**20**:143-146. DOI: 10.1007/s13233-012-0030-5

[45] Li Y, Wang H, Cao X, Yuan M, Yang M. A composite of polyelectrolyte-grafted multi-walled carbon nanotubes and in situ polymerized polyaniline for the detection of low concentration triethylamine vapour. Nanotechnology. 2008;**19**:015503. DOI: 10.1088/0957-4484/19/01/015503

[46] Porphyrins, Polymeric Porphyrins and their Metalloderivatives. Prepared

by Bee Hive Digital Concepts Cochin for Mahatma Gandhi University Kottayam

[47] Suslick K. Comprehensive Supramolecular Chemistry; Bioinorganic Systems. Vol. 5. Oxford: Elsevier; 1996

[48] Natale C, Monti D, Paolesse R. Chemical sensitivity of porphyrin assemblies. Materials Today. 2010;**13**:7-8. DOI: 10.1016/S1369-7021(10)70127-9

[49] Richardson T, Dooling C, Jones L, Brook R. Development and optimization of porphyrin gas sensing LB films. Advances in Colloid and Interface Science. 2005;**116**:81-96. DOI: 10.1016/j.cis.2005.04.009

[50] Amico A, Natale C, Paolesse R, Macagnano A, Mantini A. Metalloporphyrins as basic material for volatile sensitive sensors. Sensors and Actuators B. 2000;**65**:209-215. DOI: 10.1016/S0925-4005(99)00342-1

[51] Rushi A, Datta K, Ghosh P, Mulchandani A, Shirsat M. Selective discrimination among benzene, toluene, and xylene: Probing metalloporphyrin-functionalized single-walled carbon nanotube-based field effect transistors. Journal of Physical Chemistry C. 2014;**118**:24034-24041. DOI: 10.1021/jp504657c

[52] Gouterman M. In: Dolphin D, editor. The Porphyrins. Vol. III, Part A, Physical Chemistry. New York: Academic; 1978

[53] Buchler J. In: Dolphin D, editor. The Porphyrins. Vol. 1, Part A: Structure and Synthesis. New York: Academic; 1978

[54] Natale C, Macagnanao A, Davide F, Amico A, Paolesse R, Boschi T, et al. An electronic nose for food analysis. Sensors and Actuators B. 1997;**44**:521-526. DOI: 10.1016/S0925-4005(97)00175-5

[55] Persuad K, Dodds G. Analysis of discrimination mechanisms in the mammalian olfactory system using a model nose. Nature. 1982;**299**:352-355. DOI: 10.1038/299352a0

[56] Yang W, Xu J, Mao Y, Yang Y, Jiang Y. Detection of volatile organic compounds using langmuir–blodgett films of zinc-porphyrin and zinc-phthalocyanine. Synthesis and Reactivity in Inorganic, Metal-Organic, and Nano-Metal Chemistry. 2016;**46**:735-740. DOI: 10.1080/15533174.2014.989576

[57] Evyapan M, Dunbar A. Improving the selectivity of a free base tetraphenyl porphyrin based gas sensor for NO_2 and carboxylic acid vapors. Sensors and Actuators B. 2015;**206**:74-83. DOI: 10.1016/j.snb.2014.09.023

[58] Roales J, Pedros J, Guillén M, Cost T, Pinto S, Calvete M, et al. Optical detection of amine vapors using ZnTriad porphyrin thin films. Sensors and Actuators B. 2015;**210**:28-35. DOI: 10.1016/j.snb.2014.12.080

[59] Ahmadi E, Ramazani A, Hamdi Z, Malekzadeh A, Mohamadnia Z. 5,10,15,20-Tetrakis(4-carboxyphenyl) porphyrin covalently bound to nano-silica surface: Preparation, characterization and chemosensor application to detect TNT. Silicon. 2015;7:323-332. DOI: 10.1007/s12633-015-9304-9

[60] Gutiérrez A, Brittle S, Richardson T, Dunbar A. A proto-type sensor for volatile organic compounds based on magnesium porphyrin molecular films. Sensors and Actuators B. 2014;**202**:854-860. DOI: 10.1016/j.snb.2014.05.082

[61] Berríos E, Theriot J, Woodka M, Lewis N. Detection of ammonia, 2,4,6-trinitrotoluene, and common organic vapors using thin-film carbon black-metalloporphyrin composite chemiresistors. Sensors and Actuators

B. 2013;**188**:761-767. DOI: 10.1016/j.snb.2013.07.006

[62] Roales J, Pedrosa J, Castillero P, Cano M, Richardson T, Barranco A, et al. Selective detection of volatile organic compounds by spectral imaging of porphyrin derivatives bound to TiO_2 porous films. ACS Applied Materials & Interfaces. 2012;**4**:5147-5154. DOI: 10.1021/am3010169

[63] Penzaa M, Rossi R, Alvisi M, Signore M, Serra E, Paolesse R, et al. Metalloporphyrins-modified carbon nanotubes networked films-based chemical sensors for enhanced gas sensitivity. Sensors and Actuators B. 2010;**144**:387-394. DOI: 10.1016/j.snb.2008.12.060

[64] Penza M, Rossi R, Alvisi M, Valerini D, Serra E, Paolesse R, et al. Metalloporphyrins-functionalized carbon nanotube networked films for room-temperature vocs sensing applications. Procedia Chemistry. 2009;**1**:975-978. DOI: 10.1016/j.proche.2009.07.243

[65] Lvova L, Mastroianna M, Pomarico G, Santonico M, Pennazza G, Natale C, et al. Carbon nanotubes modified with porphyrin units for gaseous phase chemical sensing. Sensors and Actuators B. 2012;**170**:163-171. DOI: 10.1016/j.snb.2011.05.031

[66] Shirsat M, Sarkar T, Kakoullis J, Myung N, Konnanath B, Spanias A, et al. Porphyrin-functionalized single-walled carbon nanotube chemiresistive sensor arrays for VOCs. Journal of Physical Chemistry C. 2012;**116**:3845-3850. DOI: 10.1021/jp210582t

[67] Sarkar T, Srinives S, Sarkar S, Haddon R, Mulchandani A. Single-walled carbon nanotube-poly(porphyrin) hybrid for volatile organic compounds detection. Journal of Physical Chemistry C. 2014;**118**:1602-1610. DOI: 10.1021/jp409851m

[68] Rushi A, Datta K, Ghosh P, Mulchandani A, Shirsat A. Iron tetraphenyl porphyrin functionalized single wall carbon nanotubes for the detection of benzene. Materials Letters. 2013;**96**:38-41. DOI: 10.1016/j.matlet.2013.01.003

[69] Baskaran D, Mays J, Zhang X, Bratcher M. Carbon nanotubes with covalently linked porphyrin antennae: Photoinduced electron transfer. Journal of the American Chemical Society. 2005;**127**:6916-6917. DOI: 10.1021/ja0508222

[70] Guo Z, Du F, Ren D, Chen Y, Zheng J, Liuc Z, et al. Covalently porphyrin-functionalized single-walled carbon nanotubes: A Novel photoactive and optical limiting donor-acceptor nanohybrid. Journal of Materials Chemistry. 2006;**16**:3021-3030. DOI: 10.1039/B602349E

[71] Liu Z, Tian J, Guo Z, Ren D, Du F, Zheng J, et al. Enhanced optical limiting effects in porphyrin-covalently functionalized single-walled carbon nanotubes. Advanced Materials. 2008;**20**:511-515. DOI: 10.1002/adma.200702547

[72] Palacin T, Khanh H, Jousselme B, Jegou P, Filoramo A, Ehli C, et al. Efficient functionalization of carbon nanotubes with porphyrin dendrons via click chemistry. Journal of the American Chemical Society. 2009;**131**:15394-15402. DOI: 10.1021/ja906020e

[73] Cabana J, Martel R. Probing the reversibility of sidewall functionalization using carbon nanotube transistors. Journal of Materials Chemistry. 2007;**129**:2244-2245. DOI: 10.1021/ja068320r

[74] Cognet L, Tsyboulski D, Rocha J, Doyle C, Tour J, Weisman R. Stepwise quenching of exciton fluorescence in carbon nanotubes by single-molecule

reactions. Science. 2007;**316**:1465-1468. DOI: 10.1126/science.1141316

[75] Meng L, Fu C, Lu Q. Advanced technology for functionalization of carbon nanotubes. Progress in Natural Science. 2009;**19**:801-810. DOI: 10.1016/j.pnsc.2008.08.011

[76] Shin S, Lee C, So I, Jeon J, Kang T, Kee C, et al. DNA-wrapped single-walled carbon nanotube hybrid fibers for supercapacitors and artificial muscles. Advanced Materials. 2008;**20**:466-470. DOI: 10.1002/adma.200701102

[77] Mu Q, Liu W, Xing Y, Zhou H, Li Z, Zhang Y, et al. Protein binding by functionalized multiwalled carbon nanotubes is governed by the surface chemistry of both parties and the nanotube diameter. Journal of Physical Chemistry C. 2008;**112**:3300-3307. DOI: 10.1021/jp710541j

[78] Yu J, Huang K, Liu S, Tang J. Preparation and characterization of polycarbonate modified multiple-walled carbon nanotubes. Chinese Journal of Chemistry. 2008;**26**:560-563. DOI: 10.1002/cjoc.200890105

[79] Murakami H, Nomura T, Nakashima N. Noncovalent Porphyrin-functionalized single-walled carbon nanotubes in solution and the formation of porphyrin–nanotube nanocomposites. Chemical Physics Letters. 2003;**378**:481-485. DOI: 10.1016/S0009-2614(03)01329-0

[80] Zhong Q, Diev V, Roberts S, Antunez P, Brutchey R, Bradforth S, et al. Fused porphyrin-single-walled carbon nanotubes hybrids: Efficient formation and photophysical characterization. ACS Nano. 2013;7:3466-3475. DOI: 10.1021/nn400362e

[81] Ndiaye A, Bonnet P, Pauly A, Dubois M, Brunet J, Varenne C, et al. Noncovalent functionalization of single-wall carbon nanotubes for the elaboration of gas sensor dedicated to BTX type gases: The case of toluene. Journal of Physical Chemistry C. 2013;**117**:20217-20228. DOI: 10.1021/jp402787f

[82] Liu S, Moh L, Swager T. Single-walled carbon nanotube-metalloporphyrin composites in chemiresistive gas sensor arrays for volatile organic compounds. Chemistry of Materials. 2015;**27**:3560-3563. DOI: 10.1021/acs.chemmater.5b00153

[83] Rushi A, Datta K, Ghosh P, Mulchandani A, Shirsat A. Exercising substituents in porphyrins for real time selective sensing of volatile organic compounds. Sensors and Actuators B. 2018;**257**:389-397. DOI: 10.1016/j.snb.2017.10.147

Chapter 7

Behavior of X-Ray Analysis of Carbon Nanotubes

Firas Habeb Abdulrazzak, Ayad Fadel Alkiam and Falah Hasan Hussein

Abstract

Carbon nanotubes with a variety of types occupied an amazing position compared to other nanomaterials due to rarer and specific physical and chemical properties. The behavior of graphite or graphite nanotubes theoretically and experimentally encourages their use in huge applications such as industries, fields of energy, and the environment. Many attempts are being made to get more of these benefits by a better understanding of the nature of carbon nanotubes. One of the ways to achieve this aim is by enhancing the acquaintance of characterization such as spectroscopy analysis and image microscopy. In this chapter, we are concerned with using X-ray as a source to produce clear imaginations for a tubular structure. Thus, the common ways that use X-ray as a source to interact with carbon nanotubes will be reviewed with details of characterization such as XRD, XRF, XPS, and EDX techniques.

Keywords: XRD, XRF, XPS, EDX, CNT 1

1. Introduction

The behavior of light after many years of studies and experiments opens a new orientation for science and scientists taking advantage of these features. The binary behavior of light which is represented by wave nature and particle nature enabled to appear with a variety of behaviors such diffraction, interference, and refraction. **Figure 1** shows spectrum of electromagnetic radiation which includes all the radiation types.

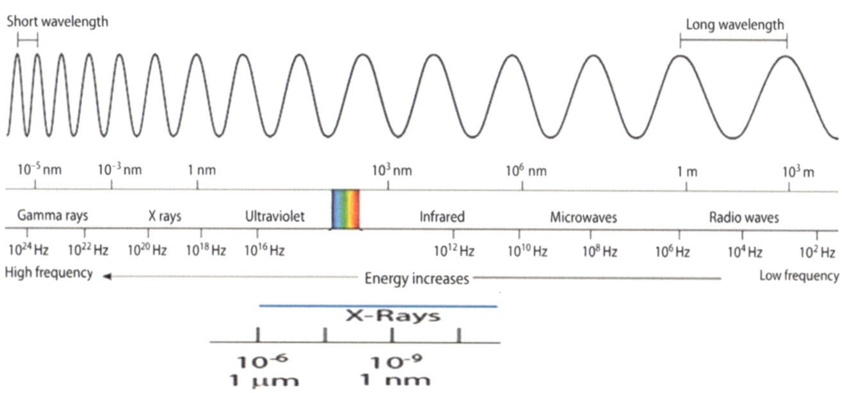

Figure 1.
The spectrum of the electromagnetic radiation with different kinds of oscillators and frequencies.

IntechOpen

Every type in spectroscopy has specific properties which urge to use it in many applications. X-rays that are mentioned in this part with the highest-energy electromagnetic waves were commonly used in hospital and engineering applications. X-rays move in a straight line without any influence with magnetic and electric field due to the negative charge on it. In this chapter, we concern toward using X-ray behavior with materials in qualitative and quantitative applications of carbon nanotubes (CNTs).

2. Carbon nanotubes (CNTs)

The term carbon nanotubes (CNTs) refers to nanomaterials with a tubular structure consisting of carbon atoms only bonded with each other by sp2 hybridization. The talks about CNTs were increasing 27 years ago after Lijima reported the first clear research [1], which represents the starting point of the huge attention on CNTs. The attentions appeared as reported literature deals with synthesis, purification, characterization, and studding physiochemical properties [2, 3]. Many results were paving the way for many successful attempts to use CNTs practically in many fields of science. The applications such as energy, construction, treatment, and environmental protection fields [4] were enhancing to increase the attention to make more clear images for CNTs. Carbon nanotubes (CNTs) were synthesized by chemical vapor deposition CVD, arc-discharge ADs, laser ablation LAs [4], and flame method FM [5]. Many types of apparatus were used to identify CNTs such scanning electron microscopy (SEM), transmission electron microscopy (TEM), helium ion microscopy (HIM), Raman microscopy, and X-ray family. In this chapter, we focus on the characterization of carbon nanotubes by using a specific technique that is used for this purpose, which depends on X-ray as a source to show the carbonic composition. The orientation of atoms within the lattice determines the peak intensity. Therefore, the X-ray behavior with materials is the fingerprint for periodic atomic arrangements in a known material. The start point with these aims to classify CNTs into two types depending on the number of sheets as shown in **Figure 2**.

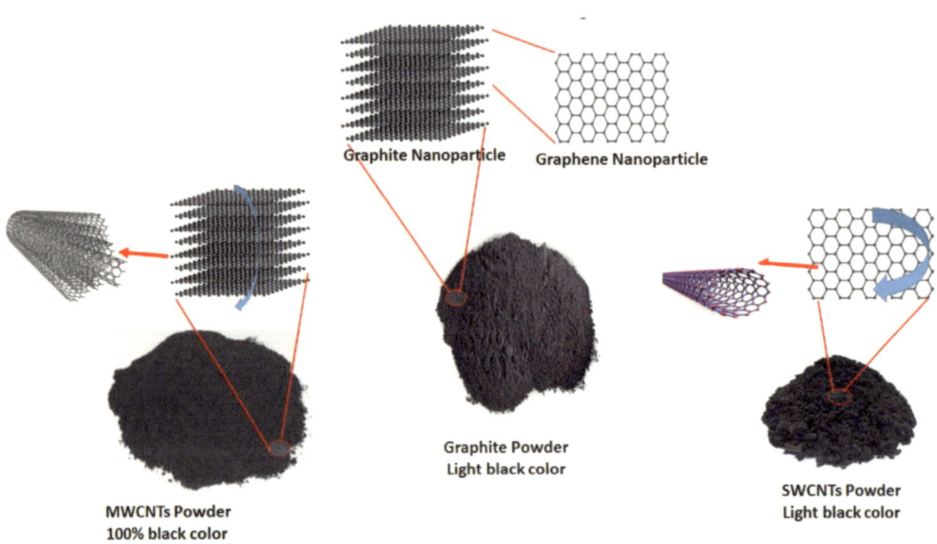

Figure 2.
Figure shows the classification of CNTs by the number of sheets.

2.1 Graphite nanotubes (GNTs)

The graphite nanotubes refer to graphite materials that show a tubular structure with nanometer dimensions, which are commonly known as multi-walled carbon nanotubes (MWCNTs).

2.2 Graphene nanotubes (gNTs)

The second type refers to graphite structure which forms nanotubes and that includes three types:

2.2.1 Single walled carbon nanotubes (SWCNTs)

There is one sheet of carbon atoms that forms SWCNTs, characterized by diameter 0.5–1.5 nm [6].

2.2.2 Double walled carbon nanotubes (DWCNTs)

Two sheets of carbon atoms are flopped with each ether to form DWCNTs, with 2.4–3.2 nm diameter [7].

2.3 Few walled carbon nanotubes (FWCNTs)

When the number of sheets is 2–6, the type is known as few walled carbon nanotubes FWCNTs. The diameter as we mentioned before was decided by the number of sheets that forms it, which ranges between 4 and 9 nm [8].

3. How X-ray is produced?

X-rays are generated in an X-ray tube responsible to generate X-rays when an electron beam is passed between a cathode part and an anode. **Figure 3** includes the essential components for an X-ray system when a stream of fast-moving accelerated electrons is transferred from the cathode to the anode. The interaction of electrons with atoms produces a great amount of energy [9] equal to 1% of the total energy while 99% appear as heat when removed from the anode. When current is passed through a conductive material, the material will heat due to the resistance in the

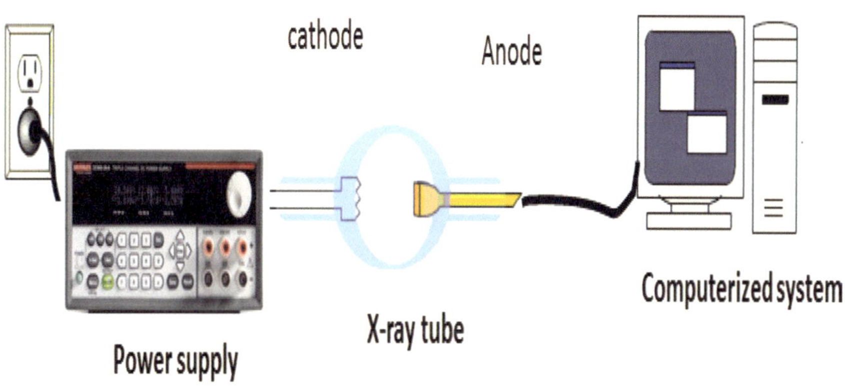

Figure 3.
Figure for essential components of X-ray generator.

wire causing excitation of the electrons and energy expansion. When the energy of the electron expands, it will return to the wire to become heat again and becomes a source of electrons [10]. The positive charge and negative charge will form a strong, attractive force to accelerate the electrons, which could increase the forces of X-ray by increasing the voltage applied to the anode.

4. Debye-Scherer equation and Bragg's law

The X-rays are commonly generated by a cathode ray tube, occupied with a specific filter to form monochromatic radiation, which concentrates toward the sample. **Figure 4** represents Bragg's law ($n\lambda = 2d \sin \Theta$) referring to the specific case of diffraction, which depends on the angles of scattering (Θ) coherent and incoherent monochromatic light (λ) from a crystal lattice consisting of n sheets. This equation ($d = (K^*\lambda)/(\beta^* \cos \theta)$) is commonly used to find partial size (d) for materials [11] when the benefit from the behavior of incident and monochromatic light (λ) with specific angle (Θ) is decided from the nature of the construction material. The value of (d) was calculated for the line broadening at half the maximum intensity (β = FWHM), while the shape factor (K) was typically equal to about 0.9, which changes with shape of the crystallite. When X-rays are incident on an atom, they make the electronic cloud move, as does any electromagnetic wave. The movement of these charges radiate wave with the same frequency, blurred slightly due to [10] a variety of effects, and this phenomenon is known as Rayleigh scattering (or elastic scattering). The scattered waves can themselves be scattered, but this secondary scattering is negligible.

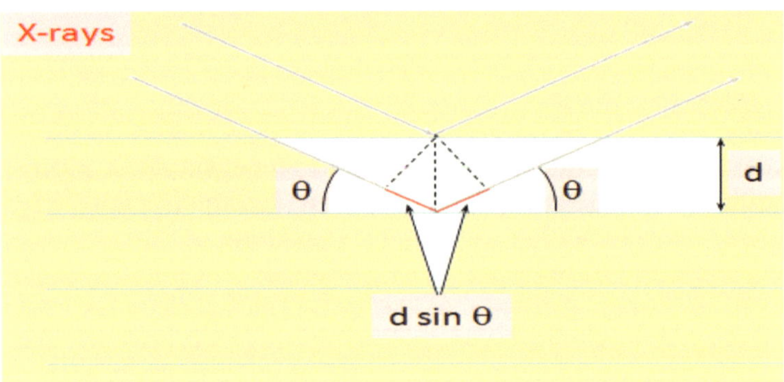

Figure 4.
The typical equation to characterize the structure of the crystal.

5. X-ray diffraction analysis (XRD)

XRD analysis is a unique method to find qualitative properties such as crystallinity, types, fingerprints, and quantitative material [11]. Graphite and graphene show great similarity in XRD with GNTs and joints, respectively, for the main peaks [12–14] at ≈24 and 43°. Most references shows these two peaks high degree of similarity, especially at the first strong peak. **Figure 5** shows the two peaks: the first on the left is of very high intensity with less width compared with the second peak at the right. The length and width of the peaks were commonly used to recognize the nature of crystal size. Generally, for carbon nanotubes, the first peak at the right refers to crystalline or amorphous nature; thus, the width increases with reducing intensity or length, which refers to exist in amorphous crystal form [15].

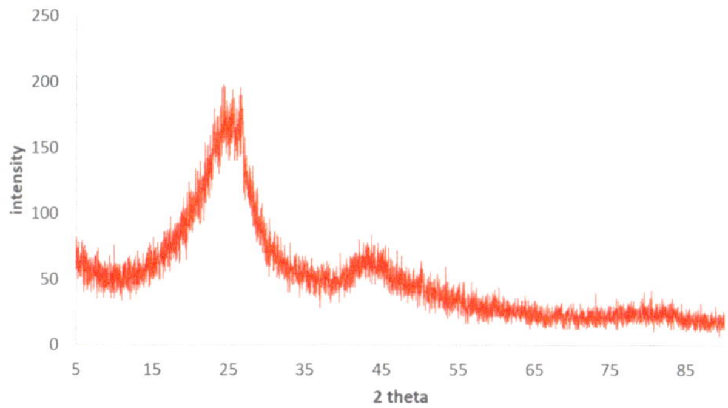

Figure 5.
Typical XRD analysis for CNTs.

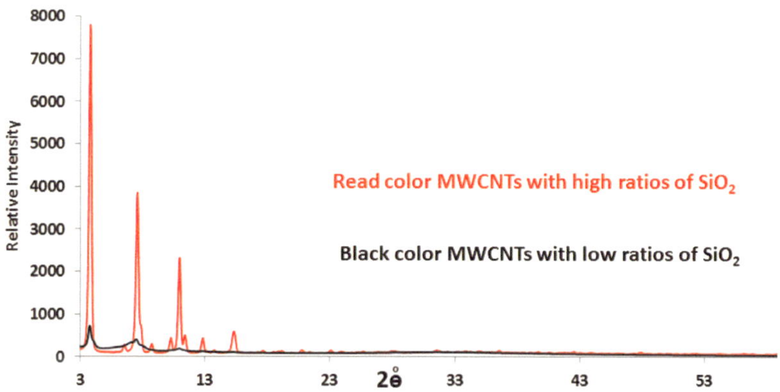

Figure 6.
XRD patterns for synthesized MWCNTs with high and low ratios of SiO_2.

Typically, the impurities influence Directly on the position of two peaks due to low intensity of the peaks compared with the impurities. For example, MWCNTs were prepared by chemical vapor deposition using 1-propanol as a source of carbon at 700°C on the surface of silica as support for precipitation as shown in **Figure 6**. The XRD analysis of synthesized MWCNTs is reported in **Figure 6**, showing the effect of amorphous carbon and Si, Na, Fe, and Mg in the diagram of XRD which removed or at least limited the appearance of the two characterized peaks of CNTs. The elements which were mentioned causing a disappearance [16] for the carbon tubular structure had spread between these components. Thus, XRD for the sample was disappearing due to the higher intensity of many crystal structures [17] of the element when reached 8000 units compared with low intensity of carbon with less than 120 units.

Figure 7 shows two notes to be taken care of when using XRD to predict the existence of CNTs: the first, the impurities inevitably caused shift for the two peaks due to real connection between tubular structure and impurities. The second refers to influence of interference, which causes removal of the second peak or complete disappearance due to the high difference between importers and carbon material. The last two points look like a photograph; picture was taken for wood including very high Trees compared to grass and few small herbs, the result or the image should be favored the trees.

The typical peak at ≈26.0° represents the characteristic graphitic peak [18] arising due to the tubular structure of the carbon atoms in the sample with (002) planes. This peak shifted to more than 28.8°, which affects [19] the ratios of impurities in the sample. The typical peaks near 43.2 and 53.74° disappeared due to

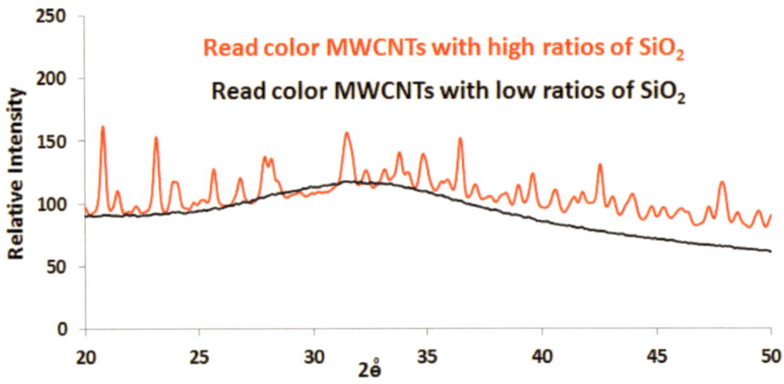

Figure 7.
XRD patterns for synthesized MWCNTs with high and low ratios of SiO_2 from 20 to 50°.

existing Si and Fe [19], which show strong peaks at ≈42 and 52°, respectively, when interfering with the weak peaks of CNTs.

5.1 MWCNTs vs. SWCNTs

This part represents one of the most important section for the science of nanomaterials generally and nanotubes particularly, which could be dependent on to identification and discrimination between SWCNTs, DWCNTs, FWCNTs, and MWCNTs. The interest in this part refers to direct application for the two roles of diffraction and the number of sheets that form the skeleton of tubular structures. This fact we referred to in the International Conference on Chemistry and Applied Research in October 29–30, 2018 Prague, Czech Republic. The fact was that there were many noises for SWCNTs as compared with MWCNTs with the two peaks, which enhance forming the noises to reduce the number of graphene sheets, as following arrangement for CNT types. Thus, the ratios of noise can be arranged as follows:

SWCNT > DWCNT > FWCNT > MWCNT

Figure 8 explains the behavior of CNTs when an increase in the number of inner shells causes more intensity for the peaks with MWCNTs than SWCNTs which show in pure and with mixture of composite. This fact was in agreement with a good

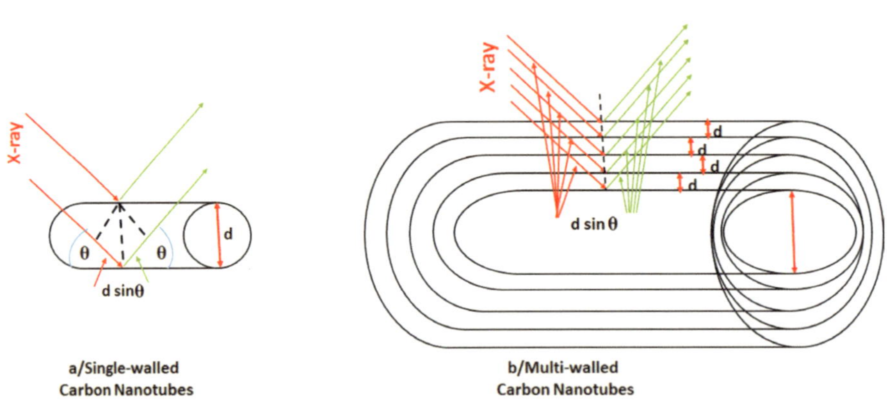

Figure 8.
Figure shows the influence of number of layers with X-ray diffraction.

deal of experimental research, such as studies by Mhamane et al., Tang et al., and Chang et al. [20–22] for graphite and graphene, showing noise with graphene and graphene oxide while reduced noise with graphite.

6. X-ray fluorescence (XRF)

The principles of this technique depend on removing the electron from lower energy level, causing an unstable state for atoms; therefore, the electron in the higher energy shell moves inside the atom. The value of the difference between higher and lower levels was emitting as specific energy to identify the fluorescing element and is proportional to the amount of element. Generally, the characteristic peaks are labeled X-rays as K, L, M, and N refer to the shells they activated or (α),

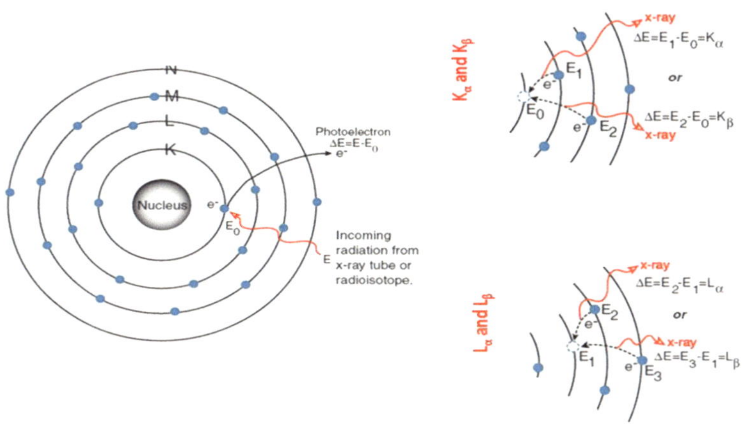

Figure 9.
A supposed mechanism for forming X-ray radiation.

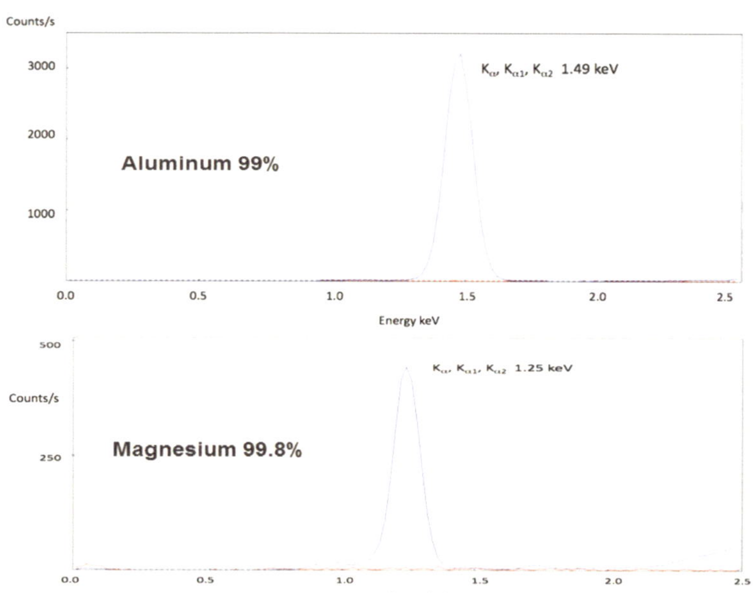

Figure 10.
XRF spectrum for aluminum and magnesium that are mostly used as a catalyst for growth of CNTs.

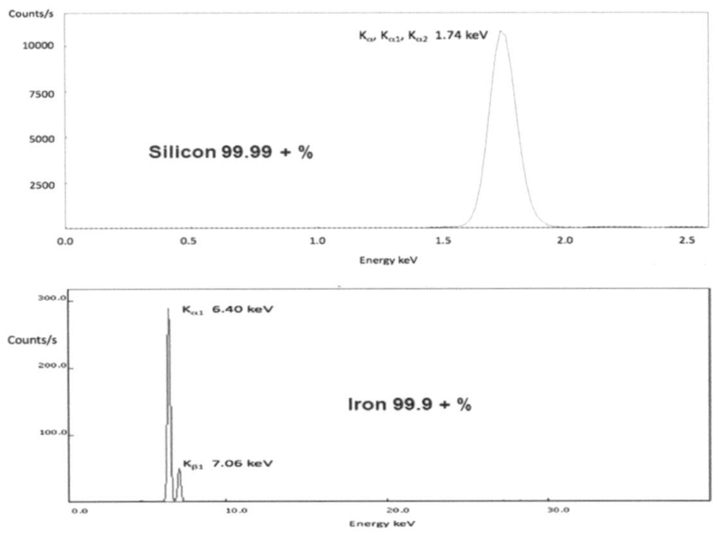

Figure 11.
XRF spectrum for silicon and iron that are mostly used as a support for catalyst CNTs.

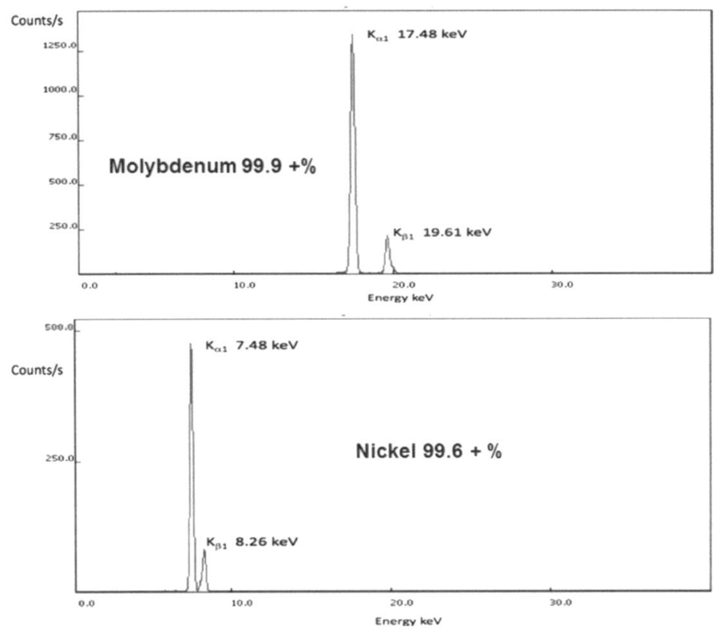

Figure 12.
XRF spectrum for nickel and molybdenum used as a catalyst for growth of CNTs.

(β), and (γ) depend on the transitions of electrons from higher shells. However, Kα2 means the transition of an electron from the L to the K shell, while Kβ1 represents the transition of an electron from the M to the K shell of multiple orbits.

Figure 9 refers to the behavior of elementary information that makes possibilities to quantify identification of metals, which are commonly used as a catalyst for precipitation of carbon on the support of the catalyst to form CNTs. The advantage of this technique is that it is rapid and it is easy to treat the sample without any complications and without being affected by chemical bonding. The disadvantages include limited abilities to identify trace contents in addition to the high radiation that may cause damage for tubular structure CNTs. Singhal et al. [23] reported the qualitative analysis when analyzing CNT-coated Al. The analysis showed three peaks at 2Θ which are 32.5, 50.8, and 144.7° corresponding to C Kα, O Kα, and Al Kα spectral lines, respectively, related to the presence of C, O, and Al elements in the composite. The quantitative analysis for synthesized composite confirms that the percentage weight of C, which corresponds to CNTs, was 3.95%. XRF can be used for studded catalyst content in CNTs during the steps of purification, but in the same time, it cannot be used to determine amorphous carbon content. [24]. **Figures 10–12** show aluminum, magnesium, silicon, iron, molybdenum, and nickel that are commonly used for synthesized carbon nanotubes as a catalyst or support for precipitation.

7. Energy-dispersive X-ray spectroscopy (EDX)

The energy dispersive X-ray analysis EDX and EDA sometimes EDS or energy dispersive X-ray Microanalysis EDXMA. This technical method depends on the interaction of X-ray radiation with a sample when convert to single that refers to the material in the chemical area and thus is used for elemental characterization. The efficiency for characterization can be related to the fundamental principle that each element has a unique atomic structure allowing a unique set of peaks in its electromagnetic emission spectrum [10]. Mostly, scanning electron microscopy (SEM) supplied with EDX is used to qualitatively and quantitatively analyze the elements present in the selected area of the SEM image to estimate of CNTs' metal contents and impurities. Together, the SEM and EDX capabilities allow irradiation by a focused electron beam, imaging secondary or backscattered electrons and an energy analysis of X-rays. Typical SEM applications include plan view and cross-sectional imaging for process development and failure analysis. EDX applications include specific defect analysis or composition analysis. **Figures 13** and **14** refer to EDX analysis for SWCNTs and MWCNTs made from Aldrich by chemical vapor deposition with high purities. When using this technique should, analysis of large-scale CNTs should be avoided since it may cause error in identification due to overlap of peaks for CNTs with many elements such as Ti Kβ, Mn Kβ, and Fe Kα [25].

Figure 15 shows an EDX spectrum of MWCNTs synthesized by the CVD technique from decomposition date palm seeds at 700°C on the boat of silicon as supported for Mo, which was used to grow MWCNTs. The spectrum shows four strong peaks, which represent carbon, silicon, Al, and Mo. The C refers to carbon materials, which mostly refer to the growth of CNTs on Mo that covers the surface of

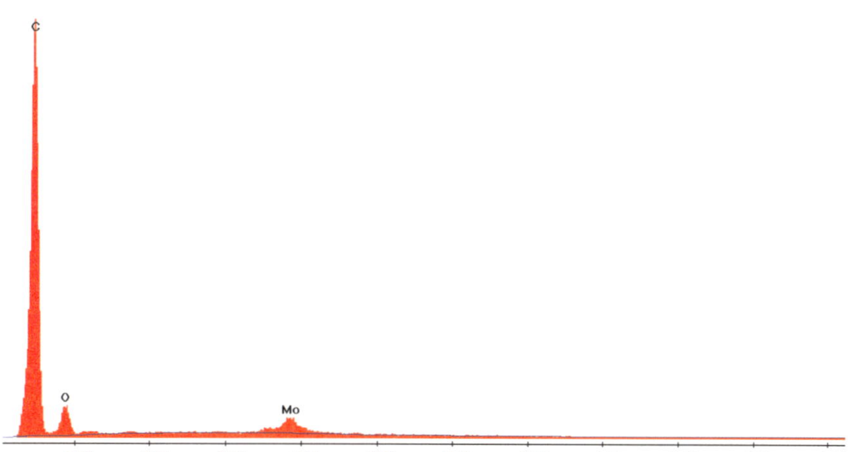

Figure 13.
EDX spectrum for Aldrich SWCNTs which were synthesized by the CVD method.

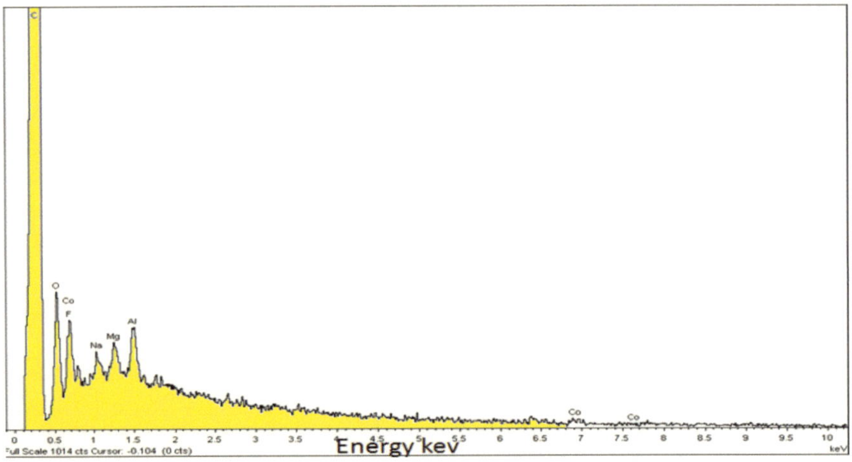

Figure 14.
EDX spectrum for Aldrich MWCNTs which were synthesized by the CVD method.

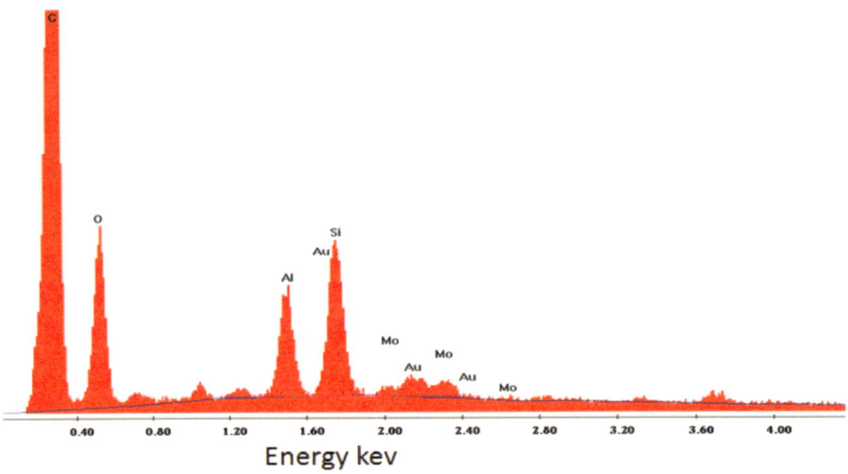

Figure 15.
EDX spectrum for Aldrich MWCNTs which were synthesized by the CVD method.

the silica boat. The O refers to the groups of oxide which are produced by a purification process by using nitric acid (5 M). **Figure 16**: EDX spectrum for MWCNTs synthesized by CVDs from decomposition date palm seeds at 700°C on the boat of silica as supported for Mo, which used to growth CNTs Shows an EDX spectrum of us received MWCNTs when synthesized from propanol at 700°C by CVDs with used silica as support and iron oxide as catalyst for growth CNTs. The EDXs in **Figure 16**, with three strong peaks, represent silicon, and two for Fe. This indicates that the growth of CNTs originated from Fe or Fe carbide particles on the silicon substrate. The peak of C is short, which refers to low precipitation on the surfaces of the boat.

The most important thing when used EDX represented by the area which needs to make quantitative and qualitative analysis due to heterogeneous distribution of the elements in analyzing samples. We mention before the aims from analysis whom responsible to decide the area of analysis. 9. X-ray photoelectron spectroscopy (XPS) is a quantitative technique that deals with measures, empirical formula, and chemical and electronic state with elemental composition in high sensitivity that exists within a material [10]. The most important nature of XPS is that it shows not only qualities of elements starting from lithium and above but also the nature of

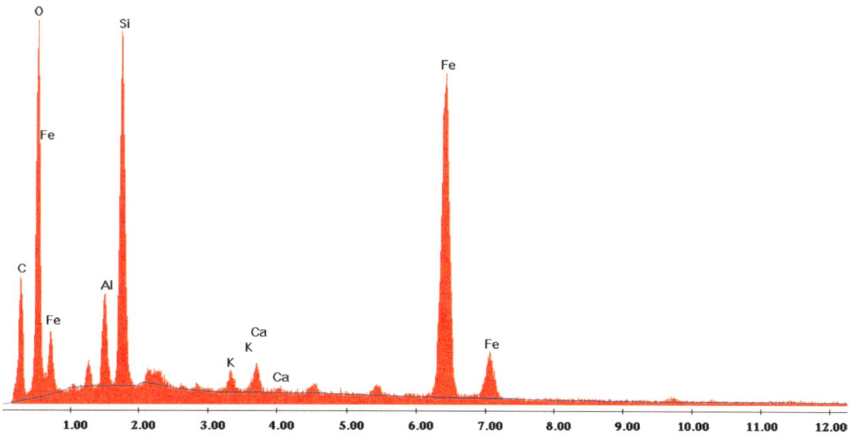

Figure 16.
EDX spectrum for MWCNTs prepared from propanol at 700°C by CVD with silica used as a support and iron oxide as a catalyst.

local bonds for all the components in a mixture. XPS spectra are obtained by irradiating thickness of analyzing material, arranges between 0 and 10 mm with a beam of X-rays to measure the number and energy of escaped electrons in high vacuum ambience. It mostly requires large amounts of sample to reach 5 mg to produce ideal analysis that commonly produces ambiguous and interpreted peaks, which are simplified by a specific computerized system. The XPS spectrum is represented by plotting the binding energy (eV) between different atoms with intensity of energy in the chemical environment of materials. The ideal detection limits for most of the elements are in the parts per thousand range while parts per million (ppm) could be made when special needs were provided such as concentration on the top surface and very long collection time. The oxidation state represents the main criterion for identifying [26–29] the elements due to influence with the local bonding environment for the atoms in molecules. In the case of carbon nanotubes, the binding energy of C1s electron is 284.6 eV; when binding with other atoms, it will be chemically shifted, which depends on the chemical state of neighboring atoms. For example, the carbon in the polymer structure shows increasing binding energy as follows [26–31]:

Carbide ($-C_2-$) < Silicone ($-Si-CH_3$) < Methylene/Methyl/Hydrocarbon ($-CH_2-CH_2- < CH_3-CH_2- < -CH=CH-$) < Amine ($-CH_2-NH_2$) < Alcohol ($-C-OH$) < Ketone ($-C=O$) < Organic Ester ($-COOR$)

The characterization peak of carbon nanotubes corresponding to C=C for graphene layer for spectra at about 284 e V.

For example, the synthesized carbon nanotubes from natural petroleum gas were tested with X-ray photoelectron spectroscopy. The nature of different groups that were covalently linked to the nanotube surface is reported in **Table 1**, where existence of C=C refers to nanotubes or graphene sheets. To have a more clear and more specific explanation, we review different groups that are covalently linked to the nanotube surface to improve C=C which refers to nanotubes by XPS analysis. The carbon nanotubes (MWCNTs) were synthesized by the chemical flame deposition method by using liquefied petroleum gas as a source of carbon on the surface of iron as supported for precipitation at 180°C. **Figure 17** shows the photoemission C1s peaks that were studies between 282 and 292 eV. A dominant peak is observed at 284 or 285 eV of the sp^2 hybridized C=C bonds in extensive conjugated systems

Functional Group		Binding Energy (eV)
hydrocarbon	C-H, C-C	285.0
amine	C-N	286.0
alcohol, ether	C-O-H, C-O-C	286.5
Cl bound to C	C-Cl	286.5
F bound to C	C-F	287.8
carbonyl	C=O	288.0

Table 1.
Common values of band that predicted existence in CNTs.

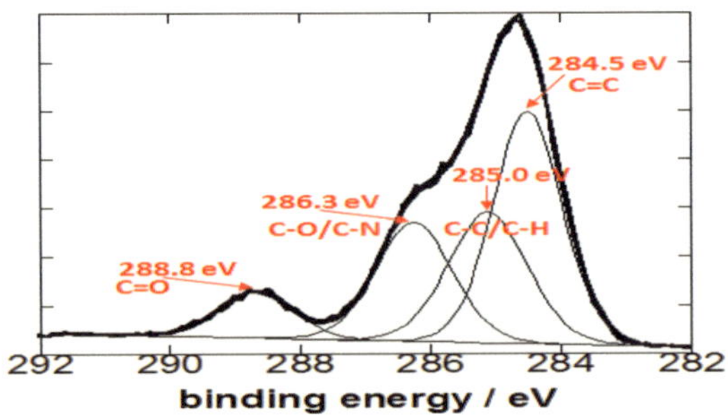

Figure 17.
XPS C 1s spectra of synthesized MWCNT between 282 and 292 eV. Arrows show possible positions of carbon peaks.

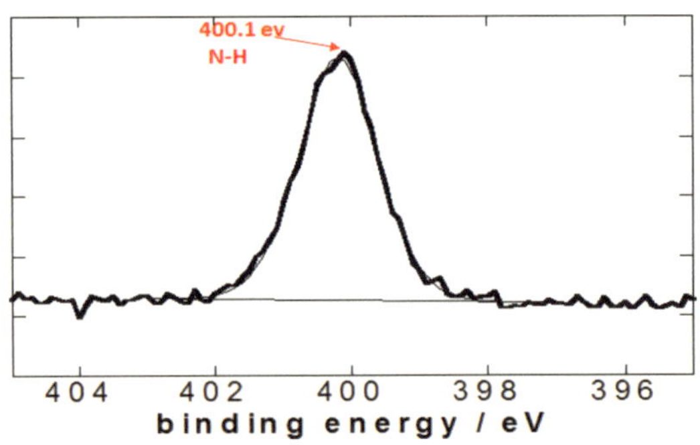

Figure 18.
XPS of N–H spectra between 396 and 404 eV. Arrows show possible position of the N–H peak.

[26, 27], which mostly refers to graphene sheets in carbon nanotubes. The same figure includes a peak at 285.0 eV, which is characteristic of sp3 hybridized C–C bonds present at defective locations [28] and tubular structure asymmetry related to C–H and C–C [29]. Two other photoemission peaks are observed at 286.3 and 288.8 eV. The first peak at 286.3 eV is attributed to carbon atoms bonded to oxygen atoms C–O which bonded with C–N. The second peak at 288.8 eV is characteristic of carbon atoms of carbonyl groups (C=O) [30]. **Figure 18** shows N 1s 400.1 eV spectra, which refer to N–H related to N–C=C groups [31].

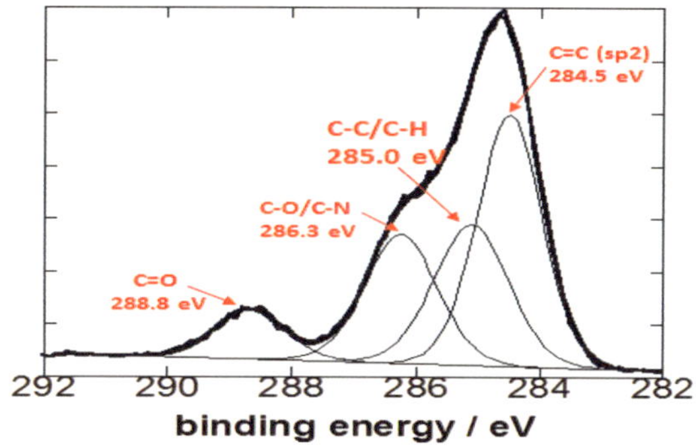

Figure 19.
XPS C 1s spectra of synthesized MWNT between 282 and 292 eV. Arrows show possible positions of carbon peaks.

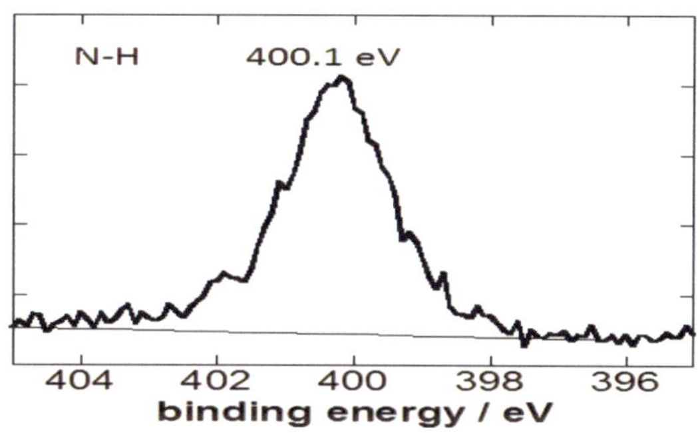

Figure 20.
XPS of N–H spectra between 396 and 404 eV.

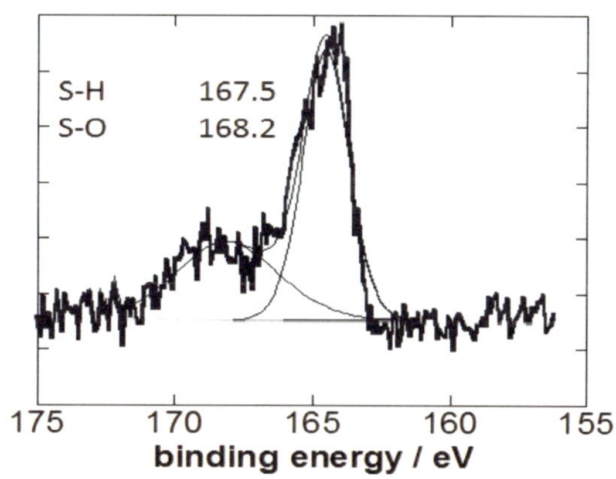

Figure 21.
XPS of S spectra between 155 and 175 eV.

Figure 19 shows the XPS spectra of the second sample with different peaks at 284.5, 285.0, 286.3, and 288.8 eV. The first peak at 284.5 eV corresponds to C=C for graphene sheets [32] while the neighboring peak at 285.0 eV corresponds to C–C and C–H. The second peak at 286.3 eV is related to C–O/C–N [33]. The third peak at 288.8 eV corresponds to C=O [30]. **Figure 20** shows the N 1 s spectra at 400.1 eV, which refers to N–H related to N–C=C groups [31].

Figure 21 refers to two overlapping peaks together at 167.5 and 168.2 eV, which correspond to S–H and S–O, respectively, while the strong peak at 164.7 eV sp2 indicates the S–C bond [34]. The second sample shows an existence of sulfur which can be related to the oil that was used to prepare this sample only.

8. Conclusion

Many techniques and methods are commonly used for characterizing and analyzing nanomaterials such as SEM, TEM, HIM, Raman microscopy, UV-visible reflection, and X-ray family. All of the techniques mentioned before and others depend on reactant materials with electrons. X-ray led to many ways of getting huge information about the composition of materials and the nature of construction. X-ray family is supplied mostly, nondestructively and fast, with the ability to provide the basic structural information that is essential for identification. In our chapter, X-ray families were used for quantification and qualification of carbon nanotubes. XRD, XRF, EDX, and XPS were used in this chapter with many synthesized CNTs to show the abilities for evaluating the purifies which are very important in applications. We anticipate that the results of this study will help to offer guidelines for understand CNTs structures by X-ray family to increase the probabilities for benefit from CNTs.

Author details

Firas Habeb Abdulrazzak[1*], Ayad Fadel Alkiam[2] and Falah Hasan Hussein[3,4]

1 Department of Chemistry, College of Education for Pure Science, Diyala University, Diyala, Iraq

2 Department of Chemistry, College of Science, Babylon University, Hilla, Iraq

3 Department of Pharmaceutics, College of Pharmacy, Babylon University, Hilla, Iraq

4 Department of Research and Studies, Al-Mustaqbal University College, Babylon, Iraq

*Address all correspondence to: firas_habeb2000@yahoo.com

IntechOpen

© 2019 The Author(s). Licensee IntechOpen. This chapter is distributed under the terms of the Creative Commons Attribution License (http://creativecommons.org/licenses/by/3.0), which permits unrestricted use, distribution, and reproduction in any medium, provided the original work is properly cited.

References

[1] Iijima S. Helical microtubules of graphitic carbon. Nature. 1991;**354**:56-58

[2] Popov V. Carbon nanotubes: Properties and application. Materials Science and Engineering R. 2004;**43**(3):61-102

[3] Yadav B, Kumar R. Structure, properties, and applications of fullerenes. International Journal of Nanotechnology and Applications. 2008;**2**(1):15-24

[4] Falah HH, Firas HA, Ayad FA. Nanomaterials: Biomedical and Environmental Applications, Chapter 1, 4: Nanomaterials: Synthesis and Characterization. 1st ed. Wiley; 2018

[5] Firas HA, Ahmad MA, Falah HH. Synthesis of multi-walled carbon nanotubes from Iraqi natural gas/CO mixture by catalytic flame fragments deposition method. Asian Journal of Chemistry. 2019;**31**(1):247-250

[6] Bansi D, Saurabh S, Shine A. Biosensors for food toxin detection: Carbon nanotubes and graphene. In: Materials Research Society Symposium Proceedings 1725. Materials Research Society; 2015

[7] Jaehyeok D, Jongsoo L. Prediction of the mechanical behavior of double walled-CNTs using a molecular mechanics-based finite element method: Effects of chirality. Computers and Structures. 2016;**169**:91-100

[8] Elsa G, Casanova O, Román M, Aguilar A, Espinosa F. Synthesis of carbon nanotubes of few walls using aliphatic alcohols as a carbon source. Materials. 2013;**6**(6):2534-2542

[9] Skinner CH. Review of soft X-ray lasers and their applications. Physics of Fluids B: Plasma Physics. 1991;**3**:2420-2430

[10] Yoon TH. Applications of soft X-ray spectromicroscopy in material and environmental sciences. Journal Applied Spectroscopy Reviews. 2009;**44**(2):91-122

[11] Ahmad M, Foroughi MR, Monish MR. Modified Scherer equation to estimate more accurately nano-crystallite size using XRD. World Journal of Nano Science and Engineering. 2012;**2**:154-160

[12] Rasel D, Abd Hamid SB, Ali ME, Seeram R, Wu Y. Carbon nanotubes characterization by X-ray powder diffraction—A review. Current Nanoscience. 2015;**11**:23-35

[13] Li ZQ, Lu CJ, Xia ZP, Luo ZZ. X-ray diffraction patterns of graphite and turbostratic carbon. Carbon. 2007;**45**(8):1686-1695

[14] Johra FT, Lee J-W, Jung W-G. Facile and safe graphene preparation on solution based platform. Journal of Industrial and Engineering Chemistry. 2014;**20**:2883-2887

[15] Longlong X, Yifu Z, Xiongzhi Z, Yu H, Xiaoyu T, Chi H, et al. Designed synthesis of tunable amorphous carbon nanotubes (a-CNTs) by a novel route and their oxidation resistance properties. Bulletin of Materials Science. 2014;**37**(6):1397-1402

[16] Tam NT, Nghia NX, Khoi PH, Minh PN. Analyzing the purity of carbon nanotubes by using different methods. Journal of the Korean Physical Society. 2008;**52**(5):1382-1385

[17] Chen XH, Chen CS, Chen Q, Cheng FG, Zhang G, Chen ZZ. Non-destructive purification of multi-walled carbon nanotubes produced by catalyzed CVD. Materials Letters. 2002;**57**(3):734-738

[18] Nishimura K, Okazaki N, Pan L, Nakayama Y. In situ study of iron catalysts

for carbon nanotube growth using X-ray diffraction analysis. Japanese Journal of Applied Physics. 2004;**43**(4A):471-474

[19] Andrew R, Jacques D, Qian D, Dickey EC. Purification and structural annealing of multi walled carbon nanotubes at graphitization temperatures. Carbon. 2001;**39**(11):1681-1687

[20] Mhamane KD, Ramadan W, Fawzy M, Abhimanyu R, Megha D, Chandra R, et al. From graphite oxide to highly water dispersible functionalized graphene by single step plant extract-induced deoxygenation. Green Chemistry. 2011;**13**:1990-1996

[21] Yu FT, Fuqiang H, Wei Z, Liu Z, Wan D. Synthesis of graphene-supported $Li_4Ti_5O_{12}$ nanosheets for high rate battery application. Journal of Materials Chemistry. 2012;**22**:11257-11260

[22] Leea CY, Baeb J-H, Kim T-Y, Chang S-H, Kim SY. Using silane-functionalized graphene oxides for enhancing the interfacial bonding strength of carbon/epoxy composites. Composites Part A: Applied Science and Manufacturing. 2015;**75**:11-17

[23] Singhal SK, Seth RK, Satish T, Rajeev C, Mathur RB. Synthesis and characterization of carbon nanotubes over iron carbide nanoparticles coated Al powder using thermal chemical vapor deposition. Applied Nanoscience. 2013;**3**:41-48

[24] Brandon C, Joshua H, Joe V, Williams S. X-ray fluorescence as a method of monitoring metal catalyst content during the purification of carbon nanotubes. Radiation Physics and Chemistry. 2012;**81**(2):131-134

[25] Goldstein J, Newbury DE, Joy DC, Lyman CE, Echlin P, Lifshin E, et al. Scanning Electron Microscopy and X-ray Microanalysis. Berlin: Springer; 2003

[26] Yang DQ, Rochett JF, Sacher E. Controlled chemical functionalization of multiwalled carbon nanotubes by kiloelectronvolt argon ion treatment and air exposure. Langmuir. 2005;**21**(18):8539-8545

[27] Li M, Boggs M, Beebe TP, Huang CP. Oxidation of single walled carbon nanotubes in dilute aqueous solutions by ozone as affected by ultrasound. Carbon. 2008;**46**:466-475

[28] Ramana GV, Kumar RN, Prabhakar KVP, Jain PK, Balaji P. Mechanical properties of multi-walled carbon nanotubes reinforced polymer. Indian Journal of Engineering and Materials Science. 2010;**17**(5):331-337

[29] Dilip CD, Veena S. Analysis of carbon nanotubes produced by pyrolysis of composite film of poly (vinyl alcohol) and modified fly ash. Materials Sciences and Applications. 2012;**3**:103-109

[30] Xia W, Wang Y, Bergstraszer R, Kundu S, Muhler M. High-density and well-aligned carbon nanotubes formed by surface decomposition of SiC. Applied Surface Science. 2007;**254**:257-261

[31] Yang K, Gu M, Guo Y, Pan X, Mu G. Effects of carbon nanotube functionalization on the mechanical and thermal properties of epoxy composites. Carbon. 2009;**47**:1723-1737

[32] Shuai W, Tang LA, Qiaoliang B, Ming L, Goh SDBM, Loh KP. Room-temperature synthesis of soluble carbon nanotubes by the sonication of graphene oxide nanosheets. Journal of the American Chemical Society. 2009;**131**(46):16832-16837

[33] Silva WM, Hélio R, Seara LM, Calado HDR, Ferlauto AS, Paniago RM, et al. Surface properties of oxidized and aminated multi-walled carbon nanotubes. Journal of the Brazilian Chemical Society. 2012;**23**(6):1078-1086

[34] Samarendra ND, Dulal D, David CF. Study of sulphur in Assam coals by X-ray photoelectron spectroscopy. Fuel. 1983;**62**(7):840-841

Chapter 8

Conductive Effect of Increased Crystallinity of Single-Walled Carbon Nanotubes as Field Emitter

Norihiro Shimoi

Abstract

Carbon nanotubes (CNTs) exhibit chemical stability, thermal conductivity, mechanical strength, and unique properties as a quasi-one-dimensional material with nanoscale needle shape. Field-emission (FE) electron sources appear to be the most promising industrial application for CNTs, and their deployment is approaching practical utilization. So far, efforts to construct an FE cathode with single-walled carbon nanotubes (SWCNTs) have only managed to average out the large FE current fluctuations in a nonhomogeneous electron emitter plane and the short emission lifetime because the crystal defects in the carbon network in CNTs prevent the realization of a stable emission current. The utilization of CNTs to obtain an effective electronic device, one with stable emission and low FE current fluctuations, relies on the high crystallization of CNTs, a task that can be fulfilled by using highly crystalline SWCNTs (hc-SWCNTs). The author could succeed in developing a model of the flow of electrons through the inside of the hc-SWCNTs and SWCNTs with crystal defects to the outside using the fluctuations of the tunneling current. Therefore, we expect that the hc-SWCNTs are used as field emitters with stable emission and low power consumption for saving energy.

Keywords: single-walled carbon nanotube, high crystallization, field emission, inelastic electron tunneling, current fluctuation

1. Introduction

In the development regime of electronic devices, carbon nanotubes (CNTs) are expected to represent a promising material because of their unique physicochemical properties—nanoscale needle shape, high chemical stability, thermal conductivity, and mechanical strength—which represent advantages for the fabrication of field emitters. The utilization of single-walled CNTs (SWCNTs) relies on their electronic properties because they can be either metallic or semimetallic, depending on the geometry of how a graphene sheet is rolled into a tube (i.e., diameter and chiral angle) [1–3]. This controllability is considered to be effective for developing electronic devices based on SWCNTs [3–5]. SWCNTs also exhibit one-dimensional confinement effects and can be used as coherent quantum wires [6–8], and the Young's modulus of SWCNTs is especially high when compared to that of other

IntechOpen

CNTs [9]. Owing to these properties, SWCNTs have been used in a wide range of applications including field emitters [10], probes in scanning microscopes [11], gas-storage materials [12], and electrode materials for secondary batteries [13], and they have been studied in a wide range of applied research fields.

However, synthesized CNTs, including SWCNTs, have crystal defects in the carbon network, and the unstable physicochemical properties make it difficult to use CNTs in electronic devices which require high reliability. Tohji et al. improved the purity and crystallinity of SWCNTs synthesized by arc discharge, and they succeeded in obtaining highly pure SWCNTs [14]. Together with Iwata and coworkers, they established a method to analyze, with high resolution, the crystallinity of highly crystalline SWCNTs (hc-SWCNTs) [15]. Based on these developments, a process for synthesizing hc-CNTs has been gradually established; however, a technique to control the crystallinity of hc-SWCNTs has yet to be demonstrated.

The application of hc-SWCNTs as field emitters is industrially promising and is approaching practical utilization. We have carried out basic research to develop field-emission (FE) devices using hc-SWCNTs [16] and succeeded, for the first time in recorded history, in employing SWCNT field emitters as the cathode of a planar lighting device [17]. These results suggest that hc-SWCNTs will be indispensable for decreasing the driving voltage for FE, increasing the emission lifetime, and improving the homogeneity of planar emission of planar lighting devices [17].

In the chapter reported here, the effect of the increased crystallinity of SWCNTs on their electrical properties was examined by comparing hc-SWCNTs and SWCNTs with crystal defects. In addition, field emitters were prepared from these SWCNTs, and the effects of the increased crystallinity of hc-SWCNTs on their electrical conductivity and FE current fluctuations were theoretically analyzed by determining their FE properties.

2. Theoretical electron tunneling model of the increased conductivity of hc-SWCNTs as field emitters

The Fowler-Nordheim (F-N) tunneling model shown in **Figure 1** is an electron tunneling model represented by Eq. (1). The F-N tunneling model is valid only when electrons in the field emitter are passing into an energy barrier quantum mechanically. We assume that electrons emit from SWCNTs to the outside in case of the FE model discussed in this chapter, and we constructed an FE model including F-N tunneling model by combining the model of inelastic and elastic tunneling electrons passing through the inside of SWCNTs to the outside.

In the quantum electron-tunneling model, the states of the electrons before and after tunneling through an energy barrier can be estimated by considering the energy transfer. According to Laks and Mills [18], the current density $\hat{J}\left(\vec{x},t\right)$ in the current-current correlation $\left\langle \hat{J}\left(\vec{x},t\right)^{*}\hat{J}\left(\vec{x},t\right)\right\rangle$ can be expressed in the following form:

$$\hat{I}(t) = \int dx_{\parallel}\hat{J}\left(\vec{x},t\right), \tag{1}$$

where $\hat{I}(t)$ is the quantum-mechanical tunneling-current operator and represents the current flow at time t. The normalization P of the wave function $\psi\left(\vec{x},t\right)$ is expressed as the probability density of electrons in a quantum barrier as follows:

$$P\left(\vec{x},t\right) = \left|\psi\left(\vec{x},t\right)\right|^2 = \psi^*\left(\vec{x},t\right)\psi\left(\vec{x},t\right). \tag{2}$$

When an electron exists in a sphere Ω,

$$\begin{aligned} &\frac{\partial}{\partial t}\int_\Omega P\left(\vec{x},t\right)d\vec{x} \\ &= \int_\Omega \left(\psi^*\frac{\delta\psi}{\delta t}\right)\mathrm{d}\vec{x} = \frac{1}{i\hbar}\int_\Omega (\psi^* H\psi - \psi H\psi^*)\mathrm{d}\vec{x} \\ &= \frac{i\hbar}{2m}\int_A (\psi^*\nabla\psi - \psi\nabla\psi^*)\mathrm{d}\vec{A} \end{aligned} \tag{3}$$

When $S\left(\vec{x},t\right) \equiv \frac{\hbar}{2im}(\psi^*\nabla\psi - \psi\nabla\psi^*)$,

$$\frac{\partial}{\partial t}\int_\Omega P\left(\vec{x},t\right)\mathrm{d}\vec{x} = -\int_\Omega \nabla S\left(\vec{x},t\right)\mathrm{d}\vec{x}. \tag{4}$$

When $\frac{\partial}{\partial t}P\left(\vec{x},t\right) + \nabla S\left(\vec{x},t\right) = 0$,

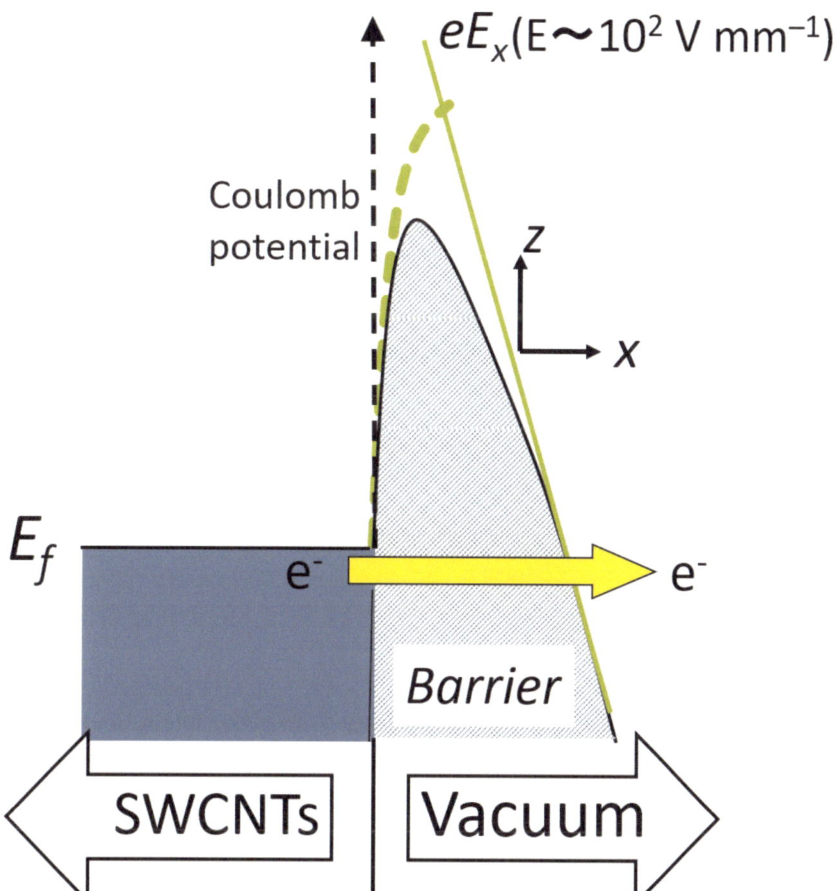

Figure 1.
Schematic diagram of Fowler-Nordheim (F-N) tunneling.

$$\hat{J}\left(\vec{x},t\right) = -e\hat{S}\left(\vec{x},t\right) = \frac{ie\hbar}{2m}\left(\psi^{*}\nabla\psi - \psi\nabla\psi^{*}\right), \tag{5}$$

where e is the electric charge. The quantum-mechanical tunneling-current operator $\hat{I}(t)$ can thus be expressed as:

$$\hat{I}(t) = \sum_{k \bullet q}\left[T_{kq}C_{q}^{r+}(t)C_{k}^{l}(t) + T_{kq}^{*}C_{k}^{l+}(t)C_{q}^{r}(t)\right], \tag{6}$$

where T_{kq} is the tunneling matrix element between state k of an electron in the SWCNT and state q of an electron in the vacuum through a tunneling energy barrier, and $C_{q}^{r+}(t)$and $C_{k}^{l}(t)$ are the creation and annihilation operators, respectively, of an electron before and after passing through an energy barrier. Using the Wentzel-Kramers-Brillouin (WKB) approximation to calculate the electron wave function, the tunneling matrix element can be expressed as [19–21]:

$$\left|T_{kq}\right| \propto \frac{\hbar^2}{2m^*}\exp\left(-\int_{z_1}^{z_r} dz\left|k_z(z)\right|\right)\exp\left(-\int_{z_1}^{z_r} dz\left|q_z(z)\right|\right), \tag{7}$$

where m^* is the effective mass of the electron in the tunneling barrier, and k_z and q_z are the z components of the electron wave vectors in the barrier perpendicular to the CNT surface:

$$k_z(z) = \frac{\sqrt{2m^*}}{\hbar}\left[V(z) - E + \frac{\hbar^2}{2m^*}k^2\right]^{\frac{1}{2}}, \tag{8}$$

$$q_z(z) = \frac{\sqrt{2m^*}}{\hbar}\left[V(z) - E + \frac{\hbar^2}{2m^*}k^2 + \hbar\omega\right]^{\frac{1}{2}}. \tag{9}$$

In Eqs. (8) and (9), $V(z)$ is the barrier potential, and z is the coordinate of the axis perpendicular to the interfaces. The average of the current-current correlation in the ith state is expressed as follows:

$$\left\langle \hat{I}^{*}(t)\hat{I}(0)\right\rangle = \frac{1}{N}\sum_{i}\left\langle i\left|\hat{I}^{*}(t)\hat{I}(0)\right|i\right\rangle. \tag{10}$$

$|I(\omega)|^2$ is the power spectrum of the current fluctuation of the tunneling current for the field-emission mechanism:

$$|I(\omega)|^2 = \int_{-\infty}^{\infty}\left\langle \hat{I}^{*}(t)\hat{I}(0)\right\rangle \exp(-i\omega t)dt. \tag{11}$$

By substituting Eqs. (6)–(9) into Eq. (11), we can finally obtain the following expression for the power spectrum of the field-emission current fluctuations:

$$|I(\omega)|^2 = \frac{e^2Am}{2\pi^2\hbar^3}\int dE \times \exp\left(-\int dz\left|k_z(z)\right|\right)\exp\left(-\int dz\left|q_z(z)\right|\right), \tag{12}$$

where A is the area of the field-emission site of the SWCNTs on a cathode, and m is the effective mass of the electron that moves parallel to the emission site in the source electrode from which tunneling electrons are injected into the barrier.

Ballistic electrons are considered to pass through CNTs without crystal defects. However, when electrons pass through CNTs with crystal defects, the ballistic property is lost; an energy barrier is believed to inhibit the conduction of these electrons. In accordance with the works of Fransen et al. [22] and Nilsson et al. [23] on the electron tunneling model, we adopted an inelastic tunneling model through an energy barrier having a rectangular waveform.

2.1 Power spectrum in the inelastic tunneling model

In the inelastic tunneling model, the energy states of electrons passing through an energy barrier can be estimated by the energy transfer before and after electron tunneling. The power spectrum with the flow of FE current for the inelastic tunneling model was proposed by Kirtley et al. [24], Bardeen [25], and Watanabe et al. [26] based on Eqs. (12) and (13):

$$|I_{\text{in}}(\omega)|^2 \sim \int dE_{\|} \int dE_k \rho_k(E_k) \int dE_q \rho_q(E_q) |M_{kq}|^2 \left[(1 - n_q{}^{\text{r}}) n_k{}^{\text{l}} - (1 - n_k{}^{\text{l}}) n_q{}^{\text{r}} \right], \tag{13}$$

where $\rho(E)$ is the state density of electrons in one-dimensional space, and $n_i{}^{\text{h}}$ (h = l (for left region) or r (for right region)) is the occupation number of electrons within state i (i = k or q).

We attempted to obtain the direct-current (DC) component of the power spectrum of the inelastic electron tunneling model using a rectangular potential barrier and Eq. (13). As mentioned above, the electrons passing through the CNTs without crystal defects are ballistic [5]. The CNTs with crystal defects are reported to have an energy bandgap that depends on the position of the defects, and electrons rarely pass through CNTs without energy loss. In order to analyze the electrical conductivity of SWCNTs, we attempted to obtain the tunneling matrix M_{kq} and the power spectrum of inelastic electron tunneling when electrons inelastically pass through the local area corresponding to crystal defects. The conductivity of electrons passing into SWCNTs with crystal defects was explained employing the electron tunneling model below.

For M_{kq} of an energy barrier having a rectangular waveform, the state of wave function transfers that in the right region (r) to the left region (l) of the energy barrier or vice versa, $\varphi^r{}_q$ is the wave function having wavenumber q localized in the right region of the barrier called as wave-B, and $\varphi^l{}_k$ is the wave function having wavenumber k localized in the left region of the barrier called as wave-A. The Fermi level of an SWCNT is represented by E_{F}. When electrons pass through the rectangular potential barrier from the left region to the right region, the wave function shifts from state k to state q, and the tunneling matrix M_{kq} is given by:

$$M_{kq} = \int dx \left\{ \frac{\hbar^2}{2m} \nabla \varphi_k^{\text{l}*} \nabla \varphi_q^{\text{r}} + V(z) \varphi_k^{\text{l}*} \varphi_q^{\text{r}} \right\}, \tag{14}$$

where $V(z)$ is the potential height of the energy barrier having a rectangular distribution on the z axis shown in **Figure 2**, and m is the effective mass of an electron.

The wave functions φ_k and φ_q localized in the left region (wave-A) and in the right region (wave-B) are represented by:

$$\begin{aligned} \varphi_k &= \frac{1}{\sqrt{S}} e^{ik_{\|}x_{\|}} \chi_k \\ \varphi_l &= \frac{1}{\sqrt{S}} e^{iq_{\|}x_{\|}} \chi_q \end{aligned}, \tag{15}$$

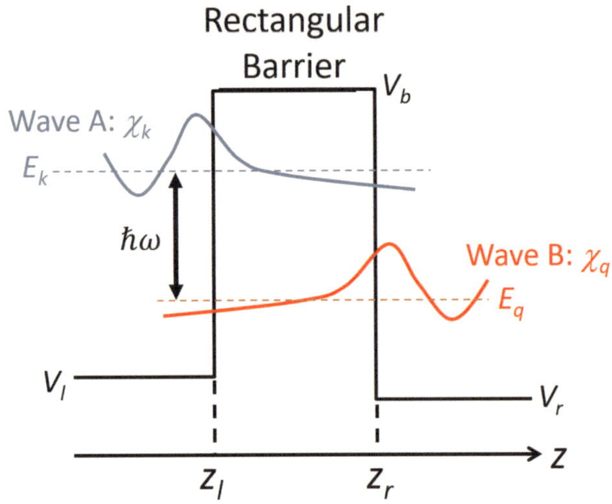

Figure 2.
Image of a rectangular energy barrier having a potential height V(z).

where S is the area in which electrons pass through an SWCNT; and $k_{||}$, $q_{||}$, and $x_{||}$ are the components, along the direction perpendicular to the z axis, of the wavenumber k, wavenumber q, and position x, respectively. The wave functions χ_k and χ_q have coefficients A_n, B_n and C_n, D_n, respectively, and are defined as:

$$\begin{aligned}\chi_k &= \sum_{n=l,\,b,\,r} \left(A_n e^{ik_n|z-z_n|} + B_j e^{-ik_n|z-z_n|}\right)\\ \chi_q &= \sum_{n=l,\,b,\,r} \left(C_n e^{iq_n|z-z_n|} + D_n e^{-iq_n|z-z_n|}\right)\end{aligned}. \tag{16}$$

The energy transfer between waves-A and -B, $\hbar\omega$, is included with the wavenumber in Eq. (16), and it corresponds to the energy loss of an electron passing through the energy barrier from the left region to the right region:

$$\begin{aligned}k_n &= \sqrt{\frac{\hbar^2}{2m}|E_z - V_n|}(n = l, b, r)\\ q_n &= \sqrt{\frac{\hbar^2}{2m}|E_z - V_n - \hbar\omega|}\ (z < z_l, z > z_r; n = l, r)\\ q_n &= \sqrt{\frac{\hbar^2}{2m}|E_z - V_n + \hbar\omega|}\ (z_l < z < z_r)\end{aligned} \tag{17}$$

Using Eqs. (16) and (17), M_{kq} can be rewritten as:

$$M_{kq} \sim \delta_{k||q||} \sum_{n=l,\,b,\,r} \left[\left(\frac{\hbar^2}{2m}k_n q_n - V_n\right)\frac{i}{k_n + q_n} \times \{A_n, B_n, C_n, D_n\}\right]. \tag{18}$$

The power spectrum of inelastic electron tunneling, $|I_{\text{in}}(\omega)|^2$, is obtained from Eqs. (13)–(18). The power spectrum at $\omega = 0$ represents the magnitude of the direct current of the power spectrum $|I_{\text{in}}(\omega)|^2$.

2.2 Power spectrum in F-N tunneling regime passing into an SWCNT with crystal defect

The thickness of the rectangular energy barrier and the energy transfer which means the energy difference between the left region and the right region are

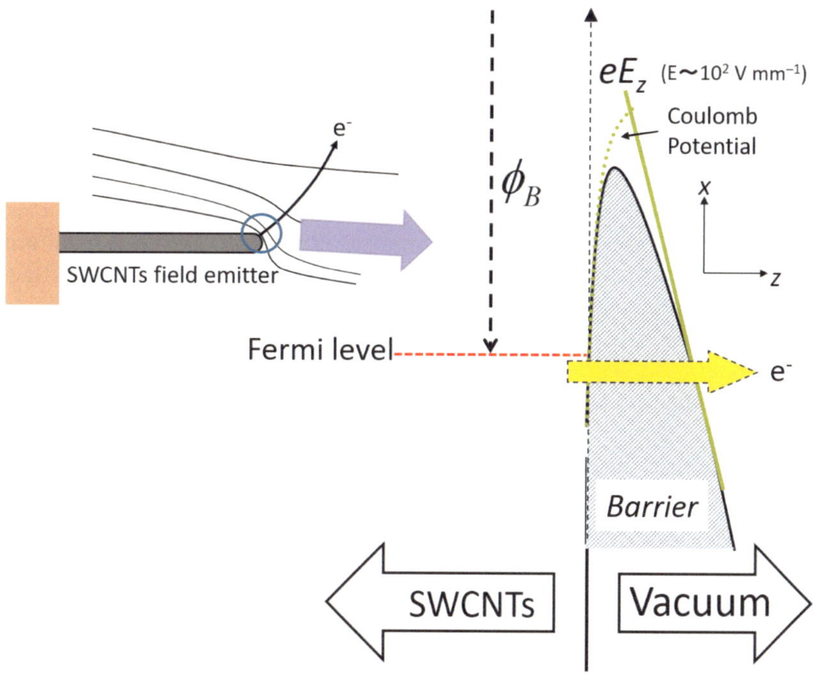

Figure 3.
Original model of field emission using an SWCNT emitter in F-N tunneling.

important factors in this simulation. Results of research on tunnel junction devices indicate that the thickness of the rectangular potential barrier for inelastic electron tunneling is only ~3 nm [20]. The energy loss caused by the tunneling of electrons through the energy barrier depends on the bandgap associated with the crystal defects [27]. The theoretical value of difference in the energy between waves A and B is calculated by fitting to be 0.027 eV, which corresponds to a temperature of 313 K for the energy loss observed when the electrons pass through SWCNTs with crystal defects.

Eq. (12) should be transformed into a more suitable form for a field-emission cathode using metallic SWCNTs. Field-emission measurements were carried out in the F-N tunneling regime, and a further simplified form is expressed as follows:

$$|I(\omega)|^2 \propto \frac{e^3 A}{4\pi^2 \hbar \phi_{\mathrm{B}}} \bullet |E|^2 \left(\frac{\hbar\omega}{e\phi_{\mathrm{B}}} + 1\right)^{-1} \times \exp\left\{-\frac{2\sqrt{2m}}{3\hbar e|E|}\left((\hbar\omega)^{\frac{3}{2}} + (e\phi_{\mathrm{B}})^{\frac{3}{2}}\right)\right\}, \quad (19)$$

where ϕ_{B}, shown in **Figure 3**, is the barrier-height potential that indicates the difference between the vacuum-level energy and the energy at the bottom of the Fermi level of metallic SWCNTs; and E is the applied electric field on the SWCNTs enhanced by the tip shape. The enhancement factor β of E can be calculated by the surface-charge method of simulation using a three-dimensional SWCNT model that protrudes from an indium-tin oxide (ITO) film [28–30]:

$$E = \beta E_0, \quad (20)$$

where E_0 represents the real electric field between the cathode and anode. Moreover, the area of the field-emission site A on the SWCNTs of the cathode was estimated from a comparison between the simulated and measured bright spots [30].

3. Experimental methods

In this study, the SWCNTs synthesized by arc discharge were annealed at a high temperature of approximately 1200 K in a low pressure of 10^{-6} Pa to obtain hc-SWCNTs. **Figure 4** shows transmission electron microscope (TEM) images of the SWCNTs before and after annealing. As shown in **Figure 4(a)**, we could find that the crystallinity of the SWCNTs after annealing was significantly improved.

The SWCNTs obtained before and after annealing were dispersed in a liquid medium with solvent including, to control the viscosity of the solution, surfactant and an ITO precursor solution, to prepare a coating film. The agitated mixture was sprayed on a silicon substrate to form an ITO film and sintered at approximately 700 K in a vacuum to remove any components in the organic solvents. In addition, to facilitate FE, the ITO film with SWCNTs was scratched with a thin needle to activate the film [16].

A diode structure fixed to the distance of 1.2 mm between a cathode of the ITO film having SWCNTs and an anode of a phosphor plate was prepared for the FE measurements; **Figure 5** shows a schematic of the FE-measurement system.

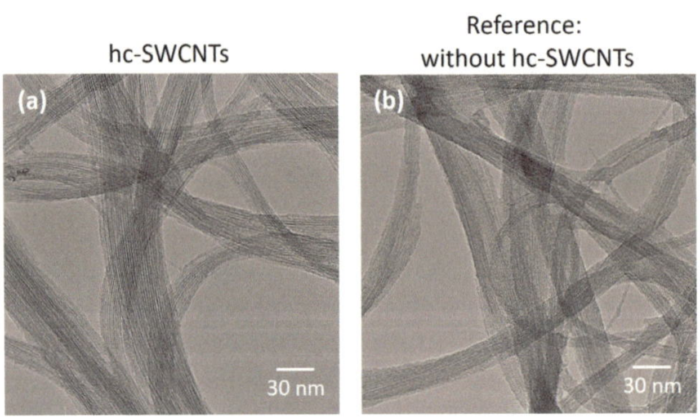

Figure 4.
TEM images with annealed and unannealed SWCNTs: (a) annealed hc-SWCNTs; (b) nonannealed SWCNTs with crystal defects.

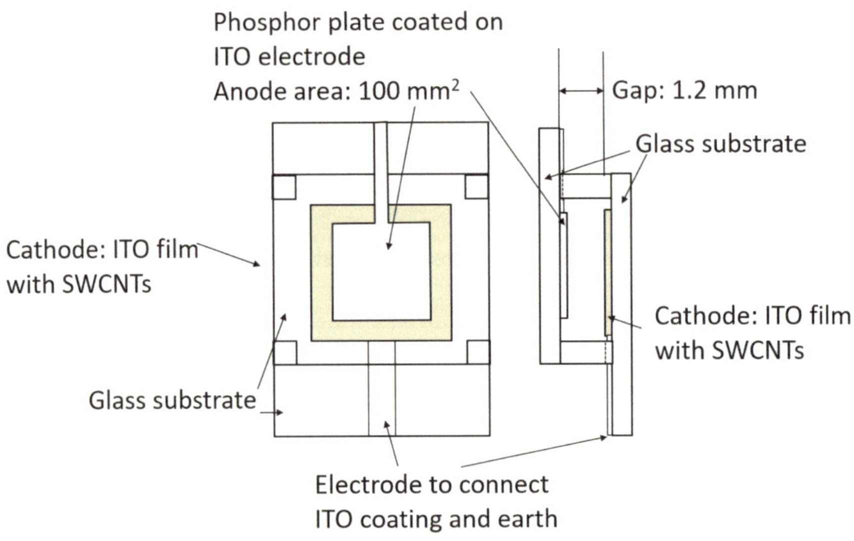

Figure 5.
Schematic diagram of the FE-measurement system with a diode structure.

The sample was set in a vacuum chamber of approx. 10^{-4} Pa. A supplied voltage with a frequency of 60 Hz was applied across the cathode and the anode to measure the FE characteristics.

4. Field-emission results and discussion

4.1 Field-emission characteristics of hc-SWCNTs

Tunneling electron microscopy and scanning electron microscopy were used to obtain images of the dispersed hc-SWCNT bundles and the ITO film which was subjected to activation treatment. As one can clearly see in the TEM image in **Figure 6(a)**, the hc-SWCNT bundles were well dispersed in the liquid medium. Meanwhile, the SWCNT bundles shown in the SEM image in **Figure 6(b)** were exposed in the scratched grooves on the activated ITO film and marked within the white circles; the dispersed SWCNTs in a scratched area are also illustrated with the red circles indicating the ends of the SWCNT bundles on the right. The number of dispersed and exposed SWCNTs per unit area could be controlled via the density of the SWCNTs. **Table 1** shows the summary of the state of the exposed SWCNTs in terms of (1) the number of SWCNTs protruding from the grooved walls of the scratched ITO film per unit area, (2) the mean diameter of the exposed SWCNT

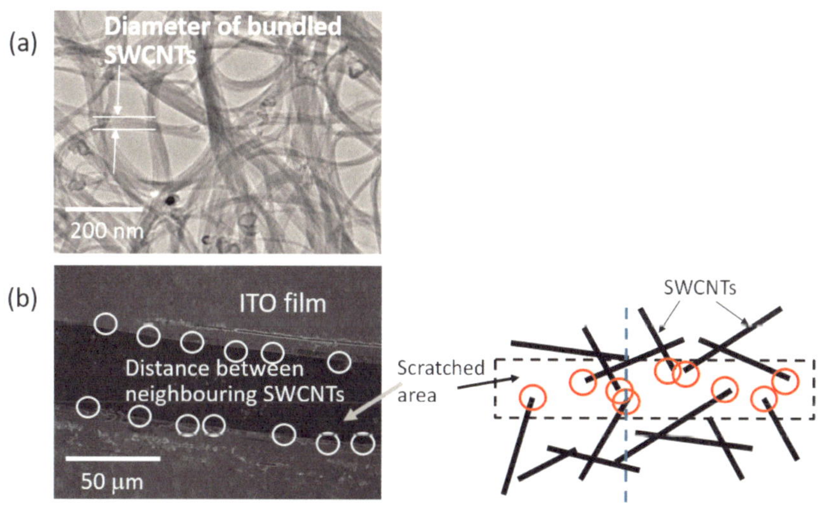

Figure 6.
(a) TEM image of dispersed hc-SWCNT bundles. (b) Left: SEM image of scratched ITO film having SWCNTs, where the white circles show the SWCNT bundles protruding from the grooved walls of the ITO film; right: schematic diagram of scratched area including dispersed SWCNTs, where the red circles show the tops of the exposed SWCNTs in the scratched area.

	hc-SWCNTs	**SWCNTs with crystal defects**
Number of SWCNTs exposed at grooved walls of ITO film after activation	Approx. 21 per 100×100 μm^2	Approx. 26
Average diameter of bundled SWCNTs (nm)	Approx. 9	Approx. 10
Average distance between neighboring bundled SWCNTs (μm)	Approx. 27	Approx. 21

Table 1.
State of homogeneity of dispersed hc-SWCNTs and SWCNTs with crystal defects.

bundles, and (3) the mean distance between neighboring SWCNT bundles for two types of SWCNTs, i.e., SWCNTs with crystal defects and hc-SWCNTs. The states of exposure of the two types of SWCNTs fabricated by our wet-coating process were similar, and they were homogeneously dispersed in the films.

Figure 7 shows the relationship between the current density and electric field for the SWCNTs, obtained using the system for FE measurements (**Figure** 5). The threshold field decreased from 2.01 V μm^{-1} for the SWCNTs with crystal defects to 1.07 V μm^{-1} for the hc-SWCNTs. As shown in **Table 1,** the dispersion density of SWCNTs and the mean diameter of SWCNT bundles in the films were similar for the two types of the SWCNTs. It means that the probability densities of SWCNTs that induced FE were similar for the two types of SWCNTs. Therefore, it can be discussed and assumed that the above improvement in electrical properties was caused by the increase in crystallinity of the SWCNTs.

Figure 8 shows the distribution of bright spots in the hc-SWCNTs and SWCNTs with crystal defects. As can be seen in **Figure** 7, the threshold and the driving field depended on the crystallinity of the SWCNTs. When the applied field was controlled so that the current density was uniform, the densities of the high- and low-brightness spots on the measured area were almost the same between the

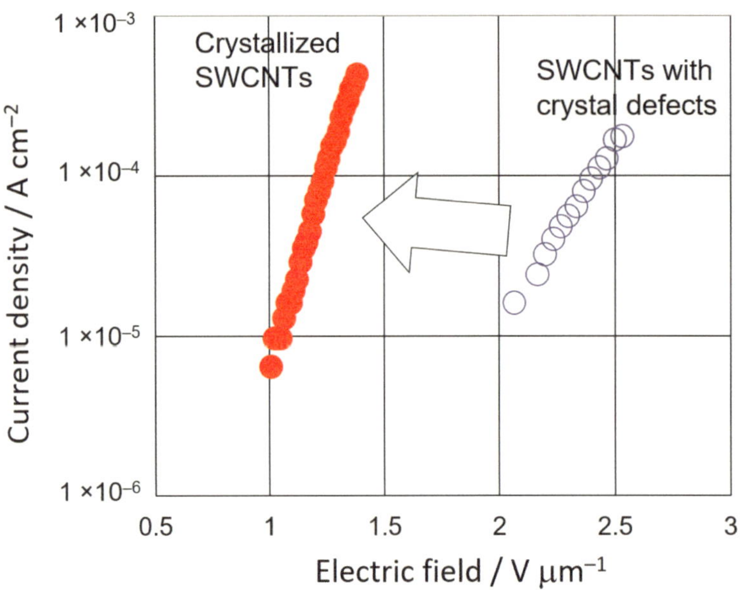

Figure 7.
Semilog plot of current density versus electric field for hc-SWCNTs and SWCNTs with crystal defects.

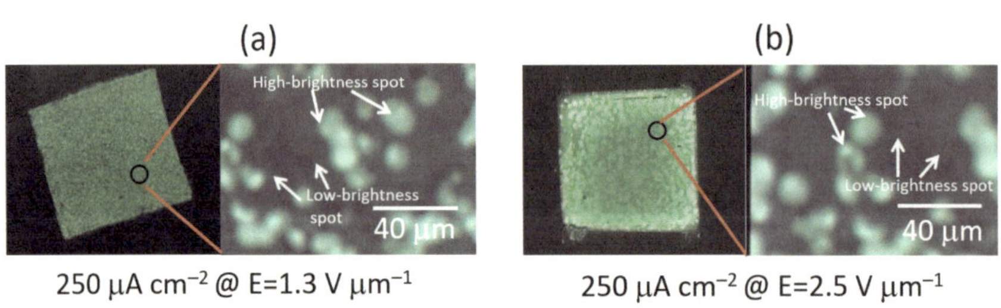

Figure 8.
Homogeneity of emission sites homogeneity in (a) hc-SWCNTs and (b) SWCNTs with crystal defects. The panels on the left show planar lighting, and the enlarged view of the circled regions in the panels on the right show the distribution of spots of cathodes employing hc-SWCNTs and SWCNTs with crystal defects.

hc-SWCNTs and the SWCNTs with crystal defects, respectively. Previous reports showed that in the SWCNTs synthesized by arc discharge, SWCNTs with metallic and semiconductive properties coexisted [31]. Therefore, the high-brightness spots can be attributed to the FE from metallic SWCNTs, and the low-brightness spots can be attributed to the FE from semiconductive SWCNTs; these phenomena and their origins are not discussed in this chapter. When the crystallinity of SWCNTs increased, the chirality of the carbon sheets making up the SWCNTs did not change, and the mixed ratio of metallic to semiconductive SWCNTs remained unchanged [32–37].

The results in **Figure 6** and **Table 1** indicate that the diameter, protruded length, and dispersion density of the SWCNT bundles protruding from the grooved ITO film were similar for the two types of the SWCNTs as FE electron sources. However, the difference in the plots of current density versus electric field of **Figure** 7 is impossible to be analyzed only by the exposed uniformity of the dispersed SWCNTs on the activated ITO film. Therefore, we surmised that the difference in the crystallinity of SWCNTs caused the difference in electric properties of SWCNTs. Fransen et al. and Nilsson et al. find that the interruption of carbon network of an SWCNT with crystal defects originates a local energy band that affects the electrical conductivity by scanning tunneling microscopy (STM) [22, 23]. On the basis of the above results, an electric-physical model in an SWCNT was developed from the inelastic electron tunneling model. Therefore, the energy difference originated by the energy bandgap yielding from the interruption of carbon network in an SWCNT with crystal defects is surmised to become an energy barrier which impedes the electrons pass through the SWCNT.

Figure 9 shows the FE properties obtained by experiment in **Figure** 7 and simulation. For the hc-SWCNTs, the results obtained by the experiment are represented by red circles and those obtained by simulation using Eq. (12) at $\omega = 0$ of the F N tunneling model without inelastic tunneling model are represented by the yellow dashed line. Moreover, the experimental results of the SWCNTs with crystal defects are represented by blue circles and those obtained by the following method are represented by the dark-red dashed line. We can find the fitting

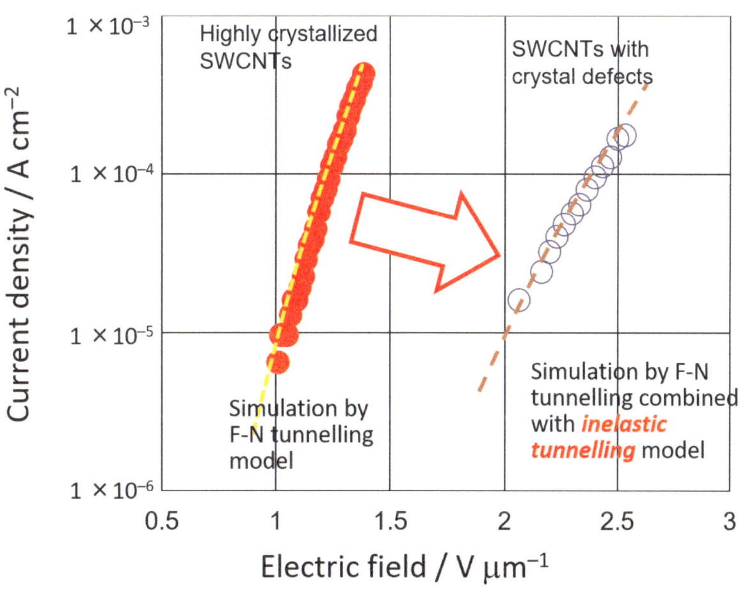

Figure 9.
Fitting of simulation results to experimental current density-electrical field characteristics.

	hc-SWCNTs	SWCNTs with crystal defects
Electron-emission site (α)	1.95×10^{-9}	
Field-enhancement factor (β)	259,500	
Effective mass of electron (m)	$1.00m_0$ ($m_0 = 9.11 \times 10^{-31}$ kg)	
Thickness of rectangular inelastic barrier	—	2.9 nm
Energy bandgap after passing into the barrier	—	0.148 eV
Energy difference between wave A and wave B	—	0.025 eV

Table 2.
Homogeneity of dispersed hc-SWCNTs and SWCNTs with crystal defects.

between experiment and simulation is in good agreement. The simulation results represented by the dark-red dashed line were obtained by combining the F-N tunneling model of electrons emitted from the end of the SWCNT with the power spectrum with inelastic tunneling model given by Eqs. (7)–(12) at $\omega = 0$. **Table 2** shows the summary of the parameters employed in the theoretical calculations on the basis of F-N tunneling and inelastic electron tunneling. The total electron emission site area (α) and the field-enhancement factor (β) for FE are important parameters in evaluating the FE characteristics. These two parameters were the same for the two types of the SWCNTs.

The inelastic electron tunneling through the SWCNTs with crystal defects, as shown in **Figure 10**, is a phenomenon based on the energy loss associated with the electron transfer in the SWCNTs. Although many researchers have reported that CNTs exhibit ballistic electrical conductivity, such property has not been demonstrated in any experiment that we know of. We can express successfully the first

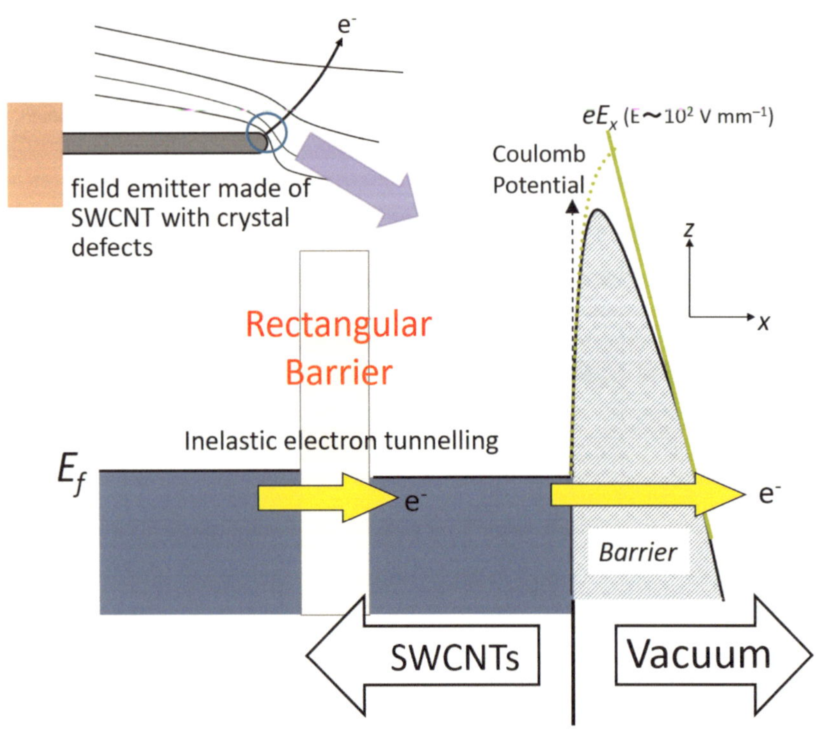

Figure 10.
Schematic of field-emission property of SWCNTs with crystal defects modeled as convoluted FN-tunneling with inelastic tunneling.

report of the development using an inelastic electron tunneling model for an SWCNT with crystal defects based on the evaluation of FE properties. The hc-SWCNTs are suitable materials for electronic devices to save energy because the energy loss associated with electron transfer in the SWCNT is suppressed.

4.2 Current fluctuations of hc-SWCNTs as field emitters: theoretical predictions versus experimental results

Figure 11 shows the relationship between the current density and electric field for the SWCNTs obtained using the FE-measurement system shown in **Figure 5**. The threshold field of the metallic SWCNTs was 4.1 V μm^{-1} at a line-current density of 0.1 mA cm^{-2}, as shown in **Figure 11**. Moreover, the planar lighting observed at 2 mA cm^{-2} was homogeneous, as shown in the inset in **Figure 11**.

Figure 12(a) shows the stability of the FE current of metallic SWCNTs when different values of DC voltage were supplied for 50 s; **Figure 12(b)** shows the power spectrum of the frequency based on the FE current fluctuations by Fourier transform calculation. The current density obtained from the sample biased using the DC power supply was measured as 1.2 mA cm^{-2} at 6 kV, 2.0 mA cm^{-2} at 6.6 kV, 4.6 mA cm^{-2} at 7.1 kV, and 10 mA cm^{-2} at 7.6 kV. The FE current fluctuations were observed to be stable owing to the desorption of gases from the SWCNT surface caused by annealing before the FE measurements. Some damped current distributions in the FE current were expected if the annealing of the sample was insufficient. However, **Figure 12(a)** shows no damping, and the SWCNT surfaces were expected to be clean for the FE measurements. The spectrum of each FE current in **Figure 12(a)** is normalized by the peak at 0 Hz, and it indicates that periodic fluctuations were recorded. A peak appears at 50 Hz, as shown in **Figure 12(b)**, and it originated from the noise of the DC power supply, which represents the frequency of the commercial power supply in the eastern region in Japan. It can be seen that the stable DC current had alternating values, especially in the

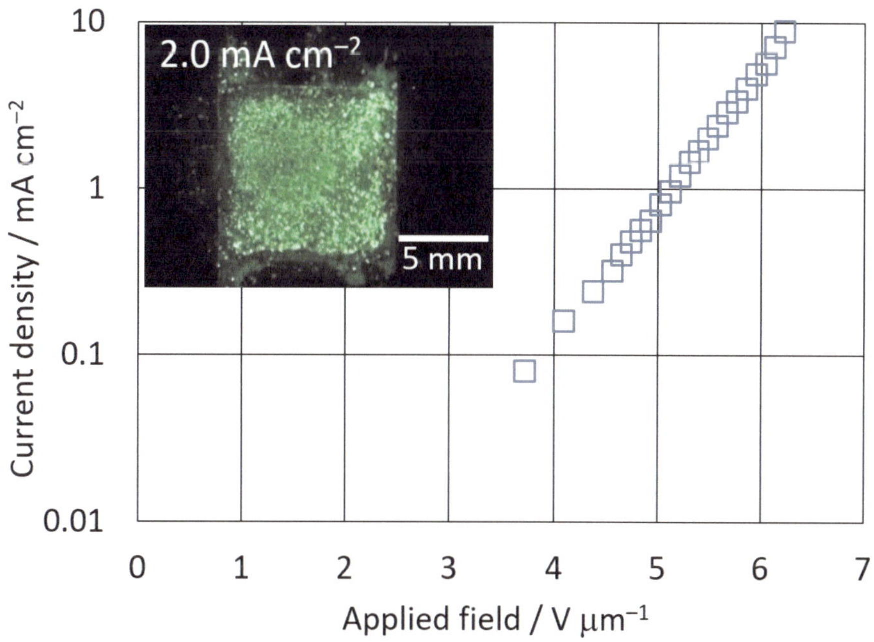

Figure 11.
Current density-electric field characteristics of metallic SWCNTs; the inset shows the case of planar emission.

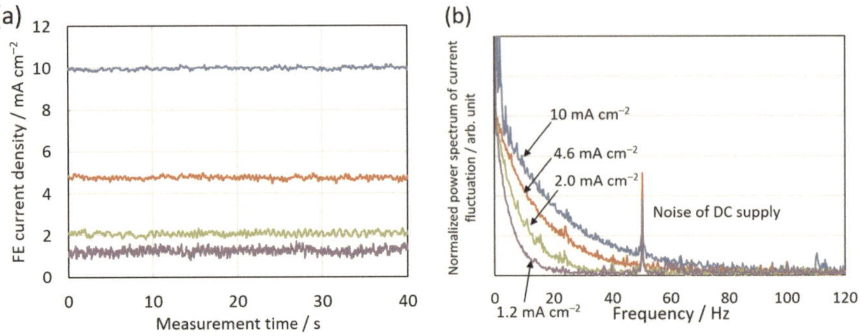

Figure 12.
(a) FE current fluctuations and (b) power spectra of FE-current fluctuations obtained by Fourier transform.

e (elementary charge)	1.60×10^{-19} C
A (area of field-emission site)	6.48×10^{-24} m^2
ϕ_B (work function of bulk carbon)	4.3 eV
$\hbar$ ($= \frac{h}{2\pi}$, h: Planck's constant)	6.58×10^{-16} eV s
β (enhancement factor)	2.56×10^3
m (free electron mass)	9.11×10^{-31} kg
A and β were obtained from the simulation reported in Ref. [28].	

Table 3.
Parameters used in the theoretical calculations.

low-frequency domain of <100 Hz. Moreover, the spectrum distribution depends on the FE current, and the contents of each spectrum increased with the DC power.

When the experimental results and the theoretical predictions explained in the previous theoretical section are compared, the parameters that determined the tunneling characteristics of the metallic hc-SWCNTs used as field emitters are listed in **Table 3**. The enhanced field E in Eq. (20) was determined from the surface potential calculations with an SWCNT model. The area of the field-emission site, A, at the SWCNT surface was calculated from the analysis that determined the FE properties. Each SWCNT model protruded from the ITO wall, which included the minimum area for calculating the electric field using the surface-charge method. The potential parameter ϕ_B was estimated as the work function value of bulk carbon.

The power spectrum $|I(\omega)|^2$ obtained from Eq. (19) simplified by the WKB approximation shows the dependence of the applied field on a cathode using hc-SWCNTs, as shown in **Figure 13**. Each calculated spectrum was normalized, and it can be seen that $|I(\omega)|^2$ increased with the applied voltage at low frequency. For the metallic hc-SWCNTs, the experimental results are shown in **Figure 12**, and those obtained by simulation are shown in **Figure 13**; the two sets of results are in good agreement. From the above comment, the experimental results could be fitted well by the composition of the inelastic tunneling model and the F-N tunneling model employing the power spectrum of current fluctuation represented by Eq. (19). These results shown in **Figures 12** and **13** indicate that the electrons passing into a barrier in the F-N model exhibited the electron flow fluctuation in the inelastic tunneling model, and the shape of the power spectrum depended on the increased component of current fluctuation obtained from FE.

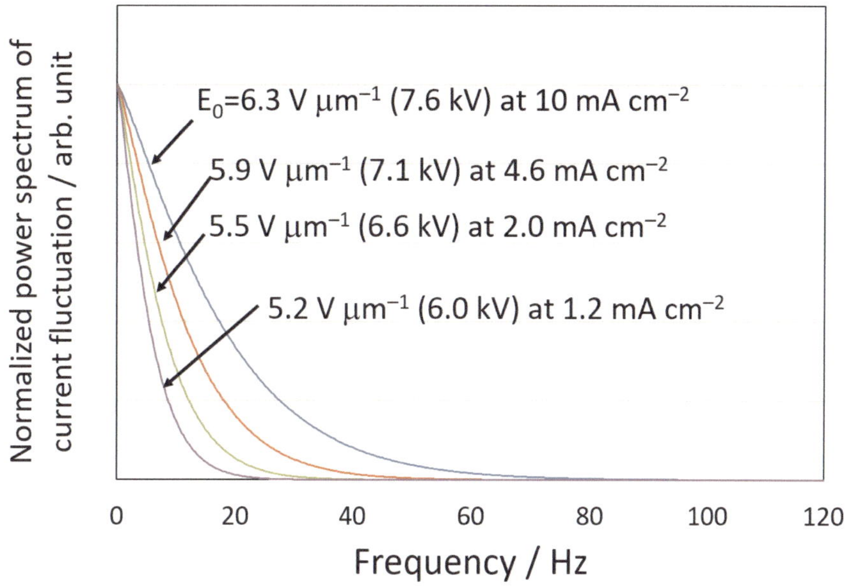

Figure 13.
Power spectrum obtained by simulation showing the dependence on applied voltage; E_0 represents the original supplied electric field between the cathode and anode.

Field emission using metallic SWCNTs was modeled as a phenomenon based on inelastic electron tunneling, and the calculated power spectra are shown in **Figure 13**. Although many researchers have reported that field emitters exhibit F-N tunneling with elastic electrical conductivity, the phenomenon has not been demonstrated in any reported experimental work. This chapter presents successfully the first report of the development of an inelastic electron-tunneling model with WKB approximation for SWCNTs from the FE properties. Moreover, the FE current from electron sources like SWCNTs contains time-depended fluctuations, and the *I-V* characteristics of the FE electron sources were expressed as the power spectrum at $\omega = 0$, which gives the magnitude of the current originating from the electrons passing through the SWCNTs, in accordance with the inelastic electron-tunneling model.

5. Conclusion

In this chapter, the author explained the conductive model of electrons flowing in the vicinity of a crystal defect that acts as a rectangular energy barrier based on inelastic electron tunneling model and the current fluctuation model for FE with hc-SWCNTs employing the F-N tunneling phenomenon.

The author could succeed in developing a model of the flow of electrons through the inside of an SWCNT to the outside using the fluctuations of the tunneling current. The electron flow model for an SWCNT with crystal defects was obtained by combining the F-N tunneling model with the power spectrum obtained using the tunneling matrix. From the previous mentioned comment, we could give a brief explanation of the effect of the increased crystallinity of SWCNTs on their electrical conductivity and describe the development of an electron flow model through the crystal defects of an SWCNT. Therefore, we expect that the hc-SWCNTs are used as field emitters with stable emission and low power consumption for saving energy.

Acknowledgements

The author kindly appreciates to support and give helpful discussions and suggestions by DOWA Holdings Co. Ltd., Japan.

Author details

Norihiro Shimoi
Graduate School of Environmental Studies, Tohoku University, Sendai, Japan

*Address all correspondence to: norihiro.shimoi.c8@tohoku.ac.jp

IntechOpen

© 2019 The Author(s). Licensee IntechOpen. This chapter is distributed under the terms of the Creative Commons Attribution License (http://creativecommons.org/licenses/by/3.0), which permits unrestricted use, distribution, and reproduction in any medium, provided the original work is properly cited.

References

[1] Hamada N, Sawada S, Oshiyama A. New one-dimensional conductors: Graphitic microtubules. Physical Review Letters. 1992;**68**:1579-1581

[2] Saito R, Fujita M, Dresselhaus G, Dresselhaus MS. Electronic structure of chiral graphene tubules. Applied Physics Letters. 1992;**60**:2204-2206

[3] Tanaka K, Okahara K, Okada M, Yamabe T. Electronic properties of bucky-tube model. Chemical Physics Letters. 1992;**191**(5):469-472

[4] Kim J-Y, Kim M, Kim HM, Joo J, Choi J-H. Electrical and optical studies of organic light emitting devices using SWCNTs-polymer nanocomposites. Optical Materials. 2002;**21**:147-151

[5] Rinzler AG, Hafner JH, Nikolaev P, Lou L, Kim SG, Tomanek D, et al. Unraveling nanotubes: Field emission from an atomic nanowire. Science. 1995; **269**:1550-1553

[6] Ebbesen TW, Lezec HJ, Hiura H, Bennett JW, Ghaemi HF, Thio T. Electrical conductivity of individual carbon nanotubes. Nature. 1997;**382**:54-56

[7] Wildoerm JWG, Venema LC, Rinzler AG, Smalley RE, Dekker C. Electronic structure of atomically resolved carbon nanotubes. Nature. 1998;**391**:59-62

[8] Odom TW, Huang JL, Kim P, Lieber CM. Atomic structure and electronic properties of single-walled carbon nanotubes. Nature. 1998;**391**:62-64

[9] Treacy MMJ, Ebbesen TW, Gibson JM. Exceptionally high Young's modulus observed for individual carbon nanotubes. Nature. 1996;**381**:678-680

[10] Saito Y, Uemura S, Hamaguchi K. Cathode ray tube lighting elements with carbon nanotube field emitters. Japanese Journal of Applied Physics. 1998;**37**:L346-L348

[11] Dai H, Hafner JH, Rinzler AG, Colbert DT, Smalley RE. Nanotubes as nanoprobes in scanning probe microscopy. Nature. 1996;**384**:147-150

[12] Dillon AC, Jones KM, Bekkedahl TA, Kiang CH, Bethune DS, Haben MJ. Storage of hydrogen in single-walled carbon nanotubes. Nature. 1997;**386**:377-379

[13] Niu C, Sichel EK, Hoch R, Moy D, Tennent H. High power electrochemical capacitors based on carbon nanotube electrodes. Applied Physics Letters. 1997;**70**:1480-1482

[14] Tohji K, Goto T, Takahashi H, Shinoda Y, Shimizu N, Jeyadevan B, et al. Purifying single-walled nanotubes. Nature. 1996;**383**(6602):679-679

[15] Iwata S, Sato Y, Nakai K, Ogura S, Okano T, Namura M, et al. Novel method to evaluate the carbon network of single-walled carbon nanotubes by hydrogen physisorption. The Journal of Physical Chemistry Letters. 2007;**111**: 14937-14941

[16] Shimoi N, Adriana LE, Tanaka Y, et al. Properties of a field emission lighting plane employing highly crystalline single-walled carbon nanotubes fabricated by simple processes. Carbon. 2013;**65**:228-235

[17] Garrido SB, Shimoi N, Abe D, Hojo T, Tanaka Y, Tohji K. Planar light source using a phosphor screen with single-walled carbon nanotubes as field emitters. Review of Scientific Instruments. 2014;**85**:104704

[18] Laks B, Mills DL. Photon emission from slightly roughened tunnel junctions. Physical Review B. 1979;**20**: 4962-4980

[19] Watanabe J, Uehara Y, Ushioda S. Multiple-scattering effect of surface-plasmon polaritons in bright emission from tunnel junction. Physical Review B. 1995;**52**:2860-2867

[20] Uehara Y, Watanabe J, Fujikawa S, Ushioda S. Light-emission mechanism of Si-MOS tunnel junctions. Physical Review B. 1995;**51**:2229-2238

[21] Hone D, Muhlschlegel B, Scalapino DJ. Theory of light emission from small particle tunnel junctions. Applied Physics Letters. 1998;**33**:203-204

[22] Fransen MJ, van Rooy TL, Kruit P. Field emission energy distributions from individual multiwalled carbon nanotubes. Appl. Sur. Sci. 1999;**146**: 312-327

[23] Nilsson L, Groening O, Emmenegger C, Kuettel O, Schaller E, Schlapbach L. Scanning field emission from patterned carbon nanotube films. Applied Physics Letters. 2000;**76**: 2071-2073

[24] Kirtley JR, Theis TN, Tsang JC, DiMaria DJ. Hot-electron picture of light emission from tunnel junctions. Physical Review B. 1983;**27**:4601-4611

[25] Bardeen J. Tunneling from a many-particle point of view. Physical Review Letters. 1961;**6**:57-59

[26] Watanabe J, Uehara Y, Murota J, Ushioda S. Light emission from Si-metal-oxide-semiconductor tunnel junctions. Japanese Journal of Applied Physics, Part 1. 1993;**32**:99-101

[27] Shimoi N. Effect of increased crystallinity of single-walled carbon nanotubes used as field emitters on their electrical properties. Journal of Applied Physics. 2015;**118**:214304

[28] Murata H, Shimoyama H, Ohye T. Computer simulation of electric field analysis for vertically aligned carbon nanotubes: I. Simulation method and computing model. Proceedings of SPIE. 2001;**4510**:156-162

[29] Shimoyama H, Murata H, Ohye T. Computer simulation of electric field analysis for vertically aligned carbon nanotubes: II. Electric field on the nanotube apex. Proceedings of SPIE. 2001;**4510**:163-171

[30] Shimoi N, Tanaka S-I. Numerical analysis of electron emission site distribution of carbon nanofibers for field emission properties. ACS Applied Materials & Interfaces. 2012;**5**:768-773

[31] Wu B, Geng D, Liu Y. Evaluation of metallic and semiconducting single-walled carbon nanotube characteristics. Nanoscale. 2011;**3**:2074

[32] Saito Y, Hamaguchi K, Nishino T, Hata K, Tohji K, Kasuya A, et al. Field emission patterns from single-walled carbon nanotubes. Japanese Journal of Applied Physics. 1997;**36**:L1340-L1342

[33] Ren ZF, Huang ZP, Xu JW,Wang JH, Bush P, Siegal MP, et al. Synthesis of large arrays of well-aligned carbon nanotubes. Science. 1998;**282**:1105-1107

[34] Tsai SH, Chao CW, Lee CL, Shih HC. Bias-enhanced nucleation and growth of the aligned carbon nanotubes with open ends under microwave plasma synthesis. Applied Physics Letters. 1999;**74**:3462-3464

[35] Kim JM, ChoiWB, Lee NS, Jung JE. Field emission from carbon nanotubes for displays. Diamond and Related Materials. 2000;**9**:1184-1189

[36] Popov VN. Carbon nanotubes: Properties and application. Materials Science and Engineering Reports. 2004; **43**:61-102

[37] Bandaru PR. Electrical properties and applications of carbon nanotube structure. Journal of Nanoscience and Nanotechnology. 2007;**7**:1239-1267

Chapter 9

Electron Transport in the Assemblies of Multiwall Carbon Nanotubes

Vladimir Samuilov, Jean Galibert and Nikolai Poklonski

Abstract

The assemblies (films) of carbon nanotubes (CNTs) possess very stable, reproducible, and extraordinary electronic properties. These films have been considered as attractive materials for various nanosensors and as electrodes of electrochemical energy storage devices, like supercapacitors, with low equivalent series resistance and highly developed internal surface. In order to develop CNT devices operating at the room temperature, it was necessary to determine the assembled films' properties, such as the mechanism of conductivity, carrier concentration, and mobility. In this study, we are focused on the assemblies (monolayers, arrays, and films) of multiwall carbon nanotubes (MWCNT). We applied a wide temperature range resistance and magnetoresistance as a tool to determine the transport characteristics of MWCNT films. The measurements of the electrical transport (temperature dependence of the resistance) in the assemblies of nanotubes were tested in the temperature range T = 1.5–300 K, and the magnetoresistance measurements were carried out in pulsed magnetic fields up to 35 tesla in the temperature range 1.5–300 K. The mechanisms responsible for the transport in these systems, including weak localization, antilocalization, Luttinger liquid, Shubnikov–de Haas oscillations, and intertube coupling, were observed.

Keywords: films of carbon nanotubes, MWCNT,
temperature dependence of resistance, magnetoresistance, localization

1. Introduction

Since their discovery [1], carbon nanotubes revealed remarkable properties, which made them as an attractive material in broad domain of applications [2–4]. In carbon nanotube electronics [5, 6], they were expected to be used as the components of nanostructured field effect transistors [5, 7–9]. Their high anisotropic shape is considered to be useful for low field emission displays. Carbon nanotubes exhibited unique mechanical properties. Their huge elastic modulus (~1 TPa) [10, 11] and their large surface area per volume made them good candidates, respectively, for composite materials [12] and lithium ion storage [13], for sustainable energy supplies, and for gas [14], chemical [15, 16], and bio- [17–21] sensors.

The electrical transport properties are the most important for a number of applications of carbon nanotubes. Usually, single-tube samples are used for electrical transport studies. The kinetic inductance of individual nanotubes contributes to

IntechOpen

the low-frequency impedance of fibers composed of SWCNTs (at cryogenic temperatures and a constant electric bias) [22, 23]. Quantum transport properties have been observed both in single-wall (SWCNT) [5, 9, 24–29] and multiwall nanotubes (MWCNT) [3, 30–35].

Quantum interference effects, such as weak localization (WL), the Aharonov–Bohm (AB) effect, the Al'tshuler–Aronov–Spivak (AAS) effect, and universal conductance fluctuations (UCF), have previously been observed in multiwalled carbon nanotubes [32]. Knowledge of phase-breaking effects in coherent transport is very important for the study of these quantum interference phenomena. The temperature (T) dependence of the phase coherence length, $L\varphi$, of MWCNTs has been reported to follow a power law dependence of $T^{-1/3}$, a characteristic of one-dimensional interference [36]. In the case of single-walled CNTs, the power law dependence also shows a $T^{-1/3}$ dependence [28]. In addition, as further evidence of low-dimensional transport, there are some experimental reports of Tomonaga–Luttinger liquid behavior in CNTs [37].

From one point of view, for the array (assemblies) of nanotube samples, the synergetic properties were expected only [38]. But, the transport in the arrays of nanotubes was found to show single nanotube properties at low temperatures due to the mostly conductive nanotubes responsible for the transport [25]. On the other hand, for specific new applications, like chemical [15, 16] and bio- [17–21] sensors, the synergetic properties of the arrays of nanotube samples are important [39]. They are based on their large surface area per volume and intertube coupling in electrical transport in the arrays of nanotubes [40].

The examples of different morphologies of the samples of arrays of nanotubes involve definitions of bundles (ropes) [32, 38, 41–43], mats [44, 45], networks [46, 47], and films [48]. Some of them were manufactured using hardly controlled deposition from an organic solvent dispersion of pristine nanotubes [49].

The Langmuir–Blodgett (LB) technique was found to be a method for depositing defect-free, molecularly ordered ultrathin films with controlled thickness and orientation. Since pristine nanotubes are non-soluble in many kinds of organic solvents, it is impossible to form LB films directly from nanotubes. The LB monolayer of SWCNTs by dispersing them in a surfactant was obtained on a solid substrate in [50–52]. However, due to their limited dispersibility, the concentration of nanotubes in the layer was quite low (<7%), and the control of the tube orientation was not achieved by this method. The LB film of SWCNTs grafted by polyethylene oxide (PEO) was realized [53]. However, due to the soft PEO film covering the top surface of SWCNT cores, the low collapse pressure (~15 mN/m) was achieved. It became clear that, for the formation of a monolayer of carbon nanotubes using the LB technique, chemical modification of nanotubes is required.

The organic functionalization of carbon nanotubes has been realized [54–56]. But, to our knowledge, there was no experimental data on electrical and magneto-transport properties' characterization of dense monolayers manufactured using LB assembling of functionalized nanotubes. This method is expected to be used for obtaining the layers of very dense arrays of nanotubes for using them for new applications in chemical and biosensors, controlled by electrical transport.

We utilized assembly of different carbon nanotubes in the layers:

- Pristine (non-modified) CNTs.
- Modified by organic chemical functionalization.
- Oxidized CNTs in the solution of sulfuric and nitric acids.

We describe the methods of modification and assembly of CNTs and the specificity of the electrical transport in the layers in respect to the technological approach utilized.

2. Pristine MWCNT (no modification/functionalization)

The purified MWCNTs were dissolved in chloroform/chlorobenzene mixture with the concentration up to 10^{-2} mg/ml of visually non-scattered solution after sonication. A droplet of this solution was spread on the water surface, and after evaporation of the solvent, an array of MWCNT was collected on the substrate with in-plane macroscopic finger-shaped electrodes. The resistance was measured as a function of temperature and magnetic field in the Laboratoire National des Champs Magnetiques Intenses de Toulouse in the range 1.8–300 K and up to 35 T. We review here the most typical features of our results obtained on these non-functionalized self-assembled arrays of MWCNT.

Between 4.2 and 70 K, the zero magnetic field resistance (**Figure 1**) exhibited a behavior as $R \propto R_0 + AT^{-\alpha}$, where $\alpha = 0.38$, while in the range 60–300 K, the resistance behaved like $R \propto R_0' + BT^{-1/3}$.

As a power law can fit the temperature dependence of the resistance, the formation of a Luttinger liquid in which the transport is managed by coherent backscattering effects could be suggested [35]. Similar power law temperature dependence with the same exponent was found earlier in SWCNT samples [25, 26].

With the magnetic field B perpendicular to the plane of our sample, a negative magnetoresistance (NMR) is observed in the whole range of magnetic fields and in the temperatures between 1.8 and 80 K, with a tendency to a saturation at high magnetic fields.

Also, the magnetoresistance shows some oscillations in the magnetic field up to 30 T, at 1.8 K, as it can be seen in **Figure 2**. At higher temperatures these oscillations tend to vanish. The oscillations can be attributed to the Shubnikov–de Haas (SdH) oscillations. They are clearly seen without background subtraction.

A maximum occurs in the resistance at magnetic field B_n when the Fermi energy E_F crosses a quantized energy level E_n in the field. In graphene [57],

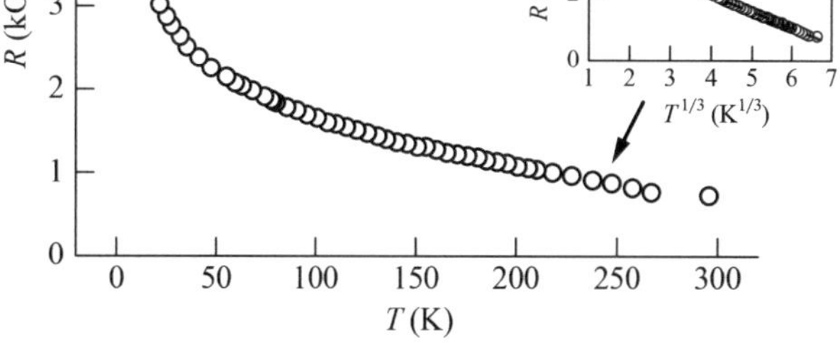

Figure 1.
Temperature dependence of the resistance of an array of pristine MWCNT sample. In the insets, the dependence R(T) is shown in the low and high temperature range.

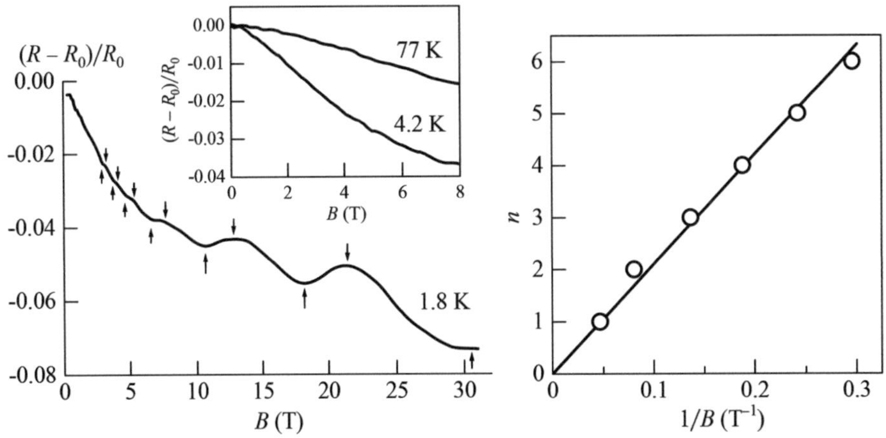

Figure 2.
Transverse magnetoresistance at 1.8 K shows some oscillations in the magnetic field up to 30 T. In the inset the magnetoresistance dependences at T = 4.2 K and at T = 77 K show no significant oscillations. The maxima of resistance oscillations are labeled with n and plotted as a function of $1/B_n$ in a Landau plot in the second inset.

$E_n(B) = \sqrt{(2n_e B v_0^2)}$ [58] and $B_n = B_0/n$, where $B_0 = E_F^2/(2ev_0\hbar^2) = \hbar k_F^2/2e$. Each maximum is labeled with n and plotted as a function of $1/B_n$ in a Landau plot (inset of **Figure 2**). The well-identified peaks (n = 1–6) define a straight line. From the slope we determine k_F and $n_s = 4k_F^2/4\pi \approx 1.9 \times 10^{12}$ cm^{-2}, which is consistent with the experiment [59].

At this stage of our investigation in these samples, we intend to suggest some explanation of the behavior of the NMR. The transverse NMR has been observed at low magnetic field with a superposition of some aperiodic oscillations consistent with the universal conductance fluctuations (UCF) [3, 30]. From a classical point of view, the interference term leading to UCF originates from adding the probability amplitudes of all paths that connect the source and drain. The NMR at low field is caused by interference contributions due to closed electron trajectories, which add up constructively at zero magnetic field; that could be considered as the definition of WL [60]. A correct treatment would need the use of the digamma functions [61, 62].

While WL is expected to decay slower with the temperature than UCF, a study of the NMR at different temperatures will allow us to discriminate from these two processes. If not, it will signify that the fluctuations could be due to the band structure of the ensemble of nanotubes and might be caused by magnetic depopulation of a one-dimensional sub-band, a phenomenon which, in extended thin films, gives rise to SdH oscillations [60].

It must be noticed that the MWCNT in our samples were not arranged as a film but rather form a "carpet" which, under some conditions, should be considered as a 2DEG. In the MWCNT films, the NMR was observed also at low magnetic fields [48], followed by a positive magnetoresistance (PMR), till 12 T. It was argued that the NMR may come from (i) quantum interference [63, 64], (ii) thermal fluctuation-induced tunneling [65], or (iii) Landau levels in disordered graphite [66]. In our case, as the NMR was observed till 25 T and at low temperature, the hypotheses (i) and (ii) can be rejected. As far as the Landau levels, or SdH oscillations [3], could be considered, we notice that the oscillations observed in our experiments appear to be mainly periodic in $1/B$ (in contradiction with which was observed in [3, 30]) and might be related to the SdH oscillations.

As a result, in this section, we report the experiments on temperature and magnetic field dependence of the resistance of self-assembled assemblies of pristine

MWCNT, in which we attribute the NMR in terms of WL and interpreted the oscillations in terms of SdH behavior.

3. Chemically functionalized multiwall carbon nanotubes

The electrical transport properties are the most important for a number of applications of carbon nanotubes. Usually, single-tube samples are used for electrical transport studies. Quantum transport properties have been obtained both in single-wall (SWCNT) [5, 9, 24–28] and multiwall nanotubes (MWCNT) [3, 30–35]. For the nanotube array samples, only mean values of the characteristic parameters were expected [38]. But, in some cases, the transport in the arrays of nanotubes was found to show single nanotube properties at low temperatures due to the mostly conductive nanotubes responsible for the transport [25]. On the other hand, for the specific applications, like chemical [15, 16] and bio- [17–21] sensors, the synergetic properties of the arrays of nanotube samples are important. They are based on their large surface area per volume and intertube coupling in electrical transport in the arrays of nanotubes [40].

We have shown experimentally that chemically functionalized multiwall carbon nanotubes could be assembled into 2D layers (dense arrays) covering large surfaces with in-plane electrodes for electrical and magnetotransport testing.

In contrast to the standard morphologies of the samples of arrays of nanotubes involving definitions of bundles (ropes), mats, networks, etc., based on hardly controlled deposition from an organic solvent dispersion of pristine nanotubes, we use the LB technique for chemically functionalized multiwall carbon nanotubes. The method we propose offers a radical departure from the existing methodology due to the possibility of covering large surfaces with dense and defect-free, molecularly thin films of carbon nanotubes.

The electrical and magnetotransport properties in the assembled monolayers of carbon nanotubes have been tested.

The organic functionalization of carbon nanotubes had been realized [54–56, 67–69]. But, to our knowledge, up to now, the experimental data on electrical and magnetotransport properties' characterization of dense monolayers manufactured using LB assembling of functionalized nanotubes is very limited. This method is expected to be used for obtaining the layers of very dense arrays of nanotubes for utilizing them for new applications in chemical and biosensors, controlled by electrical transport.

The MWCNTs produced by chemical vapor decomposition (CVD) with catalyst 5% Fe-Co/$CaCO_3$ were utilized.

Organic functionalization of MWCNTs was based on the scheme described in [69].

The purified MWCNTs were suspended in N,N-dimethylformamide [DMF] $HCON(CH_3)_2$ together with excess *p*-anisaldehyde (4-methoxybenzaldehyde) $CH_3OC_6H_4CHO$ and 3-methylhippuric acid [m-toluric acid, N-(3-methyl-benzoyl) glycine] $CH_3C_6H_4CONHCH_2CO_2H$.

Heterogeneous mixture was heated at 130°C for 3 days. The scheme of the reaction is described in **Figure 3**. After the reaction was stopped, the organic phase was separated from unreacted material by centrifugation and washing five times with chloroform ($CHCl_3$) and vacuum drying. The obtained dark solid phase was easily soluble in $CHCl_3$ up to a few mg/ml without sonication. The functionalized nanotubes were tested by HRTEM (**Figure 4**) and FTIR (**Figure 5**) methods. Due to the distinct layer on the surface of the nanotubes, observed by HRTEM (**Figure 4**), which is responsible for the absorption peaks in 1400–1500 cm^{-1} and 1800–1900 cm^{-1} on FTIR spectra (**Figures 5** and **6**), we conclude that the

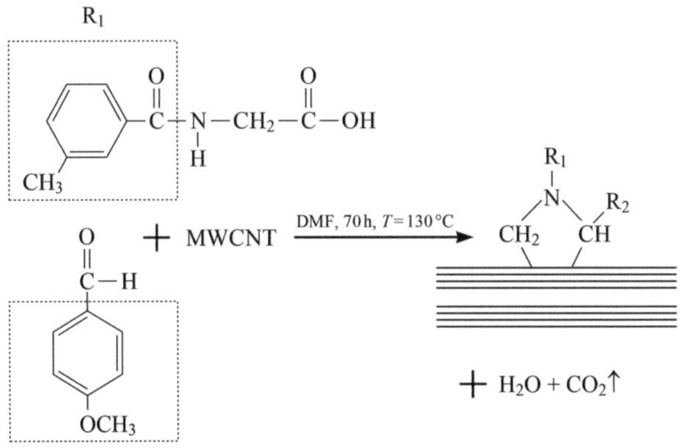

Figure 3.
The scheme of the reaction of organic functionalization of MWCNTs. The purified MWCNTs were suspended in N,N-dimethylformamide [DMF] $HCON(CH_3)_2$ together with excess p-anisaldehyde (4-methoxybenzaldehyde) $CH_3OC_6H_4CHO$ and 3-methylhippuric acid [m-toluric acid, N-(3-methyl-benzoyl) glycine] $CH_3C_6H_4CONHCH_2CO_2H$. Heterogeneous mixture was heated at 130°C for 3 days.

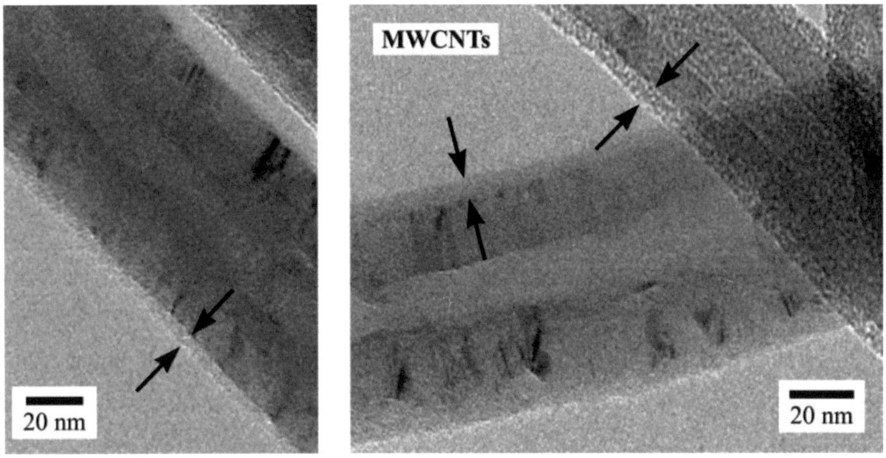

Figure 4.
High-resolution TEM image of organically functionalized MWCNT. The nanotubes are covered with the thin organic layer (shown with the arrows).

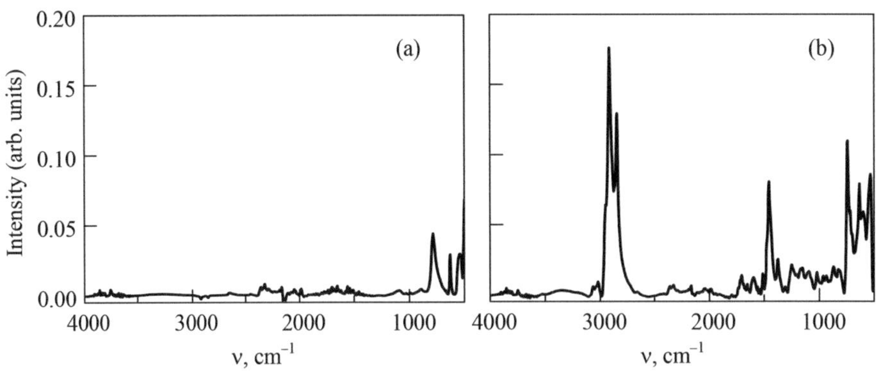

Figure 5.
FTIR spectra of MWCNTs: no functionalization (a); with functionalization (b).

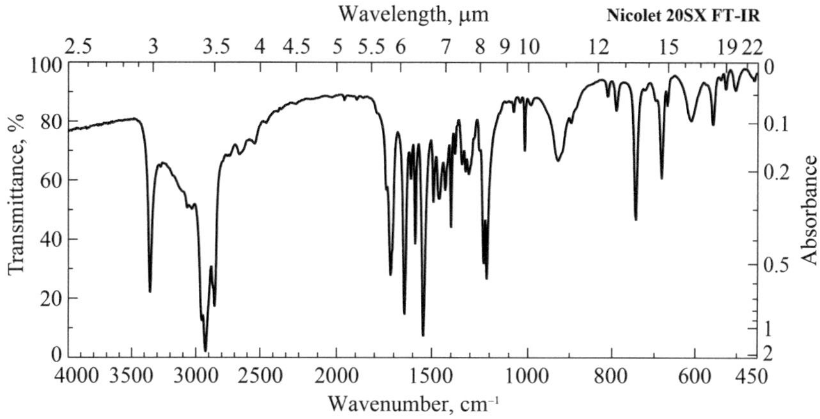

Figure 6.
FTIR spectrum 3-methylhippuric acid, condensed phase. Adapted from the Sigma-Aldrich catalog: https://www.sigmaaldrich.com/spectra/ftir/FTIR005396.PDF.

Figure 7.
Functionalized nanotubes self-assembled in dense arrays (monolayers), covering without empty space the whole surfaces of the finger-shaped electrode device.

functionalization procedure described in [69] for single-wall carbon nanotubes works for MWCNTs. An important feature of FTIR spectrum of MWCNT is the absence of -COOH group peak near 3350 cm^{-1}. This fact has proven covalent bonding of 3-methylhippuric acid and *p*-anisaldehyde with MWCNTs, and not just physical adsorption of the surfactant.

The deposition of the layers (arrays) of nanotubes on the surface of the devices with the electrodes was done by using the cell, imitating the LB trough. Once a droplet of the solution of functionalized nanotubes in chloroform was spread on the water surface, a droplet of diblock copolymer PS-PMMA solution was added in order to create surface pressure of ~9 mN/m. Functionalized nanotubes were self-assembled in dense arrays (monolayers), covering without empty space the whole surfaces of the finger-shaped electrode devices, when the monolayer was picked up from the water surface (**Figure 7**). The same layer of MWCNTs on the substrate at higher magnification with SEM is shown in **Figure 8**.

Utilizing LB method for functionalized nanotubes has shown reliable and reproducible p-A isotherms. Based on this, we expect to use LB method for covering large surfaces with dense and defect-free, molecularly ordered ultrathin films of carbon nanotubes with controlled thickness and orientation.

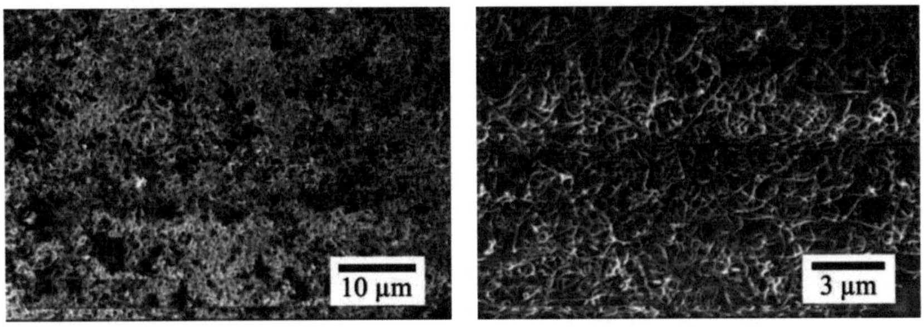

Figure 8.
SEM images of the dense arrays (monolayers) of functionalized MWCNTs, covering surfaces of the substrate.

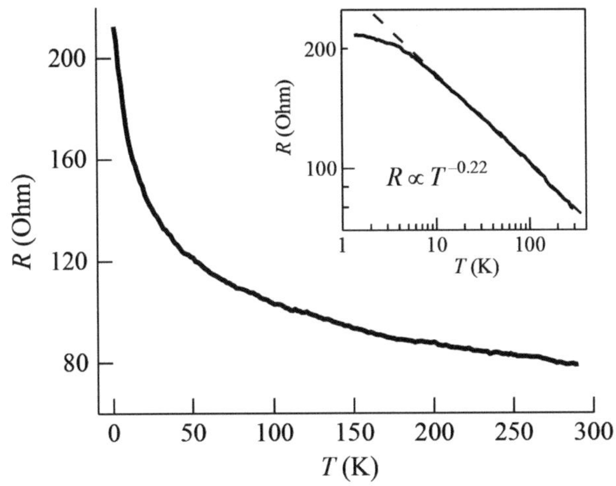

Figure 9.
The temperature dependence of the resistance of the functionalized MWCNT sample. In the inset, dependence is the same in the logR vs. logT scale. The linear dependence in the double log scale in the range 10–300 K gives a power law with an exponent −0.22 for the temperature dependence.

The electrical and magnetotransport properties of the layers (arrays) of multi-wall carbon nanotubes have been tested in the temperature range 1.8–300 K and in magnetic fields up to 35 T.

The nanotube samples on the electrodes with "finger-shape" geometry (**Figure 7**) have shown low resistance (<1 kOhm at room temperature) and a "weak" temperature dependence of the resistance in the shape of power law in the temperature range T = 4.2–300 K.

The temperature dependences of the resistance in linear and log–log scales for arrays of the functionalized nanotubes are presented in **Figure 9**. It is interesting to point out that the temperature dependences of the resistance are represented by the power law with the exponent −0.22. While a power law can fit the temperature dependence of the resistance in MWCNT, a behavior suggestive of the formation of a Luttinger liquid, in which the transport is managed by coherent backscattering effects, can be considered [35]. Similar power law temperature dependence with the same value of the exponent was found earlier in SWCNT "bundle" samples deposited on the top of the contacting electrodes [25, 26].

The saturation at low temperatures is noticeable on $R(T)$ dependence of functionalized nanotubes. This could be explained in the framework of Coulomb blockade and tunneling between tubes through thin organic layers, covering MWCNT surfaces.

The magnetoresistance measurements were carried out in pulsed magnetic fields up to 35 T in the temperature range 1.8–80 K (**Figure 10**). The magnetic field

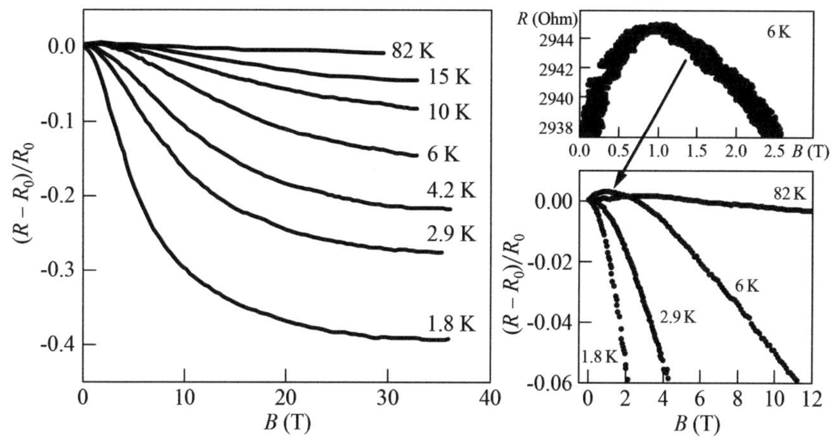

Figure 10.
The magnetoresistance of the functionalized MWCNT sample is carried out in pulsed magnetic fields up to 35 T in the temperature range 1.8–80 K.

orientation in relation to the current direction was considered "normal" to the current direction.

With the magnetic field B perpendicular to the plane of the sample, NMR was observed in the whole range of available magnetic fields between 1.8 and 80 K in the arrays of functionalized nanotubes (**Figure 10**). As already reported in section 2, the NMR was believed to be observed due to the interference contributions of the closed electron trajectories, which add up constructively at zero magnetic field; that could be considered as the definition of WL [60]. A correct treatment should be based on the use of the digamma function [61, 62].

In addition to the NMR at high fields for functionalized nanotubes, we can see a positive magnetoresistance at low fields (inset in **Figure 10**). The positive magnetoresistance increases quadratically and saturates at fields above B = 2 T. The weak antilocalization effect [70–72] in multiwall carbon nanotubes [73] appears to be responsible for the positive magnetoresistance. It is attributed to the spin-dephasing process, arising from the local interfacial fields as a genuine property of the curved multiwall tubes [73].

We have reported the experimental observation of the electrical transport properties of assembled (layers) arrays of non-functionalized and functionalized MWCNTs. The negative magnetoresistance as a characteristic of weak localization state was observed. In addition to the negative magnetoresistance at high fields for functionalized nanotubes, we observed positive magnetoresistance at low fields.

4. Oxidized carbon nanotubes

This technology introduces the formation of oxidized multiwalled carbon nanotubes. Oxidized MWCNTs become hydrophilic and soluble in water. In order to assemble uniform monolayers of MWCNTs on large surfaces we have developed, the so-called "inverted" Langmuir–Blodgett technique, the essence of which is clear from **Figure 11**. The method consists of the following major technology steps. MWCNT oxidation can be carried out by means of oxidizing liquids such as sulfuric and nitric acids.

MWCNT is dispersed in acid solution (3:1 = H_2SO_4:HNO_3) and ultra-sonicated for a few hours in a water bath at the elevated temperatures. The solution was stirred

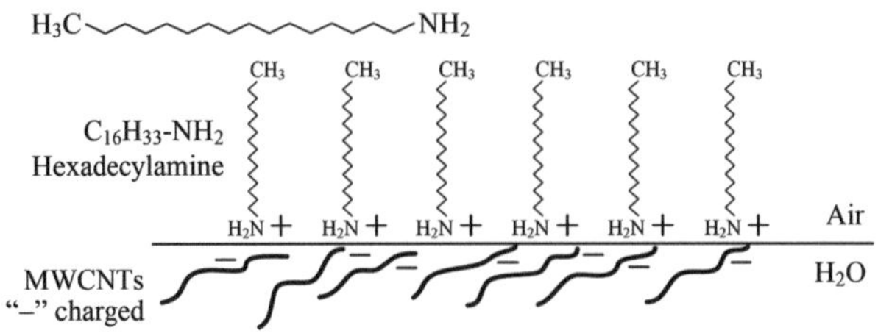

Figure 11.
Schematics of the inverted Langmuir–Blodgett technique for the monolayers of MWCNT assembly.

very well simultaneously. The mixture is filtered by using 0.1-micron polyvinylidene fluoride (PVDF) filter. The filtered oxidized MWCNT is extensively washed with distilled water until pH became neutral. The powder was dried in the oven.

This method generates oxygenated functional groups (-OH, -C=O, and -COOH).

Hydroxyl (-OH) groups are not highly reactive, but they readily form hydrogen bonds and contribute to making molecules soluble in water.

Carbonyl (-C=O) groups have one oxygen atom double-bonded to a carbon atom (symbolized as C=O). Like hydroxyl groups, carbonyl groups contribute to making CNTs water-soluble.

The carboxyl group (symbolized as -COOH) has both a carbonyl and a hydroxyl group attached to the same carbon atom, resulting in new properties— it can be ionized—releasing the H from the hydroxyl group as a free proton (H^+), with the remaining O carrying a negative charge.

MWCNT monolayers were prepared by dispersion in deionized water at the concentration of 0.04–0.05 mg/ml. The surfactant hexadecylamine is dissolved in chloroform to prepare the spreading solution at the concentration of 1 mg/ml and spread onto the surface of the oxidized MWCNT solution in the LB trough. The floating monolayer of highly polarized surfactant molecule, having positively charged hydrophilic ends, makes monolayers with negatively charged oxidized carbon nanotubes through electrostatic interaction. The hybrid monolayer compressed in the LB trough at a rate of a few mm/min. Surface pressure-area (p-A) isotherm

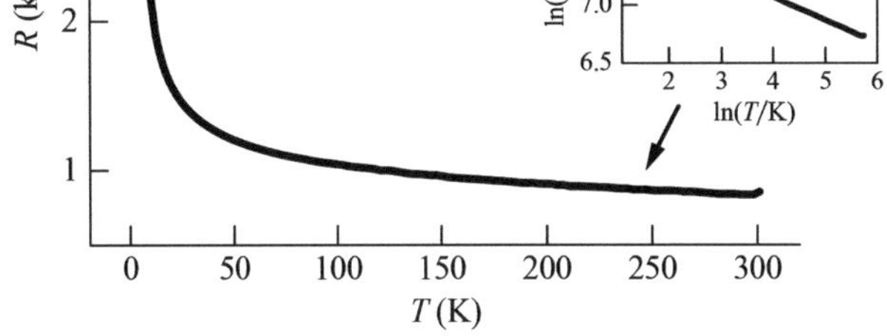

Figure 12.
The temperature dependence of the resistance of the oxidized MWCNT sample. In the insets, low temperature range dependence in the lnR vs. $T^{-1/4}$ scale shows localization regime of conductivity. The linear dependence in the double ln scale high temperature range gives a power law temperature dependence.

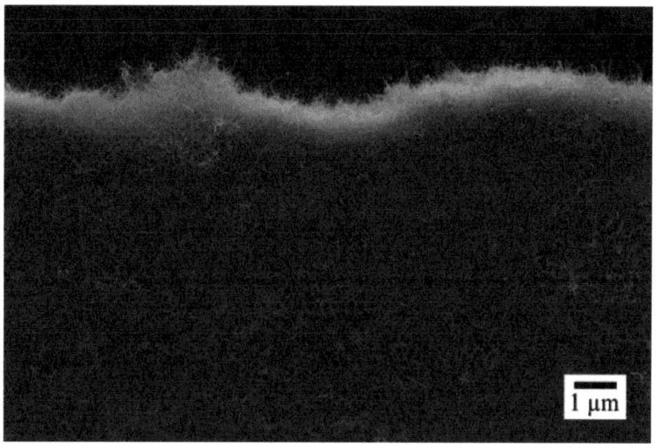

Figure 13.
SEM image of a few micrometer-thick oxidized MWCNT membrane.

curves of the hybrid monolayers are recorded during the compression. The hybrid monolayers were deposited by vertical dipping on a clean hydrophilic surface. After deposition, the surface of the transferred film was dried in a flow of N_2 gas.

The electrical transport properties of the monolayers (arrays) of the oxidized multiwall carbon nanotubes have been tested in the temperature range 4.2–300 K and represented in **Figure 12**.

The oxidized MWCNTs can be assembled in the freestanding layers of arbitrary thickness, utilizing vacuum filtering technique. The resulting MWCNT membranes are very stable and can be further utilized for sensor and energy storage applications. **Figure 13** shows the SEM image of a few micrometer-thick membrane.

5. Nano-catalysis

The common catalysis-oriented goals are to understand and predict the properties of nanosized materials and control how they facilitate chemical reactivity. Another critically important issue deals with the manufacture of nanoscale components from the bottom up and finally to integrate nanoscale components into macroscopic scale objects and catalysts for real-world uses.

We have shown a very strong catalytic activity of an outer surface of MWCNT, modified by acid oxidation. Nanocrystals of NaF were formed on the surface of MWCNT (**Figure 14**). We have also observed metal nanocrystal "decoration" on the surface of MWCNTs.

For example, electrolytic water splitting represents the most environmentally friendly alternative to generate hydrogen gas. However, the kinetics of the oxygen evolution reaction (OER) are slow and require a catalyst. Most catalysts to date have been limited to transition metal oxides or noble metals—both of which are expensive and unsustainable. The OER activity rationalized by the oxygen-containing functional groups on the surface of oxidized MWCNTs alters the electronic distribution of the surrounding carbon atoms at the MWCNT surfaces, thereby facilitating the adsorption of water oxidation intermediates. This opens the door to new applications of surface-oxidized MWCNTs for catalyzing a class of important anodic reactions in water splitting and fuel cells. Further improvements of the activity of the surface-oxidized carbon nanomaterials may enable the fine-tuning of the structure and compositions of hybrid carbon materials for specific applications.

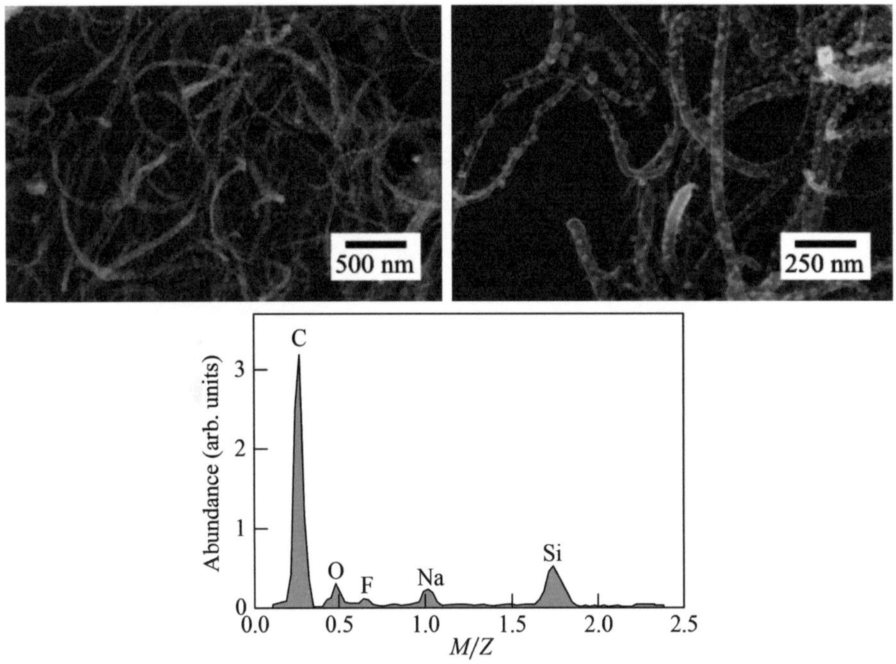

Figure 14.
Nanocrystals of NaF formed on the surface of oxidized MWCNT (top panel). Mass spectrum of NaF nanocrystals "decorated" MWCNTs (bottom panel); M/Z is the ratio of ion mass to its charge.

6. Carbon nanotube assembling and physical cross-linking

We have developed a novel physical method of stable linkage between neighboring carbon nanotubes in 2D layers (dense arrays), by intertube bridging using Ar^+-ion beam (**Figure 15**).

The carbon nanotube layer has modified using ion beam irradiation. It introduces stable link formation between neighboring carbon nanotubes, in other words, bridging or physical cross-linking.

The intertube bridging (cross-linking) of MWCNTs and SWCNTs in the arrays was observed under the Ar^+-ion irradiation. This method can be utilized for the improvement of the electrical and thermal conductivity properties of carbon nanotube layers for electron transport heat transfer applications.

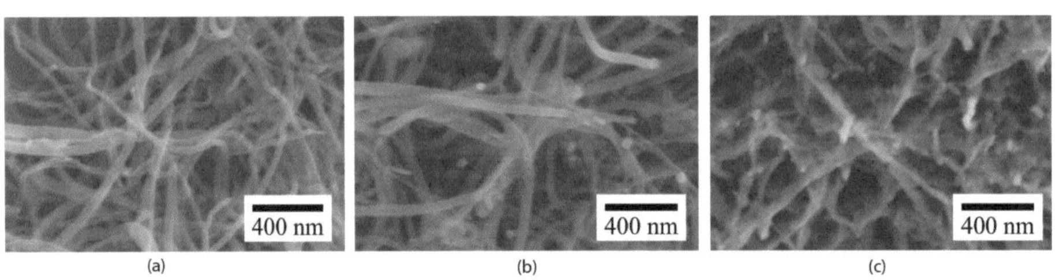

Figure 15.
(a) The image of the initial (unprocessed) array of multiwall carbon nanotubes. (b) The image of a "short-time" processing regime: 6 kV, 60 s. Just the ends of the tubes attacked by Ar^+-ion irradiation. (c) The image of a "long-time" processing regime: 6 kV, 10 min. The continuous cross-linked network of the MW carbon after the Ar^+-ion irradiation is observed.

7. Nanosensor applications of carbon nanotube films

In this chapter we would like to point out an example of application of films of MWCNT as a new icing condition resistive sensor that we have developed. These sensors are based on the adsorption of a molecular thin layer of water on the surface of carbon nanotubes and on the detection of the first-order phase transition of water molecules into ice. This transition is very well detected as a result of non-monotonous dependence of the resistance of the sensor vs. temperature in the vicinity of the freezing point due to a virtual "field effect transistor." Electronic transport in carbon nanotube films, assembled into the resistive films, was found to be extremely sensitive to the adsorption of polar H_2O molecules.

Modern sensors of icing conditions (optical or piezo devices) are based on the detection of the actual (significantly thick) layers of ice formed on the surfaces. The accumulation of the ice layer is a fast process, and detection of the massive ice formation is too late for the safety of the aircrafts.

We have developed a method of assembling MWCNT films of arbitrary thickness for sensor applications. This method involves oxidation of carbon nanotubes and does have several major advantages over the conventional methods of carbon nanotube assembling via their functionalization. The assembled carbon nanotube films are dense, homogeneous, and strong on the macro level, but internally they consist of disordered structure of self-assembled carbon nanotubes, forming conductive medium, as it is seen in **Figure 16**. Besides they are hydrophilic and adsorb water molecules strongly. Based on our experience of multiwall CNT characterization [22, 74], we were able to study them at fixed values of humidity and temperature variation.

Standard sensors of relative humidity based on multiwall carbon nanotubes were developed earlier [75, 76].

We have found that the adsorption of the water vapor at the temperatures close to freezing conditions generates a specific non-monotonous dependence of the resistance of the sensor vs. temperature.

The intensive precipitation of the water vapor, when the temperature is decreasing, results in the increase of the resistance of the nanosensor, due to the "field effect" created by the adsorbed polar water molecules on the surface of slightly charged CNT tubes. A further decrease of the temperature passing the freezing point results in the sudden drop of the resistance ("lambda-point"-type curve at the phase transitions of the first order) (see the insert in **Figure 17**) due to the water transition to nonpolar ice crystal. As a result, the "field effect" disappears, and the resistance of the carbon nanotubes decreases again.

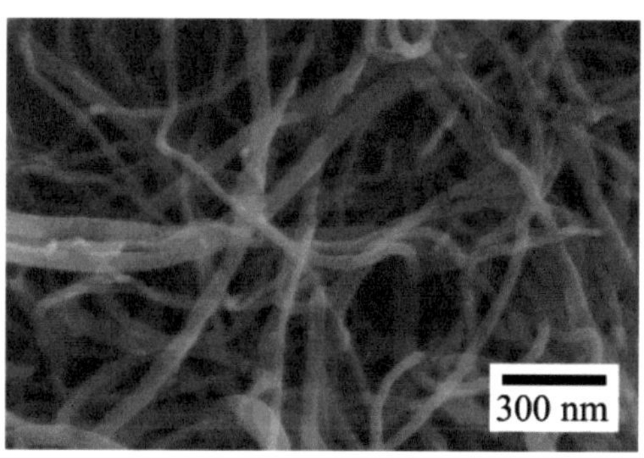

Figure 16.
SEM image of the morphology of multiwall carbon nanotube film.

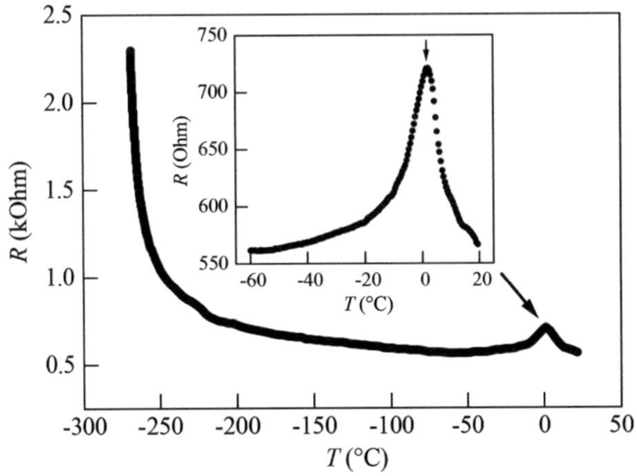

Figure 17.
Temperature dependence of the resistance of the MWCNT layer in the presence of water vapors. Note, the maximum of the resistance was found to be located in the nearest proximity of the water freezing point at T = 0°C.

In order to verify, if the positions of the peaks correspond to the humidity (dew point or frost point), we did temperature scans at different controllable values of the humidity. The dependences of the resistance of the CNT sensor measured at fast temperature scan at different humidity levels and temperature variations from approx +50°C down to −50°C are plotted in **Figure 18**. The temperature dependences of the resistance of the CNT sensors measured at the slow temperature scan (≤0.01°C/s in the cryo-cell of our own design) are shown in **Figure 19**. As can be seen in **Figures 18** and **19**, if the temperature drops down, a significant resistance increase takes place, followed by the maximum point of the resistance and sudden resistance drop due to the ice formation.

There is a significant difference in the observation of the dew points and the frost points using CNT sensors. If we observe T_{dew}, which is higher than the freezing point, we observe significant resistance increase due to condensation. Then the saturation occurs and, at the freezing point, is characterized by a sudden resistance decrease. In the meantime, if we observe frost points instead, while the temperature decreases, the condensation corresponds to the frost point.

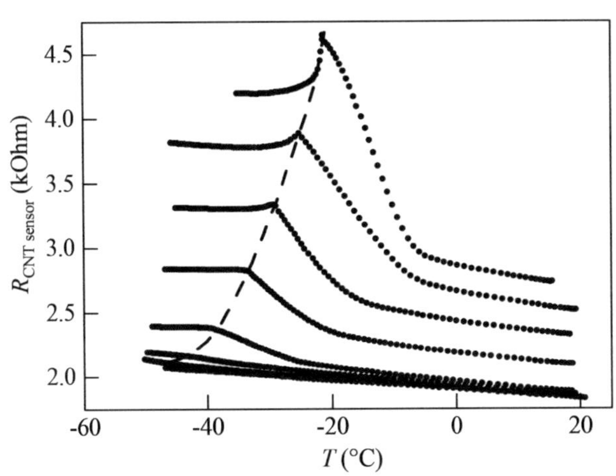

Figure 18.
The resistance of the CNT sensor vs. temperature at different humidity levels at the fast temperature scan of ~1°C/s.

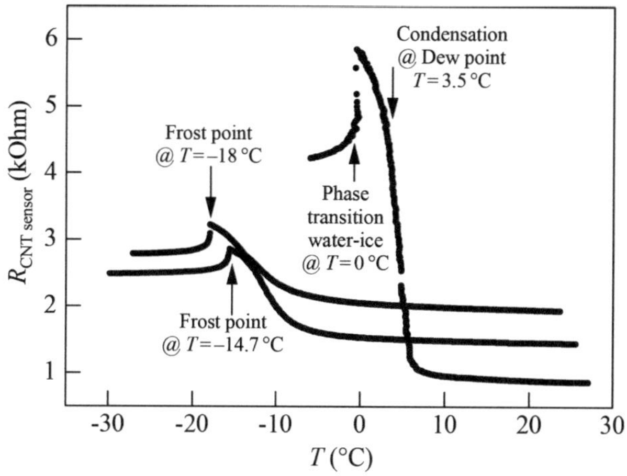

Figure 19.
Temperature dependences of the resistance of the CNT sensors measured at the slow temperature scan ≤0.01°C/s in the cryo-cell of our own design.

8. Coating carbon nanotubes with diamond-like carbon films

A commercial plasma CVD setup was used to deposit a diamond-like carbon (DLC) film over the CNT matrix (**Figure 20**).

We found out that thin (~50 nm thick) DLC films have significantly improved mechanical properties.

DLC coating on the CNT layers has a high degree of wettability. We also determined that it is adding significant reinforcement to the MWCNT matrix.

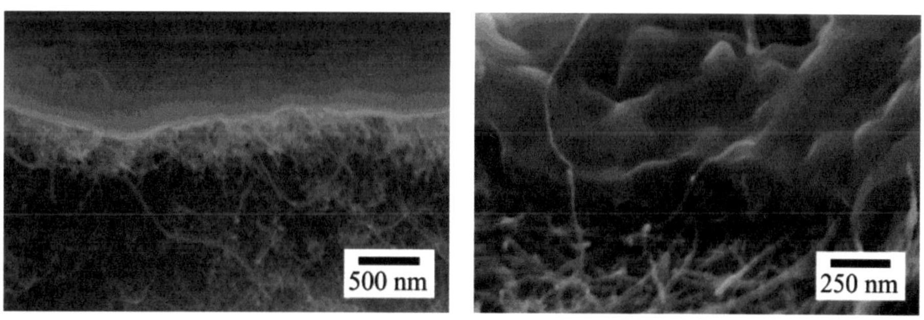

Figure 20.
SEM pictures of the DLC film deposited on the top of the MWCNT layer.

9. Conclusions

Several methods of assembling MWCNTs into monolayers and freestanding films of arbitrary thickness have been developed.

We have studied assemblies of pristine (non-modified) CNTs, CNTs modified by organic chemical functionalization, and the oxidized CNTs.

We have experimentally tested the electrical transport properties of assembled layers (films) of non-functionalized, functionalized, and oxidized MWCNTs.

The temperature dependence of the resistance and magnetoresistance of the self-assembled arrays of MWCNT, tested in the wide temperature range, were considered as a tool to determine the transport characteristics of the films, important for further applications.

The negative magnetoresistance as a characteristic of weak localization state was observed. In addition to the negative magnetoresistance at high fields for functionalized nanotubes, we observed positive magnetoresistance at low fields.

The layers of MWCNTs have been considered as attractive materials for various nanosensors and as the electrodes of electrochemical energy storage devices. We have shown an example of application of films of MWCNT as an icing condition resistive sensor. A very strong catalytic activity of an outer surface of MWCNT, modified by oxidation, has been pointed out as well.

We have developed a novel method of stable physical linkage between the neighboring carbon nanotubes in 2D layers (dense arrays) by intertube bridging using Ar^+-ion beam and the method of MWCNT coating with diamond-like carbon films as well.

Acknowledgements

V.S. is very grateful to the Ministry of Education of France for the fellowship, the National Science Foundation of Switzerland, Award #7BYPJ065694 for support, the Laboratoire National des Champs Magnetiques Intenses de Toulouse, Dr. Jean Galibert and Prof. Laszlo Forró for hospitality. He would like to express his sincere gratitude to Dr. J. Quinn for the help with SEM imaging, Dr. M. Seo for the help with TEM, Dr. Y. Seo and Dr. J. Koo for the help with MWCNT oxidation, and Dr. V. Zaitsev for the help with the organic functionalization of MWCNT. Finally, V.S. acknowledges support of the Sensor CAT and the Research Foundation of SUNY. N.P. acknowledges support of the Belarusian Research Program "Convergence-2020" and the Belarusian Republican Foundation for Fundamental Research (Grant No. F18R-253).

Author details

Vladimir Samuilov[1,2*], Jean Galibert[2] and Nikolai Poklonski[3]

1 Department of Materials Science, SUNY at SB, Stony Brook, NY, USA

2 Laboratoire National des Champs Magnetiques Intenses, LNCMI-EMFL, CNRS/INSA/UGA/UPS, Toulouse and Grenoble, France

3 Department of Physics, Belarusian State University, Minsk, Belarus

*Address all correspondence to: vladimir.samuilov@stonybrook.edu

IntechOpen

© 2019 The Author(s). Licensee IntechOpen. This chapter is distributed under the terms of the Creative Commons Attribution License (http://creativecommons.org/licenses/by/3.0), which permits unrestricted use, distribution, and reproduction in any medium, provided the original work is properly cited.

References

[1] Iijima S. Helical microtubules of graphitic carbon. Nature. 1991;**354**: 56-58. DOI: 10.1038/354056a0

[2] Dai H. Carbon nanotubes: Opportunities and challenges. Surface Science. 2002;**500**:218-241. DOI: 10.1016/S0039-6028(01)01558-8

[3] Forró L, Schönenberger C. Physical properties of multi-wall nanotubes. In: Dresselhaus MS, Dresselhaus G, Avouris P, editors. Carbon Nanotubes, Topics Applied Physics. Vol. 80. Berlin: Springer; 2001. pp. 329-391. DOI: 10.1007/3-540-39947-X_13

[4] Poklonski NA, Vyrko SA, Siahlo AI, Poklonskaya ON, Ratkevich SV, Hieu NN, et al. Synergy of physical properties of low-dimensional carbon-based systems for nanoscale device design. Materials Research Express. 2019;**6**:042002. DOI: 10.1088/2053-1591/aafb1c

[5] Avouris P. Carbon nanotube electronics. Chemical Physics. 2002;**281**:429-445. DOI: 10.1016/S0301-0104(02)00376-2

[6] Hartmann RR, Kono J, Portnoi ME. Terahertz science and technology of carbon nanomaterials. Nanotechnology. 2014;**25**:322001. DOI: 10.1088/0957-4484/25/32/322001

[7] Martel R, Schmidt T, Shea HR, Hertel T, Avouris P. Single- and multi-wall carbon nanotube field-effect transistors. Applied Physics Letters. 1998;**73**:2447-2449. DOI: 10.1063/1.122477

[8] Tans SJ, Verschueren ARM, Dekker C. Room-temperature transistor based on a single carbon nanotube. Nature. 1998;**393**:49-52. DOI: 10.1038/29954

[9] Martel R, Derycke V, Lavoie C, Appenzeller J, Chan KK, Tersoff J, et al. Ambipolar electrical transport in semiconducting single-wall carbon nanotubes. Physical Review Letters. 2001;**87**:256805. DOI: 10.1103/PhysRevLett.87.256805

[10] Goze C, Vaccarini L, Henrard L, Bernier P, Hernandez E, Rubio A. Elastic and mechanical properties of carbon nanotubes. Synthetic Metals. 1999;**103**:2500-2501. DOI: 10.1016/S0379-6779(98)01071-6

[11] Fraysse J, Minett AI, Jaschinski O, Duesberg GS, Roth S. Carbon nanotubes acting like actuators. Carbon. 2002;**40**:1735-1739. DOI: 10.1016/S0008-6223(02)00041-6

[12] Vigolo B, Pénicaud A, Coulon C, Sauder C, Pailler R, Journet C, et al. Macroscopic fibers and ribbons of oriented carbon nanotubes. Science. 2000;**290**:1331-1334. DOI: 10.1126/science.290.5495.1331

[13] Frackowiak E, Gautier S, Gaucher H, Bonnamy S, Beguin F. Electrochemical storage of lithium multiwalled carbon nanotubes. Carbon. 1999;**37**:61-69. DOI: 10.1016/S0008-6223(98)00187-0

[14] Collins PG, Bradley K, Ishigami M, Zettl A. Extreme oxygen sensitivity of electronic properties of carbon nanotubes. Science. 2000;**287**:1801-1804. DOI: 10.1126/science.287.5459.1801

[15] Varghese OK, Kichambre PD, Gong D, Ong KG, Dickey EC, Grimes CA. Gas sensing characteristics of multi-wall carbon nanotubes. Sensors and Actuators B: Chemical. 2001;**81**:32-41. DOI: 10.1016/S0925-4005(01)00923-6

[16] Kong J, Franklin NR, Zhou C, Chaplin MG, Peng S, Cho K, et al. Nanotube molecular wires as chemical sensors. Science. 2000;**287**:622-625. DOI: 10.1126/science.287.5453.622

[17] Guo Z, Sadler PJ, Tsang SC. Immobilization and visualization of DNA and proteins on carbon nanotubes. Advanced Materials. 1998;**10**:701-703. DOI: 10.1002/(SICI)1521-4095(199806)10:9<701::AID-ADMA701>3.0.CO;2-4

[18] Davis JJ, Green MLH, Hill HAO, Leung YC, Sadler PJ, Sloan J, et al. The immobilization of proteins in carbon nanotubes. Inorganica Chimica Acta. 1998;**272**:261-266. DOI: 10.1016/S0020-1693(97)05926-4

[19] Balavoine F, Schultz P, Richard C, Mallouh V, Ebbesen TW, Mioskowski C. Helical crystallization of proteins on carbon nanotubes: A first step towards the developments of new biosensors. Angewandte Chemie, International Edition. 1999;**38**:1912-1915. DOI: 10.1002/(SICI)1521-3773(19990712)38:13/14<1912::AID-ANIE1912>3.0.CO;2-2

[20] Chen RJ, Zhang Y, Wang D, Dai H. Noncovalent sidewall functionalization of single-walled carbon nanotubes for protein immobilization. Journal of the American Chemical Society. 2001;**123**:3838-3839. DOI: 10.1021/ja010172b

[21] Xue HG, Sun WL, He BJ, Shen ZQ. A new application of carbon nanotubes constructing biosensor. Chinese Chemical Letters. 2002;**13**:799-800

[22] Ksenevich V, Galibert J, Samuilov V. Charge transport in carbon nanotube films and fibers. In: Marulanda JM, editor. Carbon Nanotubes. Rijeka, Croatia: InTech; 2010. pp. 123-146. DOI: 10.5772/39422

[23] Ksenevich VK, Gorbachuk NI, Poklonski NA, Samuilov VA, Kozlov ME, Wieck AD. Impedance of single-walled carbon nanotube fibers. Fullerenes, Nanotubes, and Carbon Nanostructures. 2012;**20**:434-438. DOI: 10.1080/1536383X.2012.655562

[24] Frank S, Poncharal P, Wang ZL, de Heer WA. Carbon nanotube quantum resistors. Science. 1998;**280**:1744-1746. DOI: 10.1126/science.280.5370.1744

[25] Bockrath M, Cobden DH, Lu J, Rinzler AG, Smalley RE, Balents L, et al. Luttinger-liquid behaviour in carbon nanotubes. Nature. 1999;**397**:598-601. DOI: 10.1038/17569

[26] Nygard J, Cobden DH, Bockrath M, McEuen PL, Lindelof PE. Electrical transport measurements on single-walled carbon nanotubes. Applied Physics A: Materials Science & Processing. 1999;**69**:297-304. DOI: 10.1007/s003390051

[27] Zhou C, Kong J, Dai H. Electrical measurements of individual semiconducting single-walled carbon nanotubes of various diameter. Applied Physics Letters. 2000;**76**:1597-1599. DOI: 10.1063/1.126107

[28] Shea HR, Martel R, Avouris P. Electrical transport in rings of single-wall nanotubes: One-dimensional localization. Physical Review Letters. 2000;**84**:4441-4444. DOI: 10.1103/PhysRevLett.84.4441

[29] Laird EA, Kuemmeth F, Steele GA, Grove-Rasmussen K, Nygård J, Flensberg K, et al. Quantum transport in carbon nanotubes. Reviews of Modern Physics. 2015;**87**:703-764. DOI: 10.1103/RevModPhys.87.703

[30] Langer L, Bayot V, Grivei E, Issi J-P, Heremans JP, Olk CH, et al. Quantum transport in a multiwalled carbon nanotube. Physical Review Letters. 1996;**76**:479-482. DOI: 10.1103/PhysRevLett.76.479

[31] Bachtold A, Henny M, Terrier C, Strunk C, Schönenberger C, Salvetat J-P, et al. Contacting carbon nanotubes selectively with low-ohmic contacts four-probe electric measurement.

Applied Physics Letters. 1998;**73**:274-276. DOI: 10.1063/1.121778

[32] Bachtold A, Strunk C, Salvetat J-P, Bonard J-M, Forró L, Nussbaumer T, et al. Aharonov–Bohm oscillations in carbon nanotubes. Nature. 1999;**397**:673-675. DOI: 10.1038/17755

[33] Schönenberger C, Bachtold A, Strunk C, Salvetat J-P, Forró L. nterference and interaction in multi-wall carbon nanotubes. Applied Physics A: Materials Science & Processing. 1999;**69**:283-295. DOI: 10.1007/s003390051

[34] Lee J-O, Kim J-R, Kim J-J, Kim JH, Kim N, Park JW, et al. Magnetoresistance and differential conductance in multiwalled carbon nanotubes. Physical Review B. 2000;**61**:R16362-R16365. DOI: 10.1103/PhysRevB.61.R16362

[35] Liu K, Avouris P, Martel R, Hsu WK. Electrical transport in doped multiwalled carbon nanotubes. Physical Review B. 2001;**63**:161404. DOI: 10.1103/PhysRevB.63.161404

[36] Stojetz B, Miko C, Forró L, Strunk C. Effect of band structure on quantum interference in multiwall carbon nanotubes. Physical Review Letters. 2005;**94**:186802. DOI: 10.1103/PhysRevLett.94.186802

[37] Ishikawa S, Enomoto R, Aoki N, Umishita K, Ishibashi K, Aoyagi Y, et al. One-dimensional conduction in multi-wall carbon nano-tubes. Springer Proceedings in Physics. 2001;**87**:1661-1662

[38] Liu K, Roth S, Duesberg G, Wagenhals M, Journet C, Bernier P. Transport properties of single-walled carbon nanotubes. Synthetic Metals. 1999;**103**:2513-2514. DOI: 10.1016/S0379-6779(98)01081-9

[39] Eletskii AV, Knizhnik AA, Potapkin BV, Kenny JM. Electrical characteristics of carbon nanotube doped composites. Physics-Uspekhi. 2015;**58**:209-251. DOI: 10.3367/UFNe.0185.201503a.0225

[40] Stahl H, Appenzeller J, Martel R, Avouris P, Lengeler B. Intertube coupling in ropes of single-wall carbon nanotubes. Physical Review Letters. 2000;**85**:5186-5189

[41] Krstić V, Roth S, Burghard M. Phase breaking in three-terminal contacted single-walled carbon nanotube bundles. Physical Review B. 2000;**62**:R16352-R16355. DOI: 10.1103/PhysRevB.62.R16353

[42] Park JG, Kim GT, Park JH, Yu HY, McIntosh G, Krstic V, et al. Quantum transport in low-dimensional organic nanostructures. Thin Solid Films. 2001;**393**:161-167. DOI: 10.1016/S0040-6090(01)01064-1

[43] Fischer JE, Dai H, Thess A, Lee R, Hanjani NM, Dehaas DL, et al. Metallic resistivity in crystalline ropes of single-wall carbon nanotubes. Physical Review B. 1997;**55**:R4921-R4924. DOI: 10.1103/PhysRevB.55.R4921

[44] Kaiser AB, Park YW, Kim GT, Choi ES, Düsberg G, Roth S. Electronic transport in carbon nanotube ropes and mats. Synthetic Metals. 1999;**103**:2547-2550. DOI: 10.1016/S0379-6779(98)00222-7

[45] Kaiser AB, Düsberg G, Roth S. Heterogeneous model for conduction in carbon nanotubes. Physical Review B. 1998;**57**:1418-1421. DOI: 10.1103/PhysRevB.57.1418

[46] Kim GT, Jhang SH, Park JG, Park YW, Roth S. Non-ohmic current–voltage characteristics in single-wall carbon nanotube network. Synthetic Metals. 2001;**117**:123-126. DOI: 10.1016/S0379-6779(00)00551-8

[47] Franklin NR, Dai H. An enhanced CVD approach to extensive nanotube networks with directionality. Advanced Materials. 2000;**12**:890-894. DOI: 10.1002/1521-4095(200006)12:12 <890:AID-ADMA890>3.0.CO;2-K

[48] Baumgartner G, Carrard M, Zuppiroli L, Bacsa W, de Heer WA, Forró L. Hall effect and magnetoresistance of carbon nanotube films. Physical Review B. 1997;**55**:6704-6707. DOI: 10.1103/PhysRevB.55.6704

[49] Ausman KD, Piner R, Lurie O, Ruoff RS, Korobov M. Organic solvent dispersions of single-walled carbon nanotubes: Toward solutions of pristine nanotubes. The Journal of Physical Chemistry B. 2000;**104**:8911-8915. DOI: 10.1021/jp002555m

[50] Krstic V, Muster J, Duesberg GS, Philipp G, Burghard M, Roth S. Electrical transport in single-walled carbon nanotube bundles embedded in Langmuir–Blodgett monolayers. Synthetic Metals. 2000;**110**:245-249. DOI: 10.1016/S0379-6779(99)00302-1

[51] Krstic V, Duesberg GS, Muster J, Burghard M, Roth S. Langmuir–Blodgett films of matrix-diluted single-walled carbon nanotubes. Chemistry of Materials. 1998;**10**:2338-2340. DOI: 10.1021/cm980207f

[52] Burghard M, Krstic V, Duesberg GS, Philipp G, Muster J, Roth S. Carbon SWNTs as wires and structural templates between nanoelectrodes. Synthetic Metals. 1999;**103**:2540-2542. DOI: 10.1016/S0379-6779(99)80005-8

[53] Sano M, Kamino A, Okamura J, Shinkai S. Self-organization of PEO-*graft*-single-walled carbon nanotubes in solutions and Langmuir–Blodgett films. Langmuir. 2001;**17**:5125-5128. DOI: 10.1021/la010126p

[54] Holzinger M, Vostrovsky O, Hirsch A, Hennrich F, Kappes M, Weiss R, et al. Sidewall functionalization of carbon nanotubes. Angewandte Chemie, International Edition. 2001;**40**:4002-4005. DOI: 10.1002/1521-3773 (20011105)40:21<4002:AID-ANIE4002>3.0.CO;2-8

[55] Hirsch A. Functionalization of single-walled carbon nanotubes. Angewandte Chemie, International Edition. 2002;**41**:1853-1859. DOI: 10.1002/1521-3773(20020603) 41:11<1853::AID-ANIE1853>3.0.CO;2-N

[56] Huang W, Lin Y, Taylor S, Gaillard J, Rao AM, Sun Y-P. Sonication-assisted functionalization and solubilization of carbon nanotubes. Nano Letters. 2002;**2**:231-234

[57] Berger C, Song Z, Li X, Wu X, Brown N, Maud D, et al. Magnetotransport in high mobility epitaxial graphene. Physica Status Solidi A: Applications and Materials Science. 2007;**204**:1746-1750. DOI: 10.1002/pssa.200675352

[58] Sharapov SG, Gusynin VP, Beck H. Magnetic oscillations in planar systems with the Dirac-like spectrum of quasiparticle excitations. Physical Review B. 2004;**69**:075104. DOI: 10.1103/PhysRevB.69.075104

[59] Berger C, Song Z, Li X, Wu X, Brown N, Naud C, et al. Electronic confinement and coherence in patterned epitaxial graphene. Science. 2006;**312**:1191-1196. DOI: 10.1126/science.1125925

[60] Kawabata A. Theory of negative magnetoresistance in three-dimensional systems. Solid State Communications. 1980;**34**:431-432. DOI: 10.1016/0038-1098(80)90644-4

[61] Ousset JC, Rakoto H, Broto JM, Dupuis VV, Askenazy S, Durand J, et al. Weak-localization effects in $V_{1-x}Si_x$ amorphous alloys with high Si content.

Physical Review B. 1987;**36**:5432-5436. DOI: 10.1103/PhysRevB.36.5432

[62] Ulmet JP, Bachère L, Askenazy S, Ousset JC. Negative magnetoresistance in some dimethyltrimethylene-tetraselenafulvalenium salts: A signature of weak localization effects. Physical Review B. 1988;**38**:7782-7788. DOI: 10.1103/PhysRevB.38.7782

[63] Nguyen VI, Spivak BZ, Shklovskii BI. Tunnel hopping in disordered systems. Journal of Experimental and Theoretical Physics. 1985;**62**:1021-1029

[64] Sivan U, Entin-Wohlman O, Imry Y. Orbital magnetoconductance in the variable-range–hopping regime. Physical Review Letters. 1988;**60**:1566-1569. DOI: 10.1103/PhysRevLett.60.1566

[65] Sheng P. Fluctuation-induced tunneling conduction in disordered materials. Physical Review B. 1980;**21**:2180-2195. DOI: 10.1103/PhysRevB.21.2180

[66] Yasawa K. Negative magneto-resistance in pyrolytic carbons. Journal of the Physical Society of Japan. 1969;**26**:1407-1419. DOI: 10.1143/JPSJ.26.1407

[67] Bahr JL, Tour JM. Covalent chemistry of single-wall carbon nanotubes. Journal of Materials Chemistry. 2002;**12**:1952-1958. DOI: 10.1039/B201013P

[68] Chiu PW, Duesberg GS, Dettlaff-Weglikowska U, Roth S. Interconnection of carbon nanotubes by chemical functionalization. Applied Physics Letters. 2002;**80**:3811-3813. DOI: 10.1063/1.1480487

[69] Georgakilas V, Kordatos K, Prato M, Guldi DM, Holzinger M, Hirsch A. Organic functionalization of carbon nanotubes. Journal of the American Chemical Society. 2002;**124**:760-761. DOI: 10.1021/ja016954m

[70] Dresselhaus PD, Papavassiliou CM, Wheeler RG, Sacks RN. Observation of spin precession in GaAs inversion layers using antilocalization. Physical Review Letters. 1992;**68**:106-109. DOI: 10.1103/PhysRevLett.68.106

[71] Chen GL, Han J, Huang TT, Datta S, Janes DB. Observation of the interfacial-field-induced weak antilocalization in InAs quantum structures. Physical Review B. 1993;**47**:4084-4087. DOI: 10.1103/PhysRevB.47.4084

[72] Fujimoto A, Kobori H, Ohyama T, Ishida S, Satoh K, Kusaka T, et al. Comparison of magnetoconductance of the δ-doped layer and bulk crystal of Si:Sb in the weak localization regime. Physica B: Condensed Matter. 2001;**302-303**:7-11. DOI: 10.1016/S0921-4526(01)00399-4

[73] Liu K, Roth S, Düsberg GS, Kim GT, Popa D, Mukhopadhyay K, et al. Antilocalization in multiwalled carbon nanotubes. Physical Review B. 2000;**61**:2375-2379. DOI: 10.1103/PhysRevB.61.2375

[74] Wang G, Tan Z, Liu X, Chawda S, Koo J-S, Samuilov V, et al. Conducting MWNT/poly(vinyl acetate) composite nanofibres by electrospinning. Nanotechnology. 2006;**17**:5829-5835. DOI: 10.1088/0957-4484/17/23/019

[75] Chen W-P, Zhao Z-G, Liu X-W, Zhang Z-X, Suo C-G. A capacitive humidity sensor based on multi-wall carbon nanotubes (MWCNTs). Sensors. 2009;**9**:7431-7444. DOI: 10.3390/s90907431

[76] Lee C-Y, Lee G-B. Humidity sensors: A review. Sensor Letters. 2005;**3**:1-14. DOI: 10.1166/sl.2005.001

Chapter 10

Surface Functionalization of Carbon Nanotubes for Energy Applications

Mohamed R. Berber, Inas H. Hafez and Mohamad Y. Mustafa

Abstract

Carbon nanotubes (CNTs) are receiving a great deal of attention as a catalyst support for different energy applications, due to their high surface area and high conductivity. Recent literature studies have shown that the application of CNTs mainly depends on their surface functionalization process. Typically, pristine CNTs (as produced) have no functional groups, which is usually considered as an obstacle to their widespread application. In this chapter, we highlight the different techniques used to functionalize the surface of CNTs, including physical and chemical functionalization processes. We show the advantages and the drawbacks of the different functionalization processes. Additionally, we explain in detail the different techniques used to characterize the CNTs before and after functionalization processes. Furthermore, we focus on polymer wrapping techniques of CNTs to create active nanocomposite materials for energy applications, in particular the applications in the agriculture field to fight pollution and make farming activity easier and more efficient.

Keywords: physical functionalization, chemical functionalization, carbon nanotubes, polymers, fuel cells, energy applications for agriculture engineering

1. Introduction

The birth of nanoscience has emerged new pathways for developing new materials with new target properties for different energy applications. Among these promising materials are the carbon nanotubes (CNTs).

CNTs have been discovered by Ijima in 1991 [1]. They are made of graphene sheets which are rolled up to form nanostructured tubes (**Figure 1**). CNTs are receiving a great deal of attention for energy applications due to their high electrical conductivity, low density, large specific surface area, high chemical and thermal stability, as well as their remarkable mechanical properties [2]. CNTs are synthesized by various techniques, including chemical vapor deposition [3] (**Figure 2a**) and arc-discharge method [4] (**Figure 2b**). The as-produced (pristine) CNTs possess a hydrophobic nature due to the high interaction forces (van der Waals force and bundling) between the nanotubes (formation of aggregates). The presence of these aggregates in addition to the low solubility of CNTs shows major drawbacks for engineering useful materials based on CNTs. Thus, a functionalization process is required to overcome the CNT bundle structure which is usually obstacle their real world applications.

IntechOpen

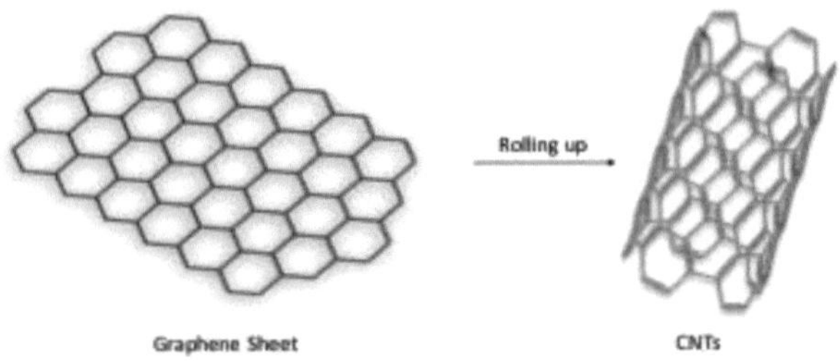

Figure 1.
CNT construction from graphene sheets.

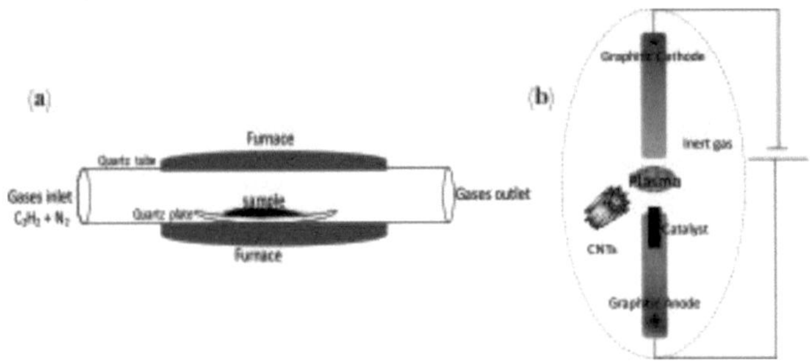

Figure 2.
Schematic illustration of (a) CVD method for preparation of CNTs and (b) arc-discharge method for preparation of CNTs.

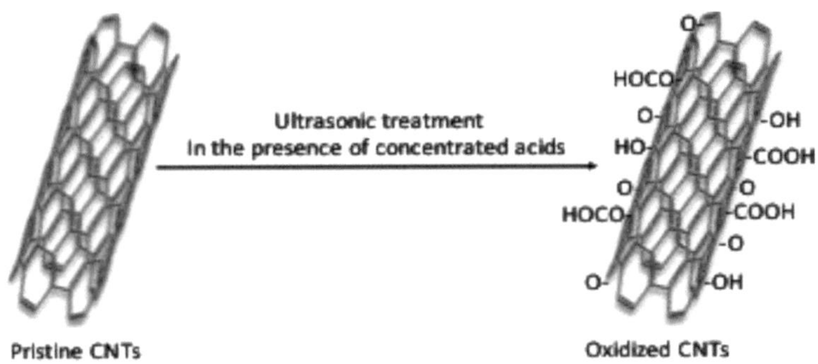

Figure 3.
Oxidation process of CNTs under harsh conditions.

Surface functionalization of CNTs is a promising technique to overcome their bundle structure and to offer a step forward for their real field application. In this chapter, we highlight the current research progress of CNT functionalization. We show the advantages and the drawbacks of the different functionalization processes. Additionally, we present and explain in details the different techniques used to characterize the CNTs. Moreover, we focus on fuel cell and supercapacitor applications of polymer functionalized CNTs. Finally, we show the future prospective of CNTs.

2. Chemical functionalization of CNTs

2.1 Activation of CNT surface via oxidation processes

CNTs are usually functionalized through a harsh oxidation process using harsh materials (e.g., nitric acid and/or sulfuric acid, permanganate using a phase transfer catalyst, or hydrofluoric acid) under ultrasonic treatment conditions [5–7]. The oxidation-functionalization process usually produces small CNT fragments with side-walls containing different oxygen functional groups (e.g., hydroxyl, carbonyl, and carboxylic acids) [8] that will be able to create covalent bonding with other groups (e.g., polymeric reagents, metal oxides, amino acids, and enzymes) (see **Figure 3**) [9].

A novel chemical activation process for CNTs is the oxidation through a silanization process. This method introduces different organo-functional groups attached to the surface of CNTs, improving the CNT chemical compatibility with specific polymers for producing new CNT-based composites. As an example, oxygen-functionalized CNTs were coupled with 3-methacryloxypropyltrimethoxysilane as an organo-functional silane agent to get double-bond-functionalized CNTs that will be used for copolymerization with vinyl monomers. The produced CNT material has been explored to achieve an improvement for CNT compatibility with other polymers and also to improve their solubility [10].

2.2 Photo-irradiation functionalization of CNTs

Photo-irradiation technique was also used to generate active species (such as nitrene) on the surface of CNTs [11]. Osmylation, for example, is one of the processes in which CNTs were photoactivated through a UV light irradiation process. Typically, CNTs are exposed to osmium tetroxide (OsO_4) under a UV light irradiation (see **Figure 4**). As a result, a photoinduced cycloaddition process of OsO_4 to the C═C bonds of CNTs was observed, leading to surface-functionalized CNTs, which facilitate the formation of an intermediate charge-transfer complex for further uses and applications [12].

2.3 Electrochemical activation of CNT surface

In a clean, nondestructive, and a controlled process, the surface of CNTs is functionalized using electrochemistry. Typically, a constant potential is applied to a CNT electrode which is immersed in a solution that contains a suitable active reagent in order to produce radical species. These radical species show a tendency to react or to polymerize, providing a surface-functionalized CNT [13].

In addition of being simple, electrochemical activation process is quite efficient in that they allow for an accurate control over the extent of film deposition thickness through the choice of suitable electrochemical conditions, i.e., duration of deposition and the magnitude of the applied potential. Moreover, by utilizing reagents containing appropriate substituents, the surface properties of the coated CNTs can be tailored [13].

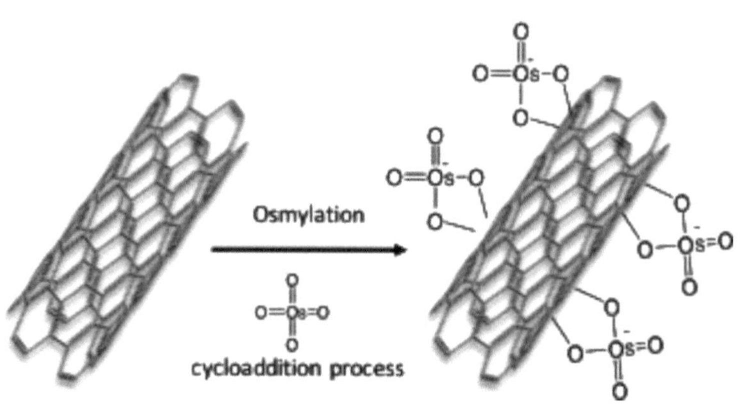

Figure 4.
Schematic illustration of cycloaddition osmylation process of CNTs using osmium tetroxide.

Using this technology, Zhang et al. [13] have electrochemically grafted CNTs in large quantities at room temperature using ionic liquids (see **Figure 5**). In their work, CNTs were considerably untangled in the ionic liquid to greatly increase the effective surface area of the electrode. Then, *N*-succinimidyl acrylate (a model monomer) was dissolved in the supporting ionic liquid and was electrochemically grafted onto CNTs. As an application example, glucose oxidase was covalently anchored on the *N*-succinimidyl acrylate/CNTs assembly, and accordingly the electrocatalytic oxidation of glucose in this assembly was investigated, showing an improvement.

In another study by Bahr et al., CNTs were functionalized via electrochemical reduction process of aryl diazonium salts. In this study, different diazonium salts were used in order to provide conducive materials for further elaboration after attachment to the CNTs [14].

2.4 Characterization of oxidized CNTs

To identify the surface functional groups of the oxidized CNTs, Fourier Transform Infrared spectroscopic analysis was applied (see **Figure 6**). The oxidized CNTs have showed the OH functionality at around 1380 cm^{-1} (bending deformation) and around 3500 cm^{-1} (stretching). The C—O (stretching) was observed at 1160 cm^{-1}, while the C = O functionality of COOH group was observed at 1730 cm^{-1}. Such absorption spectroscopic investigations have provided a strong evidence of the surface functionalization of CNTs [15].

Raman spectroscopy is also considered as a powerful technique to provide information concerning CNT purity. The Raman spectra of **Figure 7** have showed three bands at around 1340–1350, 1550–1600, and 2600–2700 cm^{-1}. These bands are assigned for structural defects of CNTs (D band), graphitized carbon atoms of CNTs, and a secondary D band (2D band), respectively. To detect the changes occurred in the structure of CNTs after functionalization process, the ratio of D and G band (I_D/I_G) was applied in order to determine the purity factor. As a result of the oxidation process of CNTs, the I_D/I_G has increased. The increasing ratio mainly depends on the oxidation parameters [16].

X-ray photoelectron spectroscopy (XPS) is a surface characterization tool to identify the surface functional groups [15]. For CNTs, it is used to detect the surface changes resulting from the oxidation process. **Figure 8** shows the XPS spectra of C1s of pristine CNTs (top spectrum) and oxidized CNTs (bottom spectrum). As seen, the spectra have revealed six peaks related to carbon satellite peaks, carbonates groups due to carbon dioxide adsorption, carbonyl groups of carboxylic acid groups and lactone, hydroxyl groups of phenolic and alcoholic moieties, sp^2 hybridized carbon. As a result of chemical oxidation of CNTs, the ratio of these carbon functional groups has changed. As reported, longer oxidation times usually lead to higher oxygen content in the oxidized form of CNTs.

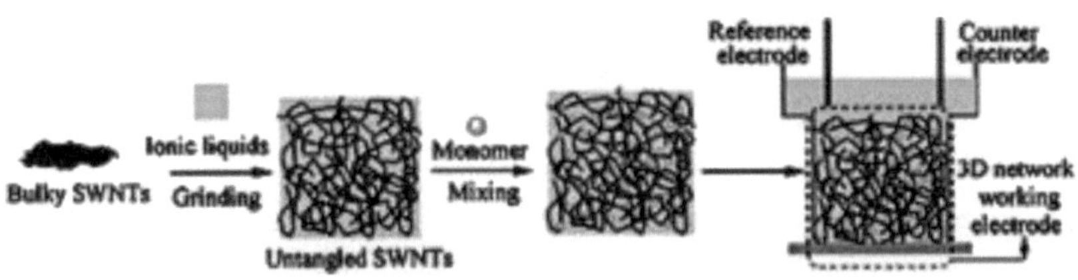

Figure 5.
Schematic illustration of electrochemical functionalization of CNTs using ionic liquids and monomer grafting. Reproduced with a permission from Ref. [13].

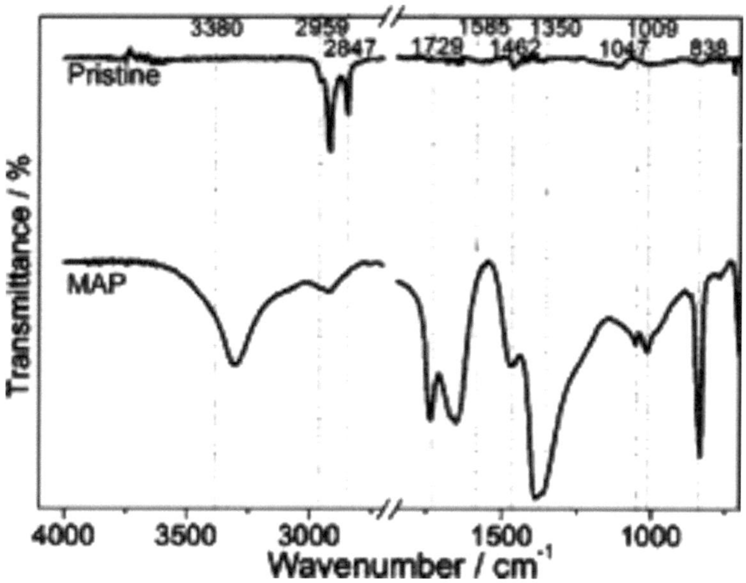

Figure 6.
Fourier Transform Infrared spectra of pristine CNTs (top spectrum) and oxidized/purified CNTs (bottom spectrum). Reproduced with a permission from Ref. [15].

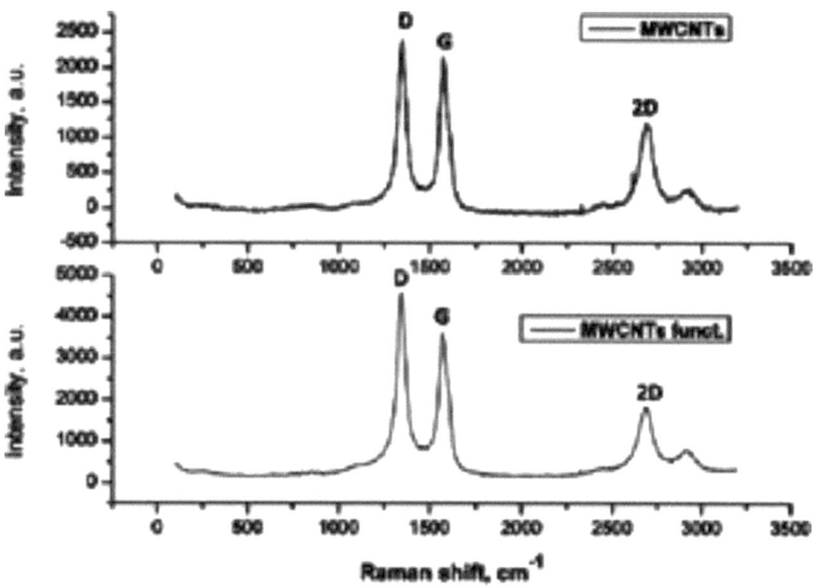

Figure 7.
Raman spectra of pristine CNTs (top spectrum) and oxidized CNTs (bottom spectrum). Reproduced with a permission from Ref. [16].

The XPS of oxygen atoms (O1s spectrum) was also an important result to provide information about the surface functionalization of CNTs. As seen from **Figure 9**, four peaks related to adsorbed water, carboxylic acid groups, hydroxyl groups, and carbonyl moieties were emerged. The ratio of these peaks has changed as a result of oxidation-functionalization process of CNTs. Hence, XPS is a powerful technique to confirm the different functionalities produced by oxidation of CNTs [17].

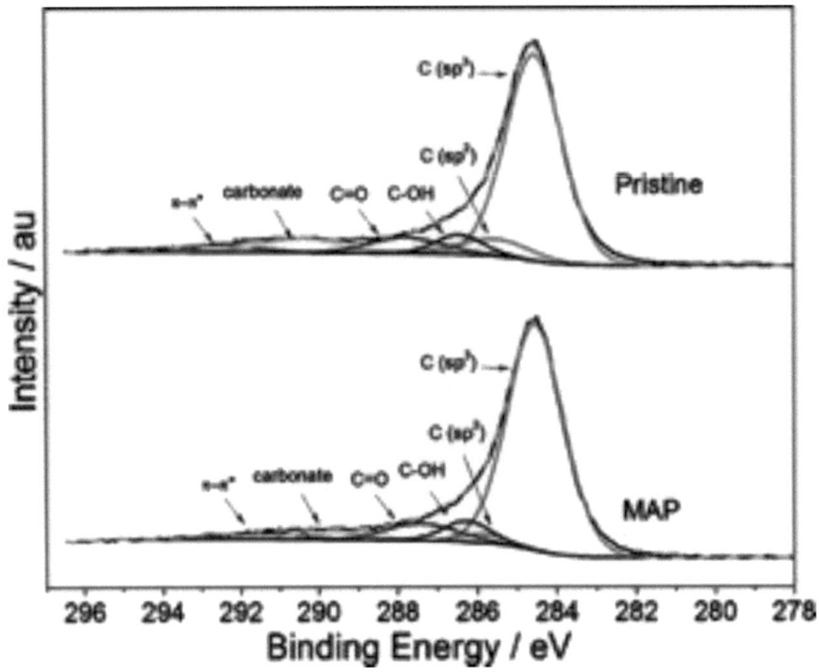

Figure 8.
XPS of C1s spectrum of pristine CNTs (top spectrum) and oxidized/purified CNTs (bottom spectrum). Reproduced with a permission from Ref. [15].

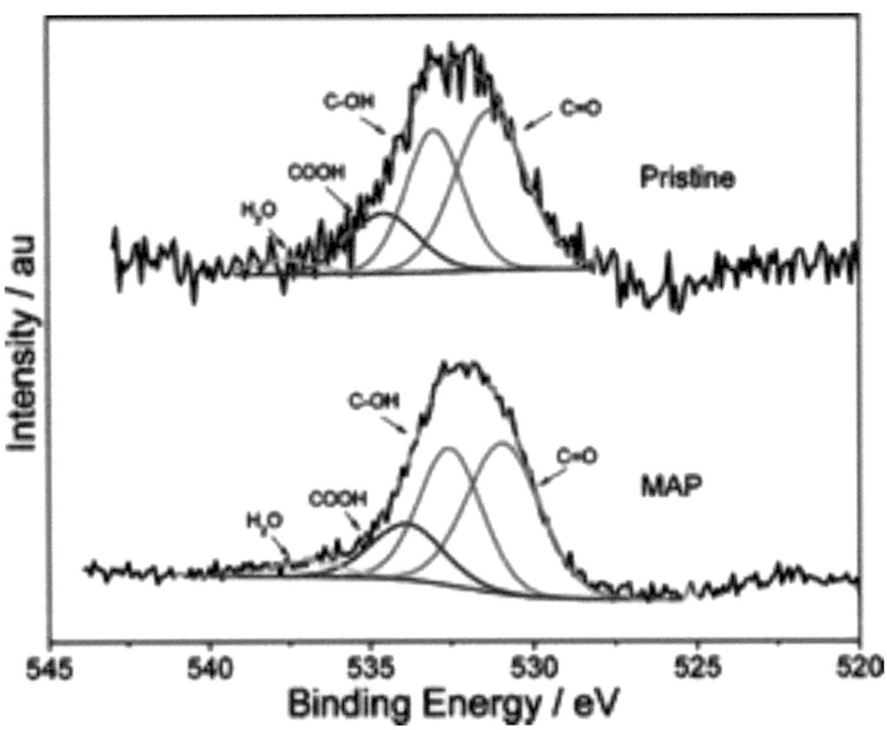

Figure 9.
XPS of O1s spectrum of pristine CNTs (top spectrum) and oxidized/purified CNTs (bottom spectrum). Reproduced with a permission from Reference.

3. Physical functionalization of CNTs

The noncovalent functionalization of CNTs has received a great deal of attention because the properties of CNTs can be tailored while maintaining their intrinsic properties. Physical functionalization of CNTs is a process in which CNTs are noncovalently activated through a π-π interaction between π bonds of sp^2-hybridized carbon atoms of CNTs and the π bonds of macromolecules, surfactants, or functional polymers that are used to activate the surface, forming supramolecular complexes. This hybrid technique usually keeps the original electronic structure and properties of CNTs.

3.1 Small molecule anchoring of CNTs

Through a nondestructive hybrid formulation process, CNTs were functionalized by a simple technique in which CNTs were assembled with a variety of small molecules. This technique has enabled a convenient process to efficiently assemble a wide variety of nanoscale particles on the surfaces of CNTs and accordingly has led to the construction of different nanoscale hetero structures with new functionalities [18].

In a study by Li et al. [18], pristine CNTs were assembled with aminopyrene molecules using the well-known π-π stacking process [19] between the pyrenyl groups of aminopyrene and the sp^2 carbon rings of the CNTs (see **Figure 10**). The resulting CNT hybrid was then used to adsorb different nanoparticle precursors through an electrostatic interaction under appropriate conditions. This unique approach has showed a wide variety of nanoparticles decorated on the surface of CNTs with high efficiency and specificity for different applications.

In a similar study, Li et al. [20] have achieved noncovalently amino-functionalized CNTs by the adsorption of amino ethanol on CNTs surface. The prepared epoxy composite of the noncovalently functionalized CNTs has showed promising mechanical properties while persevering the electrical properties of pristine CNTs.

CNTs were also physically functionalized with cytochrome to construct a new bioanalytical platform for building biosensors. The sensitivity of the resulting biosensor toward glucose has showed a high reproducibility and a successful determination of glucose in commercial beverages [21].

The amino molecules of tetrazine compounds were also attached on CNT surface by a noncovalent process. The amino groups of the functionalized CNTs have generated chemical bonds with the epoxide groups in the epoxy matrices, resulting in a homogenous dispersion of the CNTs. The electrical conductivity of the amino-functionalized CNTs/epoxide composites was significantly higher than that of the pristine CNTs/epoxide composites at the same CNT content [22].

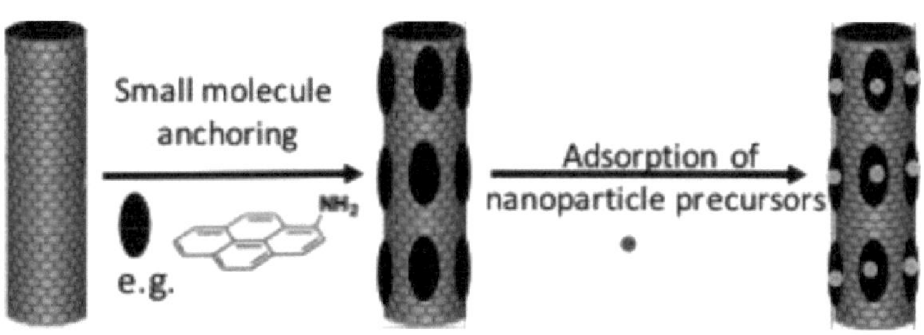

Figure 10.
Schematic illustration of small molecule anchoring onto CNTs.

3.2 Polymer anchoring of CNTs

Polymer functionalization of CNTs is generally introduced to enhance the CNT solubility and dispersion as well as synthesize new CNT composites with advanced properties. The conventional techniques used for the polymer functionalization of CNTs are the covalent attachment processes (grafting from or grafting onto). In "grafting from" technique, the polymer is added to the CNT surface through an in situ polymerization process of the previously attached monomers on the surface of CNTs in the presence of an initiator, while in "grafting onto" technique, the hole polymer is covalently added to CNTs through the surface functional groups which were previously introduced onto the CNT structure [23]. These conventional polymer functionalization techniques have been reported to be not suitable for preparing controlled and pure CNT polymer composites where the residual catalysts and the impurities removal from the reaction medium are problematic issues. Hence, advanced polymer functionalization techniques of CNTs are still required to cover the sensitive and broad application of CNTs [24].

Noncovalent functionalization of CNTs by polymer wrapping is a feasible technique to disperse CNTs, causing no changes in the electronic properties of CNTs. One of the advantages of this technique is the possibility to remove the unbound polymer while leaving the stacked polymer layer on the CNT surface [25]. **Figure 11** shows a schematic illustration of polymer wrapping of CNTs using π-conjugated polymer where a π-π stacking occurs.

Polymers such as polyvinyl alcohol and polymethyl methacrylate [26, 27] have been reported to disperse CNTs through a wrapping mechanism. The results of these studies have showed the importance of the CH-π interaction bonding for the dispersion of the nonaromatic polymers [27].

Polybenzimidazole (PBI, **Figure 12**) is one of the most promising polymers for CNT wrapping, especially for fuel cell applications since it possesses remarkable conducting properties [28]. Using PBI, we fabricated a novel electrocatalyst assembly with CNTs, for high-temperature fuel cells. The obtained PBI/CNT assembly has showed a good dispersibility (see **Figure 13**) and a high utilization efficiency of the loaded metal catalyst compared to the pristine CNTs (see **Figure 14**).

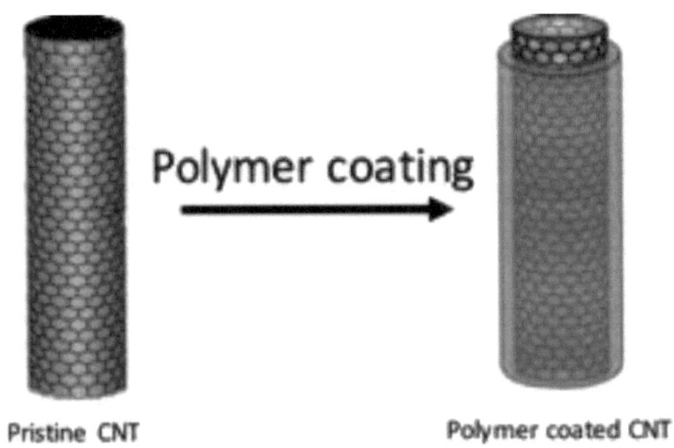

Figure 11.
Schematic illustration of polymer coating of CNTs.

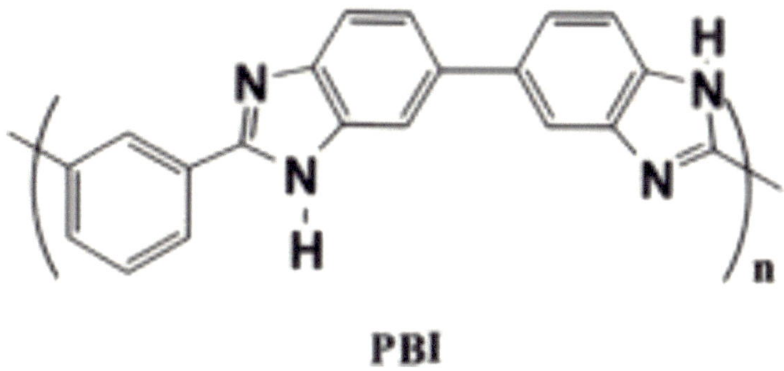

Figure 12.
The structure of polybenzimidazole polymer (PBI).

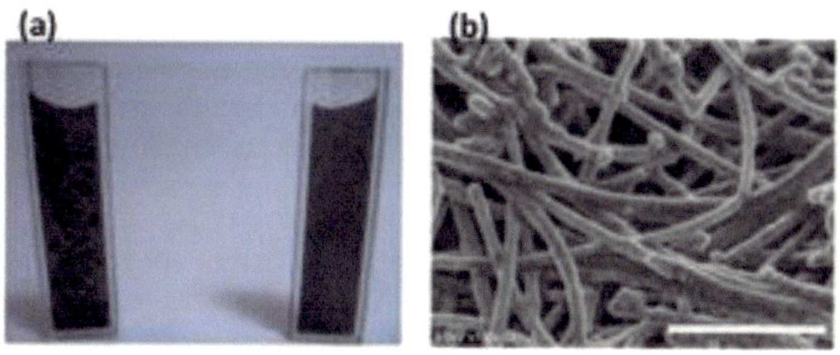

Figure 13.
(a) Photo image of CNT solution before polymer coating (left) and after polymer coating (right) and (b) TEM image of polymer-coated CNTs. Reprinted with a permission from Ref. [28].

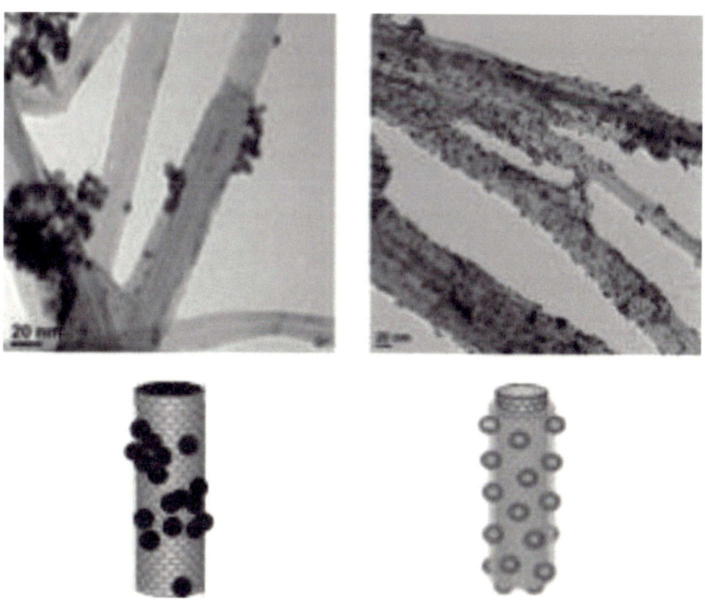

Figure 14.
TEM image of pristine CNTs loaded with metal catalyst (left) and polymer/CNT composite loaded with metal catalyst (right), reprinted with a permission from Ref. [29].

As reported in the literature, the stability of the formed polymer/CNT hybrid depends on the polymer structure [25]. The literature studies have correlated the relation between the binding strength and the length of the oligomers [30].

3.3 Applications of polymer-wrapped CNTs in fuel cells

Polymer-based fuel cell is considered as one of the best sources to realize a green energy source for many applications. It consists of two carbon-based electrodes and a conducting polymer membrane (see **Figure 15**). The carbon-based electrodes are usually composed of a carbon material decorated with a platinum metal catalyst. CNTs are considered as a good supporting material for fuel cell electrocatalysts because they possess a high electrical conductivity and a good electrochemical durability compared to other carbon-supporting materials like, for example, carbon black. The durability and cost of electrodes are the two major roadblocks to commercialize the polymer-based fuel cells. To overcome these obstacles, novel electrocatalysts with high efficiency are required.

For this target, we described a new approach to grow homogeneously platinum nanoparticles in a controllable size manner on a polymer (polybenzimidazole, PyPBI)-wrapped CNT hybrid (see **Figure 16**). A fuel cell employing this polymer-wrapped CNTs composite has exhibited eight times higher mass activity than the current state-of-the-art CNT fuel cells (see **Figure 17**) [31]. With the decrease of platinum nanosize on the polymer-wrapped CNTs, further enhancement in platinum mass activity is observed.

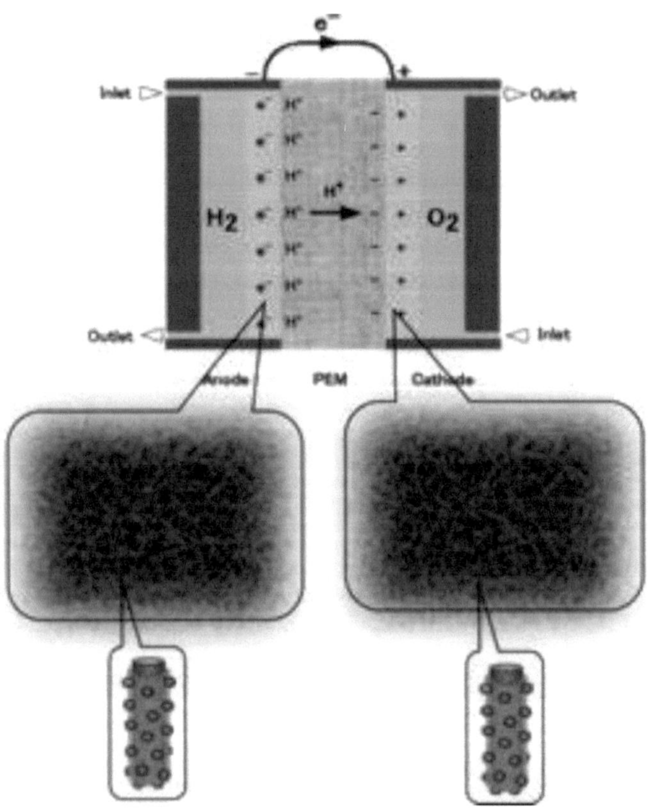

Figure 15.
Schematic illustration of polymer-based fuel cell and its components. Polymer-wrapped CNTs were used as anode and cathode electrodes.

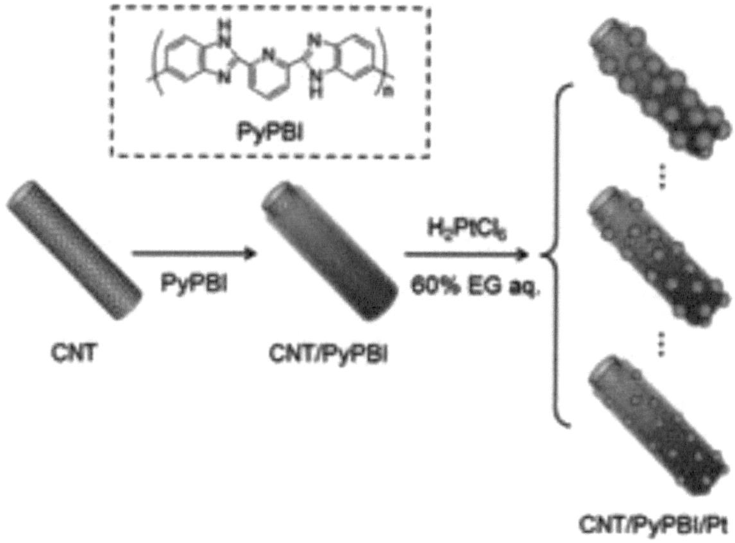

Figure 16.
Schematic illustration of polymer wrapping and platinum deposition on CNTs. The figure shows how the authors have controlled the deposition of the platinum nanoparticles. Reprinted with a permission from Ref. [31].

From the stability point of view, the assembled platinum-/polymer-wrapped CNT nanocomposite has showed a remarkable durability even at very small platinum nanosize (see **Figure 18**), thanks to the PyPBI functional polymer which offered a strong binding sites (the nitrogen atoms of the imidazole units of PyPBI) for the platinum nanoparticles. Thus, polymer wrapping technique of CNTs has offered an effective solution to use efficiently the pristine CNTs in fuel cells, keeping their high electrical properties and their stability.

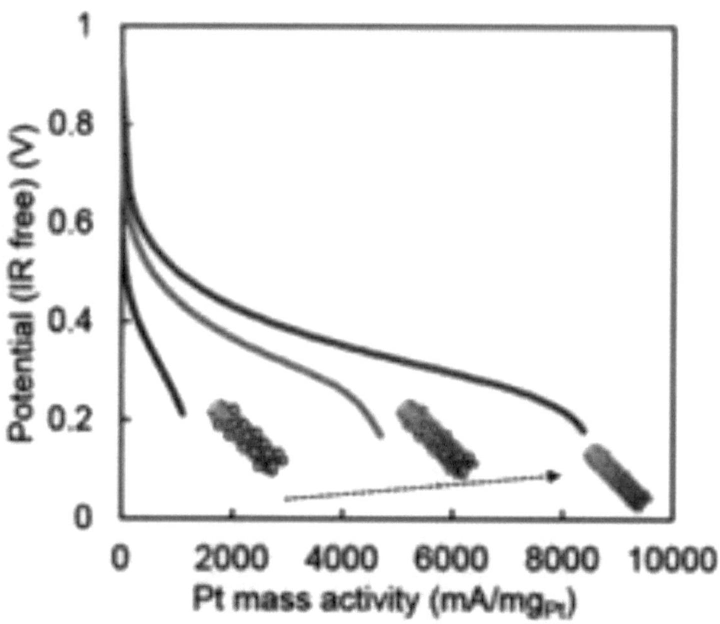

Figure 17.
Potential curves as a function of platinum (Pt) mass activity. Reprinted with a permission from Ref. [31].

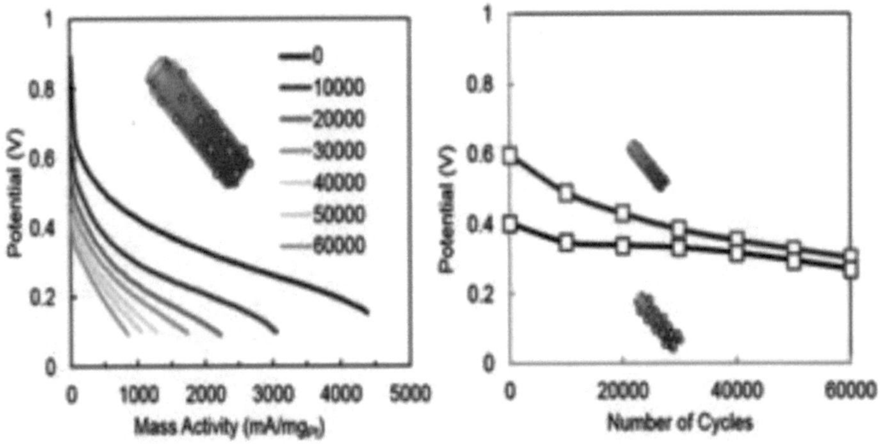

Figure 18.
Durability potential curve test using MEA10 (small platinum nanosize). The polarization curves were measured after every 1000 cycles. For convenience, the data were plotted every 10,000 cycles (c). Plots of the cell voltage at 200 mA mgPt-1 for the MEA50 and MEA10.

3.4 Fuel cell-based tractors

The agricultural sector needs to decrease its use of fossil fuels in order to diminish the global warming and to lower environmental impact. **Figure 19** shows a schematic illustration of the new fuel cell tractor built by New Holland company (http://www.newholland.com/Pages/index.html). The figure shows also the photo image of the fuel cell tack used in the tractor. The volume and the design of the fuel cell stack depend on the required power from the tractor.

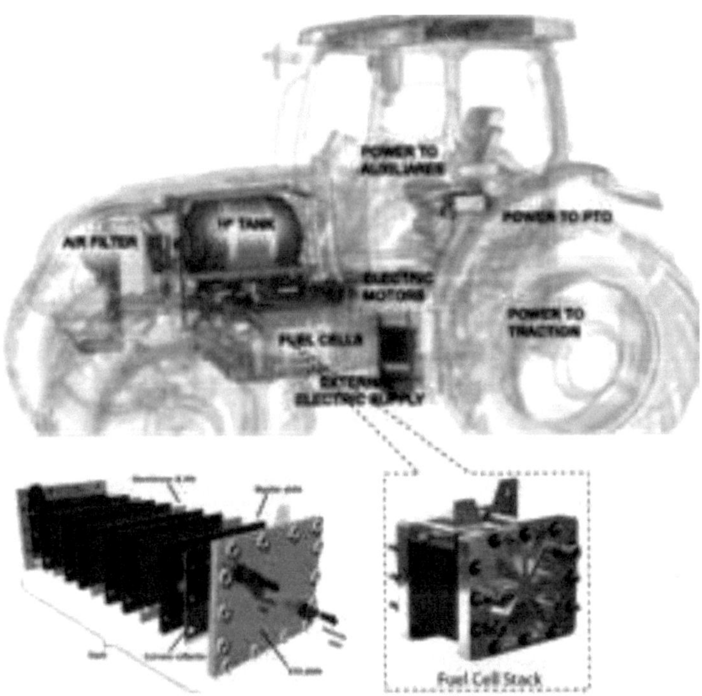

Figure 19.
Schematic illustration of fuel cell tractor built by New Holland (photo courtesy of New Holland). The fuel cell stack depends on the required power.

The current research of fuel cell tractors focuses on serving the farmer to find better, smarter, and faster ways to make the fuel cell technology simple, accessible, and easy to use. Also let farmers free from the cost of purchased fossil fuel and allow them to achieve fuel autonomy.

The new fuel cell tractors offer greater reliability than the fossil-based tractors because the use of fewer moving parts and also the replacement of the hydraulic and the mechanical systems with electrical parts.

3.5 Applications of polymer-wrapped CNTs in supercapacitor

CNTs are being tested as electrodes in supercapacitors because of their remarkable electrical and mechanical properties as well as their high specific surface area. In order to provide high-performance supercapacitor-based devices with potential applications, CNTs need to be perfectly functionalized to maintain their intrinsic properties mentioned above. The introduction of conducting polymers on the surface of CNTs is envisioned to increase the specific capacitance of the supercapacitors. Hence, a suitable engineering process is required to achieve this target, keeping the stability of the supercapacitor [32].

It has been reported that CNT composites with conducting polymers like polyaniline or polypyrrole show a high specific capacitance and a high conductivity, leading to a high power density because the entangled mesoporous network of CNTs in the polymer composite can adapt to the volume change avoiding shrinkage. Accordingly, a more stable capacitance with cycling is obtained [33]. Another important advantage of the polymer/CNT composite is the remarkable volumetric energy coming from the high density of conducting polymer/CNT composite that facilitates the accessibility of the electrode/electrolyte interface, allowing a quick charge propagation in the composite material and an efficient reversible storage of energy in the conducting polymer/CNT composite during subsequent charging/discharging cycles [34].

4. Drawbacks of chemical and physical functionalization of CNTs

As reported in the literature, covalent functionalization usually demolishes the CNT electrical properties due to the transformation of the sp^2-hybridized carbon atoms to sp^3-hybridized atoms in the CNT framework. Thus, it is not the appropriate technique to activate the structure of CNTs for electronic device applications [35].

While the physical functionalization keeps the electronic properties of the CNTs, this technique suffers from a major drawback which is the stability of the produced assembly. Typically, the molecules adsorbed on the surface of CNTs can more or less be desorbed depending on the conformation and/or the size of the attached molecules when for example the solvent is changed or the CNTs are filtered ,and then redispersed in another medium. Hence, there is a need for an advanced process to overcome the current drawbacks of the chemical and physical functionalization processes of CNTs.

5. A bridge method between physical and chemical activation processes

In a recent study, a new functionalization process of CNTs was introduced to gather the advantages of both covalent and noncovalent activation techniques, avoiding their principal drawbacks. This process is based on a controlled polymerization of hydrophobic molecules onto CNTs dispersed in micelles. This approach

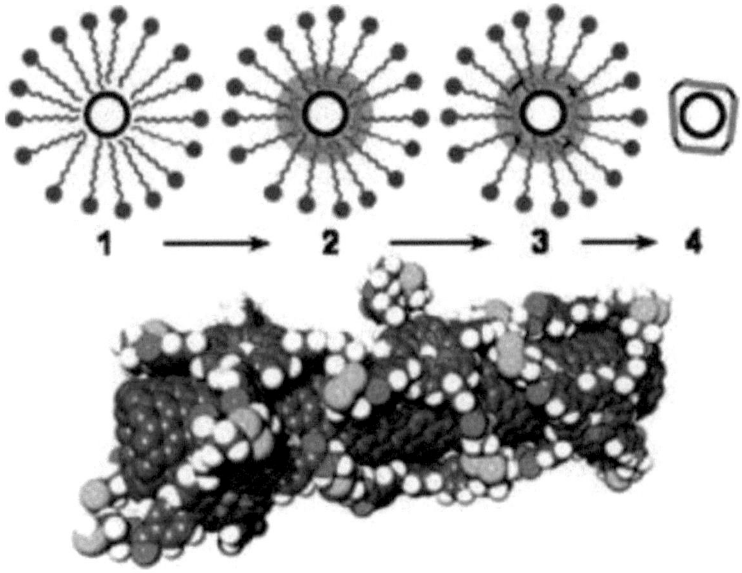

Figure 20.
Schematic illustration of the covalent/noncovalent functionalization of CNTs. (1) The CNTs dispersed in micelles, (2) dispersion of porphyrins on the CNT surfaces, (3) crosslinking of porphyrin around the CNTs, and (4) purification of CNTs via filtration and extensive washing to remove the surfactants, reagents, and unbound porphyrin moieties. Reproduced with a permission from Ref. [36].

permits to obtain CNT hybrids exhibiting high stability while preserving their π-conjugated system responsible for their outstanding optical and electrical properties. The obtained CNT hybrids can be purified, manipulated, and dispersed in various solvents without losing their functionality (see **Figure 20**) [36].

6. Conclusion and outlook

It is important to note that CNTs are now in the way of commercial production in large quantities. Also, the literature has reported many purification and functionalization processes, which maintain the original and intrinsic properties of CNTs, in addition to providing many interesting properties based on hybridization and complexation processes with other functional materials including polymers. Despite all these efforts, structural details of functional CNTs especially in the atomic level still remain to be demonstrated. In this chapter, we have highlighted the different functionalization process of CNTs including chemical, photo-irradiation, electrochemical, and physical activation processes. In addition, we have highlighted the different characterization tools used to characterize the functionalized CNTs and have indicated the drawbacks of the functionalization processes of CNTs, showing a bridge method to avoid these roadblocks that prevent the wide applications of CNTs. Finally, we have paid attention to the CNT energy applications, especially in fuel cells and supercapacitors.

To date, further work is still required to understand how the amount of the functional groups, their type, and their distribution affect the efficiency of the CNTs in its target application.

From the application point of view, CNT-based fuel cells are the future engine for the transport sector, which currently depends on the fossil fuel that causes many environmental problems. A special focus is on hydrogen- and biodiesel-based vehicles used in the agriculture field to fight pollution and make farming activity easier and more efficient. This pioneering spirit led us to look always forward in the research of fuel cells, anticipating the evolution of farming by designing effective farm vehicles.

Acknowledgements

"The publication charge for this article has been funded by a grant from the publication fund of UiT, The Arctic University of Norway".

Conflict of interest

The authors declare no "conflict of interest."

Author details

Mohamed R. Berber[1,4*], Inas H. Hafez[2] and Mohamad Y. Mustafa[3]

1 Department of Chemistry, Faculty of Science, Tanta University, Tanta, Egypt

2 Department of Natural Resources and Agricultural Engineering, Faculty of Agriculture, Damanhour University, Damanhour, Egypt

3 Institute of Buildings, Energy and Material Technology, Faculty of Engineering Science and Technology, The Arctic University of Norway, Narvik, Norway

4 Chemistry Department, College of Science, Jouf University, Sakaka, Saudi Arabia

*Address all correspondence to: mrberber@science.tanta.edu.eg

IntechOpen

© 2019 The Author(s). Licensee IntechOpen. This chapter is distributed under the terms of the Creative Commons Attribution License (http://creativecommons.org/licenses/by/3.0), which permits unrestricted use, distribution, and reproduction in any medium, provided the original work is properly cited. CC BY

References

[1] Iijima S. Helical microtubules of graphitic carbon. Nature. 1991;**354**:56

[2] Popov VN. Carbon nanotubes: Properties and application. Materials Science and Engineering: R: Reports. 2004;**43**(3):61-102

[3] Shah KA, Tali BA. Synthesis of carbon nanotubes by catalytic chemical vapour deposition: A review on carbon sources, catalysts and substrates. Materials Science in Semiconductor Processing. 2016;**41**:67-82

[4] Sharma R, Sharma AK, Sharma V. Synthesis of carbon nanotubes by arc-discharge and chemical vapor deposition method with analysis of its morphology, dispersion and functionalization characteristics. Cogent Engineering. 2015;**2**(1):1094017

[5] Gusmão R et al. Contrasts between mild and harsh oxidation of carbon nanotubes in terms of their properties and electrochemical performance. ChemElectroChem. 2016;**3**(10):1713-1719

[6] Likodimos V et al. Controlled surface functionalization of multiwall carbon nanotubes by HNO_3 hydrothermal oxidation. Carbon. 2014;**69**:311-326

[7] Fatin MF et al. Oxidation functionalization of multiwalled carbon nanotube by mild acid sonication. In: 2014 IEEE Conference on Biomedical Engineering and Sciences (IECBES). 2014

[8] Montanheiro TLdA et al. Effect of MWCNT functionalization on thermal and electrical properties of PHBV/MWCNT nanocomposites. Journal of Materials Research. 2014;**30**(1):55-65

[9] Syrgiannis Z, Melchionna M, Prato M. Covalent carbon nanotube functionalization. In: Kobayashi S, Müllen K, editors. Encyclopedia of Polymeric Nanomaterials. Berlin, Heidelberg: Springer; 2014;**2021**:1-8

[10] Dibyendu SB et al. Chemical functionalization of carbon nanotubes with 3-methacryloxy-propyltrimethoxysilane (3-MPTS). Smart Materials and Structures. 2004;**13**(5):1263

[11] Zhang Y et al. The effect of ultraviolet radiation on the mild photo-responsive surface functionalization of carbon nanotubes. Fullerenes, Nanotubes, and Carbon Nanostructures. 2017;**25**(4):273-281

[12] Balasubramanian K, Burghard M. Chemically functionalized carbon nanotubes. Small. 2005;**1**(2):180-192

[13] Zhang Y et al. Electrochemical functionalization of single-walled carbon nanotubes in large quantities at a room-temperature ionic liquid supported three-dimensional network electrode. Langmuir. 2005;**21**(11):4797-4800

[14] Bahr JL et al. Functionalization of carbon nanotubes by electrochemical reduction of aryl diazonium salts: A bucky paper electrode. Journal of the American Chemical Society. 2001;**123**(27):6536-6542

[15] Pacheco FG et al. Comparative temporal analysis of multiwalled carbon nanotube oxidation reactions: Evaluating chemical modifications on true nanotube surface. Applied Surface Science. 2015;**357**:1015-1023

[16] Dobrzańska-Danikiewicz AD et al. Characteristics of multiwalled carbon nanotubes-rhenium nanocomposites with varied rhenium mass fractions. Nanomaterials and Nanotechnology. 2017;7:1847980417707173

[17] Datsyuk V et al. Chemical oxidation of multiwalled carbon nanotubes. Carbon. 2008;**46**(6):833-840

[18] Li XL et al. Efficient synthesis of carbon nanotube–Nanoparticle hybrids. Advanced Functional Materials. 2006;**16**(18):2431-2437

[19] Calbo J et al. Understanding noncovalent interactions of small molecules with carbon nanotubes. Chemistry—A European Journal. 2017;**23**(52):12909-12916

[20] Li X et al. Non-covalent functionalization of multi walled carbon nanotubes and their application for conductive composites. Carbon. 2008;**46**(5):829-831

[21] Eguílaz M et al. Carbon nanotubes non-covalently functionalized with cytochrome c: A new bioanalytical platform for building bienzymatic biosensors. Microchemical Journal. 2016;**128**:161-165

[22] Zhang A et al. Effect of percolation on the electrical conductivity of amino molecules non-covalently coated multi-walled carbon nanotubes/epoxy composites. Applied Surface Science. 2012;**258**(22):8492-8497

[23] Tsubokawa N. Preparation and properties of polymer-grafted carbon nanotubes and nanofibers. Polymer Journal. 2005;**37**:637

[24] Hsiao A-E et al. Decoration of multi-walled carbon nanotubes by polymer wrapping and its application in MWCNT/polyethylene composites. Nanoscale Research Letters. 2012;**7**(1):240

[25] Fujigaya T, Nakashima N. Non-covalent polymer wrapping of carbon nanotubes and the role of wrapped polymers as functional dispersants. Science and Technology of Advanced Materials. 2015;**16**(2):024802

[26] Zhang X et al. Poly(vinyl alcohol)/SWNT composite film. Nano Letters. 2003;**3**(9):1285-1288

[27] Baskaran D, Mays JW, Bratcher MS. Noncovalent and nonspecific molecular interactions of polymers with multiwalled carbon nanotubes. Chemistry of Materials. 2005;**17**(13):3389-3397

[28] Berber MR et al. Remarkably durable high temperature polymer electrolyte fuel cell based on poly(vinylphosphonic acid)-doped polybenzimidazole. Scientific Reports. 2013;**3**:1764

[29] Berber MR et al. Durability analysis of polymer-coated pristine carbon nanotube-based fuel cell electrocatalysts under non-humidified conditions. Journal of Materials Chemistry A. 2014;**2**(44):19053-19059

[30] Sprafke JK et al. Noncovalent binding of carbon nanotubes by porphyrin oligomers. Angewandte Chemie International Edition. 2011;**50**(10):2313-2316

[31] Hafez IH et al. Enhancement of platinum mass activity on the surface of polymer-wrapped carbon nanotube-based fuel cell electrocatalysts. Scientific Reports. 2014;**4**:6295

[32] Chen T, Dai L. Carbon nanomaterials for high-performance supercapacitors. Materials Today. 2013;**16**(7):272-280

[33] Pan H, Li J, Feng Y. Carbon nanotubes for supercapacitor. Nanoscale Research Letters. 2010;**5**(3):654-668

[34] Lota K, Khomenko V, Frackowiak E. Capacitance properties of poly(3,4-ethylenedioxythiophene)/carbon nanotubes composites. Journal of Physics and Chemistry of Solids. 2004;**65**(2):295-301

[35] Roch A et al. Ambient effects on the electrical conductivity of carbon nanotubes. Carbon. 2015;**95**:347-353

[36] Clavé G et al. Functionalization of carbon nanotubes through polymerization in micelles: A bridge between the covalent and noncovalent methods. Chemistry of Materials. 2013;**25**(13):2700-2707

Chapter 11

Vibration Characteristics of Single-Walled Carbon Nanotubes Based on Nonlocal Elasticity Theory Using Wave Propagation Approach (WPA) Including Chirality

Muzamal Hussain and Muhammad Nawaz Naeem

Abstract

This chapter deals with the vibrational properties of single-walled carbon nanotubes (SWCNTs), based on nonlocalized theory of elasticity (NLT). The nanotube pilot control with nonlinear parameters was derived from Euler's beam theory. The wave propagation (WPA) approach was used to derive the frequency equation describing the natural frequencies of vibration in SWCNTs. Complex exponentials depend on the boundary conditions given at the edges of the carbon nanotubes used. Vibration frequency spectra were obtained and evaluated for different physical parameters such as diameter ratio for single chiral carbon nanotubes and flexural strength for chiral SWCNT. The results show that the natural frequencies are significantly reduced by increasing the nonlocal parameters, but by increasing the ratio of the diameter length (aspect ratio), the natural frequency increases. The frequency of SWCNTs is calculated with the help of MATLAB computer software. These results are compared to previously known numerical simulations.

Keywords: nonlocal, wave propagation approach, vibration, MATLAB

1. Introduction

Vibrational properties of CNTs play important critical roles in controlling the performance of various scientific and engineering fields and stability of CNT-based devices, superconductivity, and material strength analysis. New technologies and innovative improvements such as nano-probe, wood mirror, nano-electronic devices, chemical release, and drug release have been proposed. The important application of the current investigation of rotating FG-CNT is in nano-engineering structure as nano-components like sensors and actuators. The use of carbon nanotubes (CNTs) has been practiced in a variety of fields such as field emission, construction, electronics, and fashion [1]. CNT's free vibration surveys have been tested in relation to their physical properties and behavior. Much of the work was done with high rate of elasticity and characterization, a very effective Young module [2], and the bond strength between carbon atoms [3]. In the past 15 years, researchers have used different models such as the ring [4], the beam [5], the shell [6], and other continuous models [2, 7] to capture the object anew. Due to their attractive applications, dynamic features such as buckling, stability, and vibration are explored in a theoretical way and avoid potential risks for future use. Therefore, a new model is needed to capture the nanoscale structure. Researchers [8–12] have conducted investigations of higher-order elasticity theories. Other nonclassical theories of elasticity have attracted the attention of researchers such as the theory of stress [13, 14], theories of stress [15, 16], and nonlocal theory [17, 18].

In NLT, the pressure applied at a certain point depends on the stress at all points, which is quite different from conventional theory. Wang et al. [19] and Yang et al. [20] presented a survey of SWCNT based on nonlocal Timoshenko beam theory (TBM). CNT analysis has been explained by some researchers [21–25]. Bocko and Lengvarský studied the vibration frequencies of CNTs for different termination conditions and mode shapes using nonlocal elastic theory. Chawis et al. [26] analyzes vibrational behavior of SWCNTs with small-scale effects using nonlocal theory. Recently, some researchers have investigated the vibrational behavior of SWCNTs [27–29].

In addition, many researchers investigated the vibrational behavior of the above structures using different types of theories. The method chosen for studying nanoscale systems is the NLT (WPA) wave propagation method, which allows the study of fundamental frequencies of SWCNTs through combinations of different parameters. The beam model (BM) [30] is used to calculate the associated frequencies and shapes of MWCNT. Bocko and Lengvarský [31] investigate the bending free vibration of SWCNTs with four different boundary conditions. A continuum approach is used for the computation of natural frequency based on nonlocal theory of bending beam. The natural frequencies are given for several nano-parameters with two different diameters of nanotubes and continuously changed length. It is concluded that when tube length increases, the frequency decreases the nonlocal parameter and diameter. Chawis et al. [26] reported SWCNT vibration based on nonlocal theory to access the scale length. With the addition of nonlocal parameter in Euler beam theory, the governing equation is derived.

NLT-based are another option of robust research techniques of CNTs within the acceptable error range compared to previously used BMs and the subsequent other approaches [2–5, 21, 25, 32–34]. For solving ordinary differential equations (ODEs), Fourier variable separation method is used. According to the current model, the basic natural frequencies of chiral SWCNTs are calculated and obtained for various physical parameters such as the aspect ratio (length-to-diameter ratios) of SWCNTs with different nonlocal parameters and the effect of bending stiffness (rigidity) on the vibration frequencies of SWCNT. This is also our motivation for doing the present work.

2. Governing equation of motion

In classical theory, the physical mass acts as a local action. In conventional theory, the stress produced at a point is influenced by stress at the point. According to Eringen [35], in nonlocal theory, the stress applied at a given point depends on the stress at all points [1], which is quite different from conventional theory. Under this assumption, the nonlocal relationship based on homogeneous isotropic beams can be expressed as

$$\varepsilon_{xx} - \dot{\varpi}^2 \kappa_j{}^2 \frac{\partial^2 \varepsilon_{xx}}{\partial x^2} = E\upsilon_{xx} \tag{1}$$

The factor $p = \dot{\varpi}^2 \kappa_j{}^2$ is termed as the small-scale effect, where $\dot{\varpi}$ and κ_j are defined as the material constant and lattice spacing length or internal characteristic length.

Equation (1) can be written as

$$\varepsilon_{xx} - p \frac{\partial^2 \varepsilon_{xx}}{\partial x^2} = E\upsilon_{xx} \tag{2}$$

where ε_{xx}, υ_{xx}, and E are, respectively, the normal pressure, normal stress, and modulus of the child. In general, the parameter is called a nonlocal parameter, and in classical theory, this parameter is used to investigate the vibration, pinching, and bending problems of the beam [36, 37]. According to Euler's theory [36] equation can be written as

$$\alpha(x) p \frac{\partial^2 s}{\partial t^2} + \frac{\partial^2}{\partial x^2}\left[EJ(x)\frac{\partial^2 s}{\partial t^2}\right] = 0 \tag{3}$$

where α and J are the mass per unit length and moment of inertia of CNT, respectively. By using the Fourier method of variable separation, the two leading differential equation systems are normalized (ODE). In this system, an equation involving the space variable x and other equations is connected to the time variable

$$\mathrm{S}(x,t) = \beta(x)T(t) \tag{4}$$

$$\alpha p \frac{\partial^2}{\partial t^2}\beta(x)T(t) + \frac{\partial^2}{\partial x^2}\left[EJ\frac{\partial^2}{\partial x^2}\beta(x)T(t)\right] = 0 \tag{5}$$

$$\alpha p \beta(x)\frac{d^2T}{dt^2} + EJT(t)\frac{d^4\beta}{dx^4} = 0 \tag{6}$$

$$EJT(t)\frac{d^4\beta}{dx^4} = -\alpha p \beta(x)\frac{d^2T}{dt^2} \tag{7}$$

For harmonic response

$$T(t) = cos\omega t \text{ or } sin\omega t \text{ or } e^{i\omega t} \tag{8}$$

$$\alpha p \beta\left(-\omega^2 cos\omega t\right) + EJcos\omega t\frac{d^4\beta}{dx^4} = 0 \tag{9}$$

$$\frac{d^4\beta}{dx^4} - \frac{\alpha p \omega^2}{EJ}\beta(x) = 0 \tag{10}$$

$$\frac{d^4\beta}{dx^4} - \mu^4\beta(x) = 0 \tag{11}$$

Here $\beta(x)$ denotes the mode shape (eigenshape).

For parameter μ

$$\mu^4 = \frac{\alpha p \omega^2}{EJ} \tag{12}$$

where the general solution of fourth-order ODEs is

$$\beta(x) = \gamma_1 sin\mu x + \gamma_2 cos\mu x + \gamma_3 sin\mu x + \gamma_4 cosh\mu x \tag{13}$$

where $\gamma_1, \gamma_2, \gamma_3$ and γ_4 are the unknown constants.
Equation (11) becomes

$$\beta^{iv}(x) - \mu^4 \beta(x) = 0 \tag{14}$$

3. Application of WPA

For the solution of CNT problem, an analytical technique wave propagation approach is evoked. A simple approach called the propagation waveform (WPA) was developed by Zhang et al. [38]. Thus, a simple and effective technique is applied as the wave propagation method [34, 39, 40] used for problem-solving in the form of differential equations. Prior to this, many techniques have been sequentially used to study the vibration of CNTs [41–43]. Previously, the current approach was used continuously to study the vibration of carbon nanotubes [27, 29, 44–47].

$$\beta(x) = e^{-iq_m} \tag{15}$$

where q_m denoted the wave number in axial direction and used for support conditions of SWCNTs [29]. $\omega = 2\pi f$ is the angular frequency.

$$\beta^{iv}(x) = q^4{}_m e^{-iq_m x} \tag{16}$$

After putting these values in Eq. (14), we get

$$q^4{}_m e^{-iq_m x} - \mu^4 e^{iq_m x} = 0 \tag{17}$$

$$\mu^4 = q^4{}_m \tag{18}$$

By using Eq. (12), we can write as

$$\frac{\alpha p \omega^2}{EJ} = q^4{}_m \tag{19}$$

4. Nonlocal boundary conditions

In the article, the vibration of chiral SWCNTs with CC boundary conditions was investigated. In addition, the network interpretive directives (m, n) for chiral CNTs can be expressed as (m, n) for n m correspondingly as shown in **Figure 1**. General boundary conditions C—C and C—F are considered for the system; then it is used in order to find the frequency equation of SWCNTs and eigenfrequencies of different indices for chiral SWCNTs such as (12, 5), (22, 7), and (25, 10).

Support conditions in the form of frequency.

From Eq. (19)

$$\text{For clamped} - \text{clamped}, \frac{\alpha p \omega^2}{EJ} = \left(\frac{(2n+1)\pi}{2L}\right)^4 \tag{20}$$

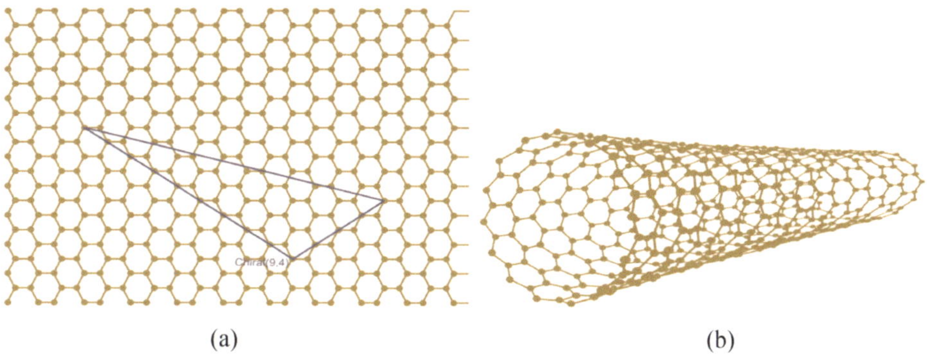

Figure 1.
(a) Graphene sheet. (b) SWCNT.

where $q_m = \frac{(2n+1)\pi}{2L}$ (CC boundary condition).

$$\text{For clamped} - \text{free}\ \frac{\alpha p\omega^2}{EJ} = \left(\frac{(2n-1)\pi}{2L}\right)^4 \tag{21}$$

where $q_m = \frac{(2n-1)\pi}{2L}$ (CF boundary condition).

5. Results and discussion

The fundamental natural frequencies (FNF) f (Hz) of SWCNTs obtained from nonlocal theory (NLT) based on wave propagation approach (WPA) with C-C and C-F boundary conditions are presented. A comparison of nondimensionalized natural frequencies $\Delta = \omega R\sqrt{\rho/E}$ of SWCNT is presented in **Table 1**. It is noted that from **Table 1**, the frequency value of present model have the small values as the values followed by the Alibeigloo and Shaban [48] shows a frequency difference between these studies. It can be seen that the error percentage is negligible, hence showing high rate of convergence. The results of nondimensional frequency are computed for two different values of n = 1, 2 with circumferential wave number (m = 0, 1, 2, 3, 4, 5) as shown in **Table 1**.

In order to analyze the effect of nonlocal parameter and bending rigidity on the vibration of chiral SWCNTs for different scales, the effects of nonlocal parameters on natural frequencies are illustrated in **Figures 2–5**. The natural frequencies are reduced by increasing nonlocal parameters (p = 0.5, 1, 1.5, 2). The results for the

m	*n* = 1		*n* = 2	
	Alibeigloo and Shaban [48]	**Present**	**Alibeigloo and Shaban [48]**	**Present**
0	0.97087	0.97063	0.99351	0.99289
1	0.59721	0.59698	0.88357	0.88301
2	0.34025	0.34019	0.68072	0.68013
3	0.20145	0.20099	0.50059	0.5003
4	0.12886	0.12872	0.36918	0.36897
5	0.09105	0.9087	0.27671	0.27662

Table 1.
Comparison of nondimensional frequencies $\Delta = \omega R\sqrt{\rho/E}$ (L/R = 1, n = 1).

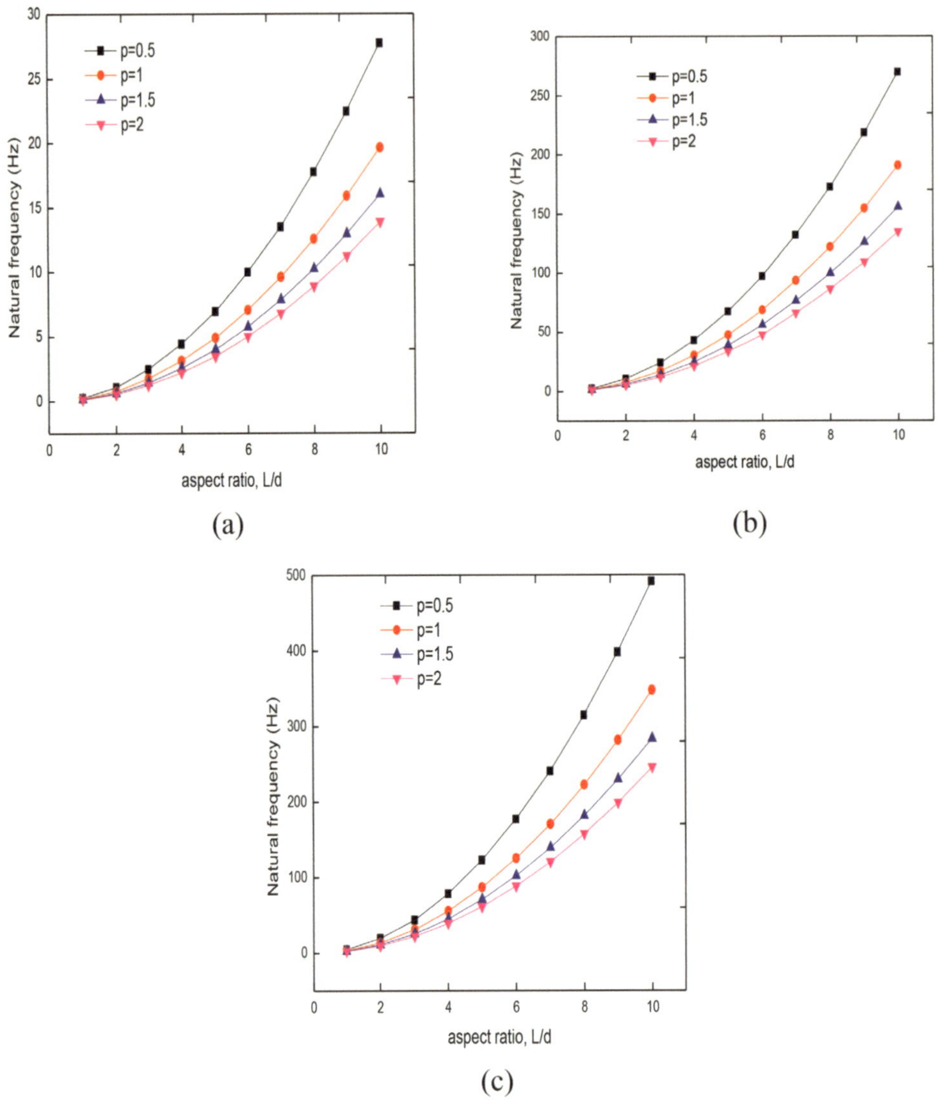

Figure 2.
FNFs versus aspect ratio for CC chiral SWCNTs (a) (12, 5), (b) (22, 7), and (c) (25, 10) with different nonlocal parameter p.

chiral clamped SWCNT with the indices (12, 5), (22, 7), and (25, 10) are shown in **Figure 2**. For the effect of nonlocal parameters with chiral index C—C(=12, 5) at $L/d(=1\sim10)$, the first 10 frequencies at p = 0.5 are 0.2773, 1.1094, 2.4961, 4.4376, 6.9337, 9.9846, 13.501, 17.7503, 22.4653, and 27.7349. When p = 2, then the frequency peaks are 0.1387, 0.5547, 1.2481, 2.2188, 3.4669, 4.9923, 6.7951, 8.8752, 11.2326, and 13.8675. Now, for C—C(=22, 7), with the same parameters, the first 10 frequencies at p = 0.5 and 2 are 2.6958, 10.7833, 26.2624, 43.1331, 67.3955, 97.0496, 132.0953, 172.5326, 218.3616, and 269.5822 and 1.3472, 5.3886, 12.1244, 21.5566, 33.6788, 48.4975, 66.0105, 86.2178, 109.1808, and 134.7152, respectively. Now, for C—C(=25, 10), with the same parameters, the first 10 frequencies at p = 0.5 and 2 are 4.9142, 19.6569, 44.2280, 78.6276, 122.8556, 176.9120, 240.7969, 314.5103, 398.0521, and 491.4223 and 2.4571, 9.8284, 22.1140, 39.3138, 61.4278, 88.4560, 120.3985, 15.2551, 199.0260, and 245.7112, respectively. It can be seen that FNF is reduced by increasing nonlocal parameters (p = 0.5, 1, 1.5, 2). To illustrate the effect

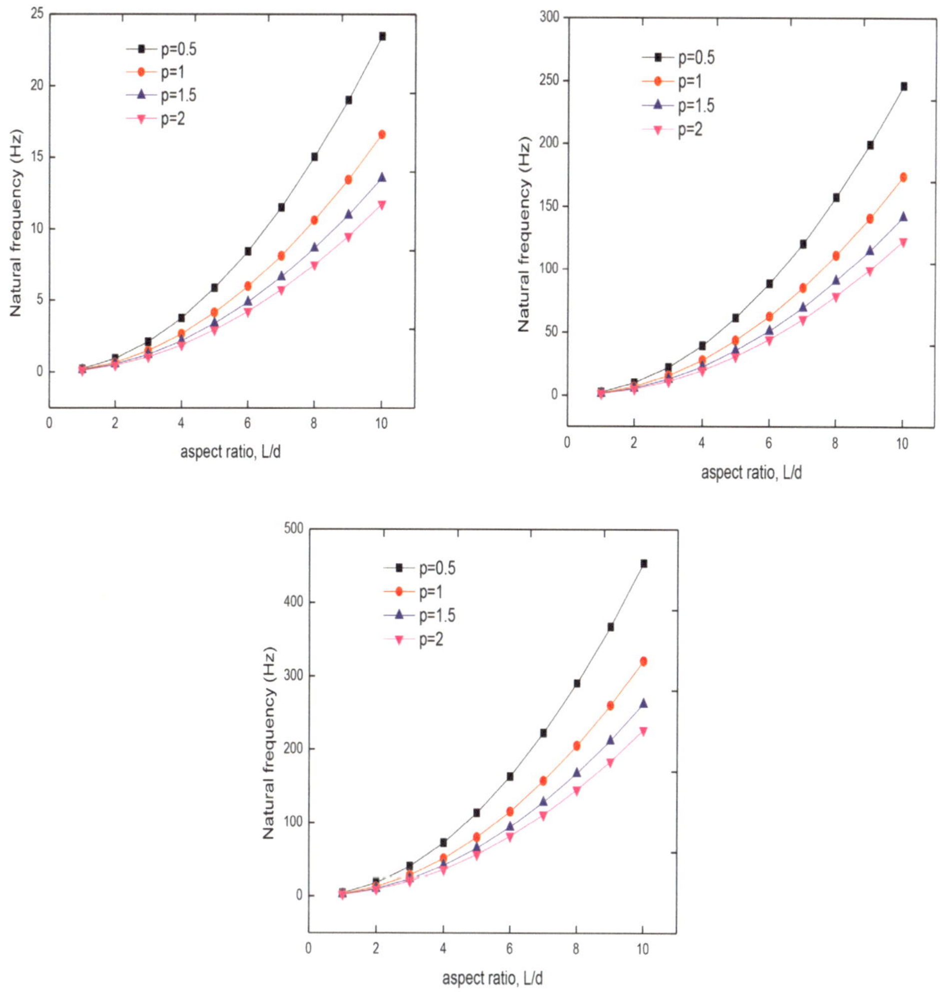

Figure 3.
FNFs versus aspect ratio for CF chiral SWCNTs (a) (12, 5), (b) (22, 7), and (c) (25, 10) with different nonlocal parameter p.

of different nonlocal parameters on natural frequencies for chiral SWCNTs with indices (12, 5), (22, 7), and (25, 10) based on NLT as shown in **Figures 2** and **3**. It is remarkable that from **Figures 2–5**, the FNF values of the chiral CC tubes are certainly higher than the chiral CF values of the SWCNTs.

Figures 4 and **5** show the FNFs against aspect ratio with varying bending strength index (EI). They refer to instances when EI changes from $5.1122e^{-9}$ to $7.2629e^{-9}$nm with nonlocal parameters $p = 1$. These figures show the natural frequency behavior of the calculated SWCNT system under bending rigidity (EI) parameters. It is considered that with the increase of the bending rigidity ($EI = 5.1122e^{-9}$ to $7.2617e^{-9}$ nm), the fundamental natural frequencies increase, and with the increase of aspect ratio, the frequencies also increase as C—C = (12, 5) f (Hz): 0.1961–0.2337 [C—F (12, 5) f (Hz): 0.1660–0.1978)] and C—C = (22, 7) f (Hz): 1.9062–2.2719 [CF (22, 7) f (Hz):1.7406–2.0745)] and C—C = (25, 10) f (Hz): 3.4749–4.1415 [C—F (25, 10) f (Hz):3.2077–3.8230)] at $L/d = 1$. The fundamental natural frequencies at $L/d = 10$ are as C—C = (12, 5) f (Hz): 27.7349–23.3737 [C—F (12, 5) f (Hz:16.5992–19.7835)] and CC = (22, 7) f (Hz): 190.6234–227.1912 [C—F (22, 7) f (Hz):174.0556–207.4452)] and C—C = (25, 10) f (Hz): 347.4881–414.1476

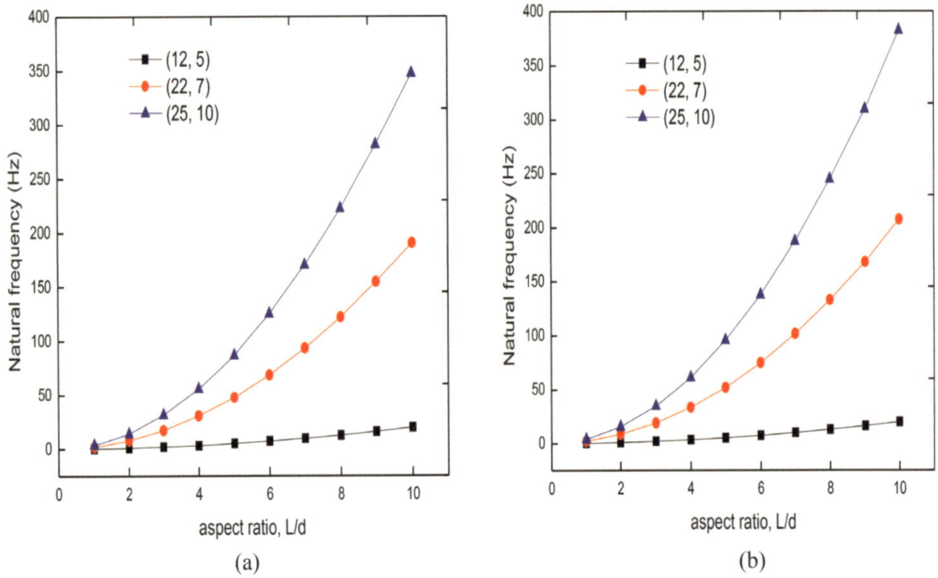

Figure 4.
FNFs with aspect ratio for CC chiral SWCNTs (12, 5), (22, 7), and (25, 10) with nonlocal parameter p = 1 and bending rigidity (a) EI = 5.1122e^{-9}nm and (b) EI = 7.2629e^{-9}nm.

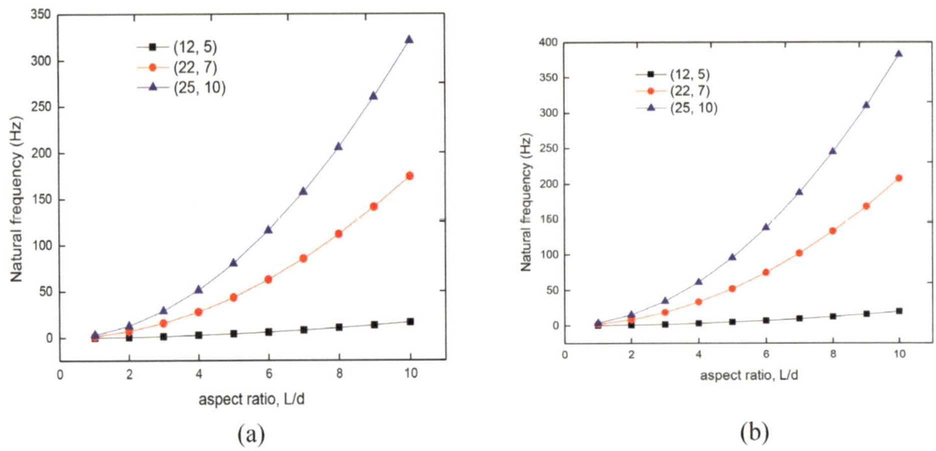

Figure 5.
FNFs with aspect ratio for CF chiral SWCNTs (12, 5), (22, 7), and (25, 10) with nonlocal parameter p = 1 and bending rigidity (a) EI = 5.1122e^{-9}nm and (b) EI = 7.2629e^{-9}nm.

[C—F (25, 10) *f* (Hz): 320.7685–382.3023)]. These relate to the case where *EI* varies from 5.1122*e*$^{-9}$ to 7.2617*e*$^{-9}$ nm and the nonlocal parameter $p = 1$. A trend of increasing frequencies of indices with bending rigidity is as (25, 10) > (22, 7) > (12, 5).

The tendency to increase the frequency of indices with bending stiffness is (25, 10) > (22, 7) > (12, 5). **Figure 5** shows that FNF, calculated by NLT, is based on WPA, with (12, 5), (22, 7), and (25, 10) CF chiral SWCNT, respectively. It was observed that FNF increased with increasing *EI* (hardness) and its value increased with increasing *L/d*. From our results, we can easily conclude that the climbing frequencies for bending the hardness of the curves (12, 5), (22, 7), and (25, 10) are as follows: (12, 5) < (22, 7) < (25, 10).

6. Conclusion

In this study, the influence of boundary conditions on the vibration of single-walled carbon nanotubes was analyzed in chiral fashion with indices (12, 5), (22, 7), and (25, 10), respectively. An attempt of nonlocal elasticity theory models has been employed to study the vibration characteristics of SWCNTs analytically, and the WPA is exploited to develop the ODE of the vibrations of the SWCNTs. The influences of different boundary conditions and bending rigidity of chiral SWCNTs against aspect ratio have investigated. As can be seen from these, by increasing the aspect ratio of the carbon nanotube, fundamental natural frequency increases. In addition, as can be seen, increasing the bending rigidity results in the increase of the fundamental frequencies. The frequencies of CC end condition are higher than CF end condition for all computations in this chapter.

Author details

Muzamal Hussain* and Muhammad Nawaz Naeem
Department of Mathematics, Government College University Faisalabad, Faisalabad, Pakistan

*Address all correspondence to: muzamal45@gmail.com

IntechOpen

© 2019 The Author(s). Licensee IntechOpen. This chapter is distributed under the terms of the Creative Commons Attribution License (http://creativecommons.org/licenses/by/3.0), which permits unrestricted use, distribution, and reproduction in any medium, provided the original work is properly cited.

References

[1] Iijima S. Helical microtubules of graphitic carbon. Nature. 1991;**345**: 56-58

[2] Li C, Chou TW. A structural mechanics approach for the analysis of carbon nanotubes. International Journal of Solids and Structures. 2003;**40**(10): 2487-2499

[3] Sakhaee-Pour A, Ahmadian MT, Vafai A. Vibrational analysis of single-walled carbon nanotubes using beam element. Thin-Walled Structures. 2009; **47**(6):646-652

[4] Vodenitcharova T, Zhang LC. Effective wall thickness of a single-walled carbon nanotube. Physical Review B. 2003;**68**(16):165401

[5] Hsu JC, Chang RP, Chang WJ. Resonance frequency of chiral single-walled carbon nanotubes using Timoshenko beam theory. Physics Letters A. 2008;**372**(16):2757-2759

[6] Yakobson BI, Brabec CJ, Bernholc J. Nanomechanics of carbon tubes: Instabilities beyond linear response. Physical Review Letters. 1996;**76**(14): 2511-2514

[7] Cao G, Chen X, Kysar JW. Strain sensing of carbon nanotubes: Numerical analysis of the vibrational frequency of deformed single-wall carbon nanotubes. Physical Review B. 2005;**72**(19):195412

[8] Murmu T, Pradhan SC. Thermo-mechanical vibration of a single-walled carbon nanotube embedded in an elastic medium based on nonlocal elasticity theory. Computational Materials Science. 2009;**46**:854-859

[9] Demir Ç, Civalek O. Nonlocal finite element formulation for vibration. International Journal of Engineering and Applied Sciences (IJEAS). 2016;**8**: 109-117

[10] Civalek O, Demir C, Akgoz B. Static analysis of single walled carbon nanotubes (SWCNT) based on Eringen's nonlocal elasticity theory. International Journal of Engineering and Applied Sciences (IJEAS). 2009;**2**:47-56

[11] Narendar S, Gopalakrishnan S. Critical buckling temperature of single-walled carbon nanotubes embedded in a one-parameter elastic medium based on nonlocal continuum mechanics. Physica E: Low-dimensional Systems and Nanostructures. 2011;**43**:1185-1191

[12] Yayli MO. Torsion of nonlocal bars with equilateral triangle cross sections. Journal of Computational and Theoretical Nanoscience. 2013;**10**: 376-379

[13] Toupin RA. Theory of elasticity with couple stresses. Archive for Rational Mechanics and Analysis. 1964;**17**:85-112

[14] Mindlin RD, Tiersten HF. Effects of couple-stresses in linear elasticity. Archive for Rational Mechanics and Analysis. 1962;**11**:415-448

[15] Fleck NA, Hutchinson JW. A phenomenological theory for strain gradient effects in plasticity. Journal of the Mechanics and Physics of Solids. 1993;**41**:1825-1857

[16] Fleck NA, Hutchinson JW. A reformulation of strain gradient plasticity. Journal of the Mechanics and Physics of Solids. 2001;**49**:2245-2271

[17] Eringen AC. On differential equations of nonlocal elasticity and solutions of screw dislocation and surface-waves. Journal of Applied Physics. 1983;**54**:4703-4710

[18] Eringen AC, Edelen DGB. On nonlocal elasticity. International Journal of Engineering Science. 1972;**10**:233-248

[19] Wang B, Deng ZC, Zhang K. Nonlinear vibration of embedded single-walled carbon nanotube with geometrical imperfection under harmonic load based on nonlocal Timoshenko beam theory. Journal of Applied Mathematics and Mechanics. 2013;**34**:269-280

[20] Yang J, Ke LL, Kitipornchai S. Nonlinear free vibration of single-walled carbon nanotubes using nonlocal Timoshenko beam theory. Physica E: Low-Dimensional System and Nanotechnology. 2010;**42**:1727-1735

[21] Azrar A, Azrar L, Aljinaidi AA. Length scale effect analysis on vibration behavior of single-walled carbon nanotubes with arbitrary boundary conditions. Revue de Mécanique Appliqué et Théorique. 2011;**2**:475-484

[22] Azrar A, Azrar L, Aljinaidi AA, Hamadiche M. Dynamics instability analysis of multi-walled carbon nanotubes conveying fluid. Advanced Materials Research. 2013;**682**:153-160

[23] Ansari R, Rouhi H, Sahmani S. Calibration of the analytical nonlocal shell model for vibrations of double-walled carbon nanotubes with arbitrary boundary conditions using molecular dynamics. International Journal of Mechanical Sciences. 2011;**53**: 786-792

[24] Ansari R, Sahmani S, Rouhi H. Rayleigh-Ritz axial buckling analysis of single-walled carbon nanotubes with different boundary conditions. Physics Letters A. 2011;**375**:1255-1263

[25] Swain A, Roy T, Nanda BK. Vibration behavior of single-walled carbon nanotube using finite element. International Journal on Theoretical and Applied Research in Mechanical Engineering. 2013;**2**:129-133

[26] Chawis T, Somchai C, Li T. Non-local theory for free vibration of single-walled carbon nanotubes. Advanced Materials Research. 2013;**747**:257-260

[27] Hussain M, Naeem MN, Shahzad A, He M. Vibrational behavior of single-walled carbon nanotubes based on cylindrical shell model using wave propagation approach. AIP Advances. 2017;**7**(4):045114

[28] Rouhi H, Bazdid-Vahdati M, Ansari R. Rayleigh-Ritz vibrational analysis of multi-walled carbon nanotubes based on the non-local Flügge shell theory. Journal of Composites. 2015:750392. http://dx.doi.org/10.1155/2015/750392

[29] Hussain M, Naeem MN. Vibration analysis of single-walled carbon nanotubes using wave propagation approach. Mechanical Sciences. 2017; **8**(1):155-164

[30] Yoon J, Ru CQ, Mioduchowski A. Noncoaxial resonance of an isolated multiwall carbon nanotube. Physical Review B. 2002;**66**(23):233402

[31] Bocko J, Lengvarský P. Vibration of single-walled carbon nanotubes by using nonlocal theory. American Journal of Mechanical Engineering. 2014;**2**:195-198

[32] Duan WH, Wang CM, Zhang YY. Calibration of nonlocal scaling effect parameter for free vibration of carbon nanotubes by molecular dynamics. Journal of Applied Physics. 2007;**101**(2):024305

[33] Elishakoff I, Pentaras D. Fundamental natural frequencies of double-walled carbon nanotubes. Journal of Sound and Vibration. 2009; **322**(4):652-664

[34] Natsuki T, Endo M, Tsuda H. Vibration analysis of embedded carbon nanotubes using wave propagation approach. Journal of Applied Physics. 2006;**99**(3):034311

[35] Eringen AC. Nonlocal Continuum Field Theories. Springer Science & Business Media; 2002

[36] Thongyothee C, Chucheepsakul S. Finite element modeling of van der Waals interaction for elastic stability of multi-walled carbon nanotubes. Advanced Materials Research. 2008;**55**: 525-528

[37] Reddy JN, Pang SD. Nonlocal continuum theories of beams for the analysis of carbon nanotubes. Journal of Applied Physics. 2008;**103**(2):023511

[38] Zhang XM, Liu GR, Lam KY. Vibration analysis of thin cylindrical shells using wave propagation approach. Journal of Sound and Vibration. 2001; **239**(3):397-403

[39] Wang L, Guo W, Hu H. Group velocity of wave propagation in carbon nanotubes. Proceedings of the Royal Society of London A: Mathematical, Physical and Engineering Sciences. 2008;**464**(2094):1423-1438

[40] Liew KM, Wang Q. Analysis of wave propagation in carbon nanotubes via elastic shell theories. International Journal of Engineering Science. 2007; **45**(2):227-241

[41] Hussain M, Naeem M, Shahzad A, He M. Vibration characteristics of fluid-filled functionally graded cylindrical material with ring supports. In: Computational Fluid Dynamics. Rijeka, Croatia: InTech Open; 2018. DOI: 10.5772/intechopen.72172. ISBN: 978-953-51-5706-9

[42] Vibration characteristics of zigzag and chiral FGM rotating carbon nanotubes sandwich with ring supports. Journal of Mechanical Engineering Science, Part C. 2019

[43] Hussain M, Naeem MN. Effects of ring supports on vibration of armchair and zigzag FGM rotating carbon nanotubes using Galerkin's method. Composites Part B Engineering. 2019;**163**:548-561. DOI: 10.1016/j.compositesb.2018.12.144

[44] Hussain M, Naeem M. Effect of various edge conditions on free vibration characteristics of rectangular plates. In: Advance Testing and Engineering. Rijeka, Croatia: InTech Open; 2018. ISBN: 978-953-51-6706-8

[45] Hussain M, Naeem M. Vibration of single-walled carbon nanotubes based on Donnell shell theory using wave propagation approach. In: Novel Nanomaterials—Synthesis and Applications. Rijeka, Croatia: InTech Open; 2018. DOI: 10.5772/intechopen.73503. ISBN: 978-953-51-5896-7

[46] Hussain M, Naeem MN, Isvandzibaei MR. Effect of Winkler and Pasternak elastic foundation on the vibration of rotating functionally graded material cylindrical shell. Proceedings of the Institution of Mechanical Engineers, Part C: Journal of Mechanical Engineering Science. 2018;**232**(24): 4564-4577. DOI: 0954406217753459

[47] Hussain M, Naeem MN, Shahzad A, He M, Habib S. Vibrations of rotating cylindrical shells with FGM using wave propagation approach. Proceedings of the Institution of Mechanical Engineers, Part C: Journal of Mechanical Engineering Science. 2018;**232**(23): 4342-4356

[48] Alibeigloo A, Shaban M. Free vibration analysis of carbon nanotubes by using three-dimensional theory of elasticity. Acta Mechanica. 2013;**224**(7): 1415-1427

Section 4

Risk Management and Limitation of Carbon Nanotubes

Chapter 12

Cytotoxicity Evaluation of Carbon Nanotubes for Biomedical and Tissue Engineering Applications

Narsimha Mamidi

Abstract

Carbon nanotubes (CNTs) are one of the most studied allotropes of carbon nanomaterials. The exceptional chemical and physical properties of CNTs make them potential candidates for several applications such as electrical, gene therapy, biosensors, and drug delivery applications. However, the toxicity of CNTs has been a major concern for their use in tissue engineering and biomedical applications. In this chapter, we present an overview of carbon nanotubes in biomedical and tissue engineering applications. We discussed various factors including impurities, length, agglomeration, and size of CNTs that cause toxicity of CNTs. Further, other toxic methods are also examined, and possible ways to overcome these challenges have been discussed.

Keywords: carbon nanotubes (CNTs), biomedical and tissue engineering, cytotoxicity, agglomeration, size, length

1. Introduction

Amalgamation of nanotechnology with biomedical and tissue engineering offers an admirable opportunity for developing great nanomaterials that would significantly improve treatment and diagnosis of diseases [1]. It is also anticipated that the development and use of nanomaterials at industrial scale would be the driving forces for the emerging industries and economies. Carbon nanotubes are novel carbon nanomaterials, and they have attracted a wide range of applications due to their inimitable properties. Particularly, CNTs have the potential to modernize biomedical and tissue engineering because of their impeccable chemical, electrical, thermal, structural, and mechanical properties, which have made them as an area of great research interest [1]. CNTs exhibit semiconducting, metallic, and superconducting electron transport properties, and they also display high elastic modulus compared to all other nanomaterials. Numerous research studies have been conducted on the applications of CNTs in the biomedical and tissue engineering fields. Most specifically, CNTs have been used in a variety of applications such as diagnostic tools, biosensors, nanofluidic systems, radiation oncology, quantum dots, drug delivery, nanorobots, and nanosensors [2–4]. However, low dispersibility, toxicity, and solubility of unfunctionalized multi-walled carbon nanotubes (MWCNTs) have been the main concern for their potential use in biomedical and tissue engineering applications. Therefore, the interaction of CNTs with biological systems is very complex and unpredictable. Biological properties, performance, and behavior of CNTs have

IntechOpen

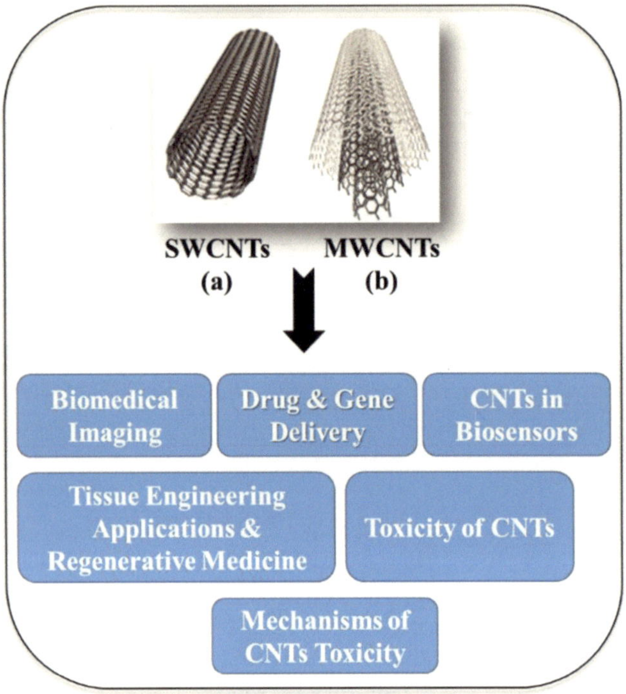

Figure 1.
(a and b) Structures of carbon nanotubes and schematic representation of the issues addressed in this chapter.

to be thoroughly understood. It is reported that CNTs exhibit different levels of toxicity based on their manufacturing method, shape, surface-area-to-volume ratio, concentration, size, composition, functional groups, applied dosages, and extent of oxidation [5–8]. In addition, CNTs have the ability to damage the cell membrane and DNA due to their high hydrophobicity. CNTs can also extend their toxicity through protein synthesis, oxidative stress, mitochondrial activities, modifications, apoptosis, necrosis, as well as intramolecular metabolic paths [8].

This chapter discusses the critical roles of CNTs in biomedical and tissue engineering applications. It explains synthetic methods and recent advances in the application of CNTs in bioimaging, drug delivery, biosensing, and tissue engineering applications. In the end, this chapter also encapsulates the surface chemistry, shape, size, and the route of synthesis of CNTs which can affect their toxicity levels (**Figure 1**). At the end the mechanism responsible for CNS toxicity and potential remedies to overcome their drawbacks has also been discussed.

2. Synthesis of carbon nanotubes

CNTs exist approximately 1 nm in diameter and 1–100 μm in length and formed by cylinder-shaped graphite layers [8]. CNTs are mainly divided into two types: single-walled carbon nanotubes (SWCNTs) and multi-walled carbon nanotubes (MWCNTs). SWCNTs are formed by a single layer of graphene, and MWCNTs are formed by multiple layers of graphene (**Figure 1a** and **b**). In addition, a variety of nanomaterials are reported including fullerenes, nanohorns, carbon nanobuds, carbon nanoporous, carbon nanopeapods, and carbon nano-onions [2, 8, 9]. Mainly three different techniques are used to fabricate CNTs such as the arc discharge technique, the chemical vapor deposition (CVD), and the laser ablation technique [2].

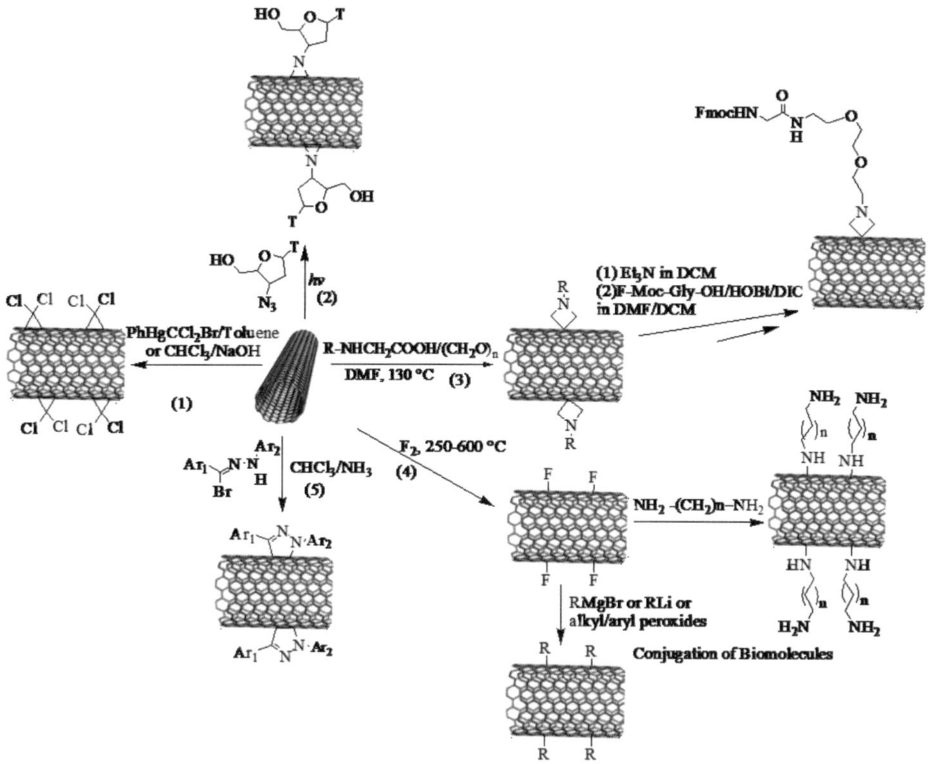

Figure 2.
Different covalent functionalization strategies of CNTs: (1) cycloaddition with dichlorocarbene, (2) photoinduced production of reactive nitrenes, (3) 1,3 dipolar cycloaddition of azomethine ylides, (4) fluorination of nanotubes and defunctionalization followed by derivatization reactions, and (5) 1,3 dipolar cycloaddition of nitrile amines.

First time in 1991, Ijima used the arc discharge technique to fabricate MWCNTs; later, metal catalyst was used in the same technique to provide the first SWCNTs [10, 11]. After that, the laser ablation technique was used by Thess to produce aligned SWCNTs [12]. Yacamán developed the catalytic growth of MWCNTs using CVD technique [12]. Further, cobalt (Co) catalyzed low-pressure chemical vapor deposition which was used on silicon oxide as a silicon substrate to produce Co-MWCNTs films and used as sensors to detect carbon dioxide (CO_2) [13].

Pristine or unfunctionalized CNTs are highly hydrophobic and insoluble in an aqueous solution and organic solvents that cause great limitation for their tissue engineering and biomedical applications. Therefore, it is important to functionalize CNTs to make them hydrophilic and amalgamate into numerous organic solvents and biological systems [4]. Three main methods have been reported to modify CNT structures, namely, (a) the covalent functionalization (**Figure 2**), (b) the noncovalent adsorption of many biomolecules, and (c) the endohedral filling of their inner empty cavity [14].

3. CNTs in biomedical applications

CNTs show numerous unique properties that make them promising carbon nanomaterials for a wide range of biomedical applications. For this, the surface of CNTs can be functionalized with suitably biocompatible moieties. These moieties can interact with cell membrane receptors that can guide their cell internalization. These receptor-mediated methods can support in drug loading, inflammation, and minimizing toxicity [15].

3.1 Drug and gene delivery

CNTs have been used in several drug delivery systems for the treatment of many diseases. Anticancer drug-loaded CNTs have attracted much attention mainly in two strategies such as selective targeting and controlled release of drugs [15, 16]. It is reported that CNT-based anticancer drug was developed that could control the multidrug-resistant cancer cells without affecting cell cycles and proliferation [17]. Further, an anticancer drug, tamoxifen-loaded peptide-modified SWCNT, was developed that showed a high level of antitumor effect and proficient tumor targeting. SWCNTs successfully delivered acetylcholine (Ach) into the mice brain to control Alzheimer's disease [18]. Anticancer drug-appended MWCNTs have been used for effective breast cancer treatment and intranuclear drug delivery [19].

In addition, functionalized CNTs have shown other advantages through covalent conjugation methods. For instance, functionalized SWCNTs with ester and amide enabled to sustain drug delivery and improved the solubility in aqueous and organic solvents [20]. Liposomes were covalently attached to MWCNTs that facilitated the delivery of large doses of the drug [21]. Peptide-based MWCNTs were also developed to deliver therapeutics into the target mitochondria to treat genetic disorders [22]. Some other functionalized MWCNTs have been used in the central nerve system (CNS) through neural tissue cell interactions [23]. On the other hand, noncovalent functionalized CNTs have been used in a wide range of medical applications that release the drugs in tumor environments at low pH and kill the cancer cells [24]. Polyethyleneimines bearing noncovalent functionalized SWCNTs were developed for effective gene delivery [25]. In addition, biomolecules including genes, DNA and siRNA, also can be loaded into CNTs. CNTs were functionalized with poly(lactic-co-glycolic) (PLGA) to deliver proapoptotic protein caspase-3 (CP3) into osteocarcinoma cells [26]. CNTs have also been used in in vivo gene silencing without any toxicity and induction of immune response. Polyethylenimine-functionalized CNTs successfully delivered siRNA into the HeLa-S3 cells [27].

3.2 Biomedical imaging

Biomedical imaging is a powerful tool that can provide high-resolution imaging of cells, organs, tissues, and even the complete body of animals or humans. Due to unique physicochemical properties, CNTs have been used in different biomedical imaging technologies [28]. There are three major methods such as fluorescence emission, photoacoustic imaging, and magnetic resonance imaging that can be executed on CNTs to detect the nanotubes in live cells [29]. It is reported that CNTs can be directed using external magnetic source toward a specific organ [30]. Surface modification, as well as the addition of elements to CNT structure, can provide a new perspective of the analysis of their performance. For example, by assembling different nanoparticles including quantum dots, gold nanoparticles, upconversion nanoparticles, iron oxide nanoparticles, PET imaging nanoprobes into CNTs enhancing their properties and wider of their applications [31]. Thus, biomedical imaging is a potential and easy platform to accomplish imaging of living cells in tissue engineering and biomedical engineering.

3.3 CNT-based biosensor

CNTs are highly effective sensing elements for biosensors due to their excellent electrical, tensile, and electrochemical properties; high surface area; and high exposure sensitivity to various biomolecules [32]. Specifically, their high surface-area-to-volume ratio has made them a potent tool to acquire fast biological species detection and

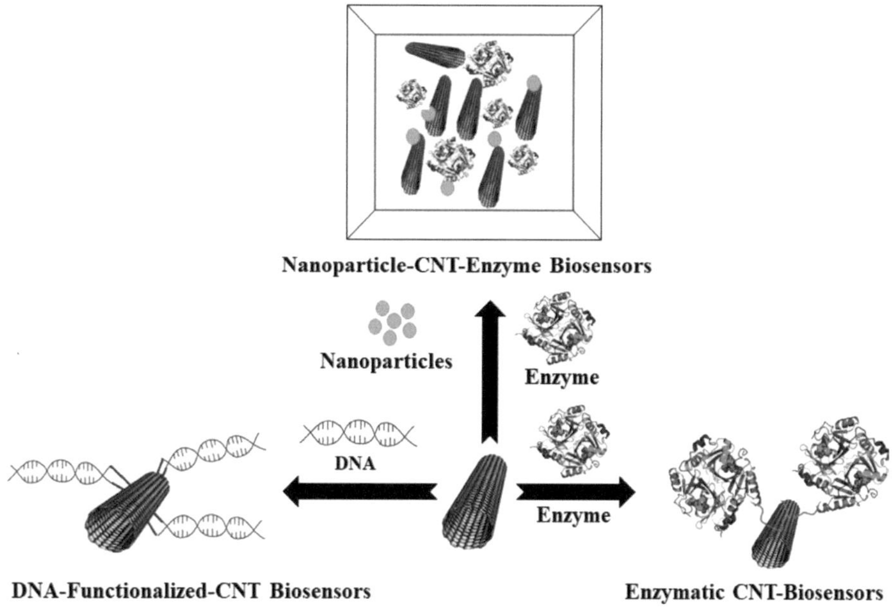

Figure 3.
Schematic illustration of CNT-based biosensors.

CNT-based biosensors which are frequently used in ultrasensitive biosensing applications [32]. In addition, enzyme biosensors are also one of the most important and commonly used biosensors (**Figure 3**). For example, durable and stable tyrosinase biosensor was developed from functionalized MWCNTs, 1-butyl-3-methylimidazolium chloride (IL), and tyrosinase (Tyr) within a dihexadecyl phosphate (DHP) film, and improved response signal was observed [33]. 3α-Hydroxysteroid dehydrogenase was incorporated into CNTs/IL/NAD+ composite electrode to develop a biosensor that detected androsterone. Further, CNT-based enzyme biosensors were also used in the detection of glucose in the blood, and it is one of the potential applications of enzyme biosensors [34]. DNA biosensors are other famous CNT-based biosensors, and they have been used in medical diagnostics, forensic science, and several other applications. Single-stranded DNA (ssDNA) or double-stranded DNA (dsDNA) is the major sensing element of DNA biosensors (**Figure 3**). It is reported that ssDNA is highly adsorptive to CNTs rather than dsDNA. For example, SWCNT-FET-based electronic DNA biosensor was developed and used in chip-on system applications [35]. Ultrasensitive DNA biosensor was developed to control DNA methyltransferase (DNA MTase) from MWCNT signal amplification and fluorescence polarization detection [36]. In addition, a glassy carbon-based sensitive DNA biosensor was fabricated to detect DNA sequencing using polydopamine (PDA), MWCNTs, and gold nanoparticles [36]. Thus, the inimitable properties of CNTs have made them a powerful tool for biosensing applications.

4. Tissue engineering applications

Tissue engineering is a new approach to fabricate artificial tissues for graft replacement and tissue models for in vitro diseases and drug discovery [37]. Maintaining proper electrical, mechanical, and biological properties of CNT-based biomaterials is a challenge in tissue engineering [38]. For example, CNTs have been used in a variety of tissue engineering applications such as enhancing electrical and mechanical properties of scaffolds, tracking of cells, sensing the cell microenvironments, and delivery of appropriate chemical and biological agents [39]. Development of CNTs bearing scaffolds for in vitro nerve generation opened a new route for neural tissue engineering. In fact, CNT-based scaffolds have been used to improve cardiac tissue growth, bone, and neural growth [40]. In cardiac and nerve growth, SWCNTs greatly improved the electrical properties of scaffolds, whereas in bone growth, SWCNTs enhanced the attraction of calcium cations [41]. Some scaffolds were developed and used for bone formation via functionalization of MWCNTs with fibroblast growth factors [42]. In addition, MWCNT-gelatin nanofiber scaffolds were used for myoblast (C_2C_{12}) growth, and MWCNTs have improved tensile properties of the fiber scaffolds [43]. Thus, CNTs are used as a potential component for many biomaterials in tissue engineering applications.

5. Toxicity of CNTs

Many toxicological investigations of CNTs have been done both in vivo and in vitro; however, they are inconsistent to each other due to the variabilities in the type of functionalization, the synthetic method, and the dose of CNTs. In addition, a different type of cell viable indicator dye also contributed different cytotoxicity results. Some of the common indicator dyes are alamarBlue, 3-(4,5-dimethylthiazol-2-yl)-2,5-diphenyltetrazolium bromide (MTT), Coomassie Brilliant Blue, neutral blue, and water-soluble tetrazolium salt (WST-1) [44]. Several factors that contribute to the toxicity of CNTs include metal impurities, diameter, length, type of carbon nanotubes, surface functionalization, and existence of dispersant [36].

Toxicological effect of CNTs has been studied; particularly, commercial and acid-purified MWCNTs and SWCNTs have shown a significant effect on toxicity levels. This study explained that commercial CNTs showed increased reactive oxygen species (ROS) that enhanced oxidative stress and decrease mitochondrial membrane potential. In contrast, acid-purified SWCNTs with less metal impurity showed less toxicity effect [45]. However, MWCNTs exhibited some toxicity at high concentration due to metal impurities [46]. Further, iron-contaminated MWCNTs presented improved CD8+ levels and CD4+/CD8+ ratio of peripheral T cell in mice models as well as increased ROS [46].

The length of CNTs also showed a great impact on the toxicity of CNTs, and smaller CNTs are less toxic than long CNTs. Smaller CNTs can easily penetrate into cell membrane as well as the success of their cellular internalization, whereas long CNTs cause biopersistence or retaining [47]. Another study demonstrated that short CNTs were cleared from the pleural cavity, whereas long CNTs were retained [48]. Similarly, the retention of long CNTs (825 nm) has exhibited induced inflammation than small CNTs (220 nm) [48]. In addition, long MWCNTs (5–15 μm of length and 20–60 nm of diameter) showed more genotoxicity in alveolar carcinoma epithelial cells (A549) than smaller MWCNTs (1–2 μm of length and 60–100 nm of diameter) [49].

Based on the structural characteristic, carbon nanotubes are divided into SWCNTs and MWCNTs. These CNTs differ in length, structure, and chemical surface. SWCNTs have a smaller diameter (0.6–2.4 nm), whereas MWCNTs have a larger diameter (2.5–100 nm) [6]. SWCNTs have a higher surface area that helps to form a bundle of CNTs. MWCNTs have a lower tendency to form a bundle due to lower surface area and their side walls [44]. Similarly, smaller size and length of SWCNTs form less aggregation of 5–30 μm that can easily phagocytosed, whereas longer MWCNTs generate larger aggregation around 300 μm, which cannot be phagocytosed [50]. The fibrous surface of MWCNTs and SWCNTs showed different mechanisms in the plasma membrane. MWCNTs caused toxicity due to the plasma membrane damage and aberrant phagocytosis, while SWCNTs initiated oxidative damage to cells [51]. As

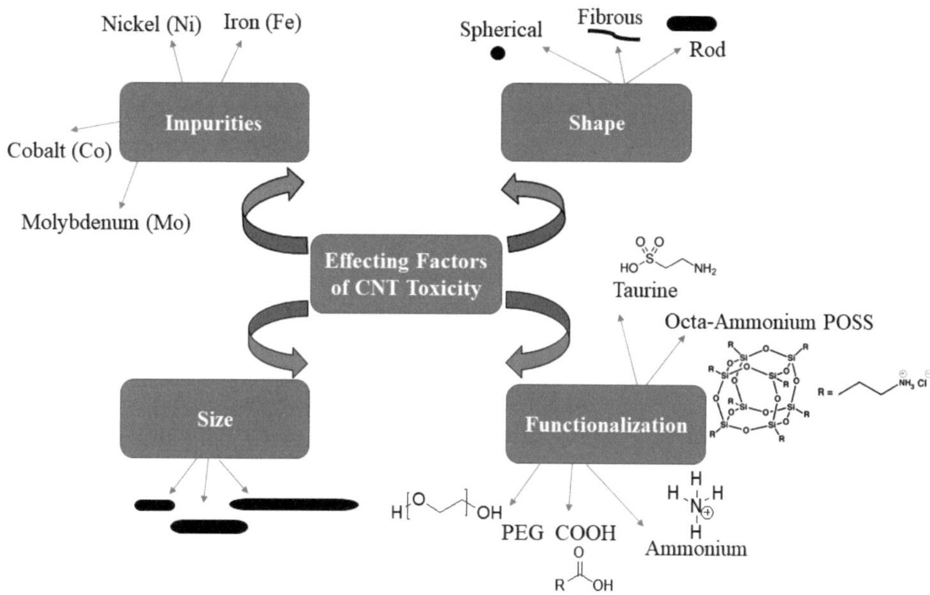

Figure 4.
Schematic representation of toxic factors of CNTs. It shows size, impurities, shape, and functionalization which are the major factors to cause CNT toxicity.

a result, based on the various characteristics (size, aggregation, and surface state) of SWCNTs and MWCNTs, they exhibit a different level of toxicity.

The surface structure of CNTs can be modified by functionalization by introducing several functional groups. The functionalization enhances solubility, biocompatibility, dispersibility, and agglomeration of CNTs. The functionalization may be either covalent bonding or noncovalent binding [52]. Functionalization enables conjugation of various groups with CNTs that help in cell receptor binding as well as cellular processing and elimination. The use of biomolecules (proteins and antibodies) in conjugation strengthens the specific binding of CNTs to targeted biomolecules. In fact, functionalized MWCNTs with 220 kDa lectin protein exhibited apoptosis and reduced toxicity in J774A macrophage [53]. On the other hand, functionalized SWCNTs-COOH showed higher toxicity in the HUVEC cell line than pristine SWCNTs [54].

The degree of toxicity of CNTs has a great influence on the type of physicochemical properties, size, shape, impurities, and functionalization of CNTs (**Figure 4**) [55]. Further, CNT-based polymer nanocomposites also developed and cytotoxicity was assessed with human cells. For instance, unfunctionalized MWCNTs were reinforced with ultrahigh molecular weight polyethylene (UMWPE), and biocompatibility was evaluated with human fibroblasts [5]. The results revealed that MWCNTs exhibited a positive influence on fibroblast cells. Therefore, several studies reported that cell membrane injury, oxidative stress, and genotoxicity are the possible mechanisms of CNT toxicity [36].

6. Conclusions

Carbon nanotubes are a new class of carbon nanomaterials that have a potential in biomedical and tissue engineering applications including drug delivery, biosensors, biomedical imaging, and artificial tissue scaffolds. In order to decrease cytotoxicity and increase biocompatibility and physicochemical properties, CNTs can be functionalized with various biomolecules either covalently or noncovalently.

Several investigations have made to study the relationship between physical and cytotoxicity properties of short-walled or multi-walled nanotubes. In vitro and in vivo biocompatibility of MWCNTs and SWCNTs have been effectively influenced by their functionalization, diameter, and length. It is also clear that the synthesis method and metal impurities of CNTs can influence the cell viability and interaction of CNTs with cells. Although many investigations have been conducted for the toxicological characterizations of CNTs in biomedical and tissue engineering, however, a complete understanding of internalization, cellular uptake, and gene expression changes linked with CNTs has still remained elusive. This understanding would be required for the future use of CNTs in biomedical and tissue engineering applications.

Acknowledgements

The authors acknowledge the financial support from the Consejo Nacional de Ciencia y Tecnología de México (CONACyT) and Tec de Monterrey, Monterrey Campus, NL, Mexico.

Conflict of interest

The authors declare no conflict of interest.

Author details

Narsimha Mamidi
School of Engineering and Science, Tecnologico de Monterrey, Nuevo León, México

*Address all correspondence to: narsimhachem06@gmail.com

IntechOpen

© 2019 The Author(s). Licensee IntechOpen. This chapter is distributed under the terms of the Creative Commons Attribution License (http://creativecommons.org/licenses/by/3.0), which permits unrestricted use, distribution, and reproduction in any medium, provided the original work is properly cited.

References

[1] Simon J, Flahaut E, Golzio M. Overview of carbon nanotubes for biomedical applications. Materials. 2019;**12**(4):624. ISSN 1996-1944

[2] Sinha N, Yeow JT. Carbon nanotubes for biomedical applications. IEEE Transactions on NanoBioscience. 2005;**4**:180-195

[3] Beg S, Rizwan M, Sheikh AM, Hasnain MS, Anwer K, Kohli K. Advancement in carbon nanotubes: Basics, biomedical applications, and toxicity. The Journal of Pharmacy and Pharmacology. 2011;**63**:141-163

[4] Bianco A, Kostarelos K, Partidos CD, Prato M. Biomedical applications of functionalised carbon nanotubes. Chemical Communications. 2005;**5**:571-577

[5] Mamidi N, Leija HM, Diabb JM, Lopez Romo I, Hernandez D, Castrejón JV, et al. Cytotoxicity evaluation of unfunctionalized multiwall carbon nanotubes-ultrahigh molecular weight polyethylene nanocomposites. Journal of Biomedical Materials Research. Part A. 2017;**105**:3042-3049

[6] Foldvari M, Bagonluri M. Carbon nanotubes as functional excipients for nanomedicines: II. Drug delivery and biocompatibility issues. Nanomedicine. 2008;**4**:183-200

[7] Vardharajula S, Ali SZ, Tiwari PM, Eroğlu E, Vig K, Dennis VA, et al. Functionalized carbon nanotubes: Biomedical applications. International Journal of Nanomedicine. 2012;**7**:5361-5374

[8] Mehra NK, Jain K, Jain NK. Pharmaceutical and biomedical applications of surface engineered carbon nanotubes. Drug Discovery Today. 2015;**20**:750-759

[9] Sano N, Wang H, Chhowalla M, Alexandrou I, Amaratunga GAJ. Synthesis of carbon 'onions' in water. Nature. 2001;**414**:506-507. DOI: 10.1038/35107141

[10] Iijimaa S. Helical microtubules of graphitic carbon. Nature. 1991;**354**:56-58

[11] Iijima S, Ichihashi T. Single-shell carbon nanotubes of 1-nm diameter. Nature. 1993;**363**:603-605

[12] José-Yacamán M, Miki-Yoshida M, Rendon L, Santiesteban J. Catalytic growth of carbon microtubules with fullerene structure. Applied Physics Letters. 1993;**62**:202-204

[13] Khan ZH, Ansari MS, Salah NA, Memic A, Habib S, Shahawi M. COBALT catalyzed-multi-walled carbon nanotubes film sensor for carbon mono-oxide gas. Digest Journal of Nanomaterials and Biostructures. 2011;**6**:1947-1956

[14] Tasis D, Tagmatarchis N, Bianco A, Prato M. Chemistry of carbon nanotubes. Chemical Reviews. 2006;**106**:1105-1136

[15] Chen J, Chen S, Zhao X, Kuznetsova LV, Wong SS, Ojima I. Functionalized single-walled carbon nanotubes as rationally designed vehicles for tumor-targeted drug delivery. Journal of the American Chemical Society. 2008;**130**:16778-16785

[16] Zhang X, Meng L, Lu Q, Fei Z, Dyson PJ. Targeted delivery and controlled release of doxorubicin to cancer cells using modified single wall carbon nanotubes. Biomaterials. 2009;**30**:6041-6047

[17] Cheng J, Meziani MJ, Sun YP, Cheng SH. Poly(ethylene glycol)-conjugated multi-walled carbon

nanotubes as an efficient drug carrier for overcoming multidrug resistance. Toxicology and Applied Pharmacology. 2011;**250**:184-193

[18] Yang Z, Zhang Y, Yang Y, Sun L, Han D, Li H, et al. Pharmacological and toxicological target organelles and safe use of single-walled carbon nanotubes as drug carriers in treating Alzheimer disease. Nanomedicine. 2010;**6**:427-441

[19] Das M, Singh RP, Datir SR, Jain S. Intranuclear drug delivery and effective in vivo cancer therapy via estradiol-PEG appended multiwalled carbon nanotubes. Molecular Pharmaceutics. 2013;**10**:3404-3416

[20] Khazaei A, Rad MN, Borazjani MK. Organic functionalization of single-walled carbon nanotubes (SWCNTs) with some chemotherapeutic agents as a potential method for drug delivery. International Journal of Nanomedicine. 2010;**5**:639-645

[21] Karchemski F, Zucker D, Barenholz Y, Regev O. Carbon nanotubes-liposomes conjugate as a platform for drug delivery into cells. Journal of Controlled Release. 2012;**160**:339-345

[22] Battigelli A, Russier J, Venturelli E, Fabbro C, Petronilli V, Bernardi P, et al. Peptide-based carbon nanotubes for mitochondrial targeting. Nanoscale. 2013;**5**:9110-9117

[23] Bardi G, Nunes A, Gherardini L, Bates K, Al-Jamal KT, Gaillard C, et al. Functionalized carbon nanotubes in the brain: Cellular internalization and neuroinflammatory responses. PLoS One. 2013;**8**:e80964

[24] Petrov P, Stassin F, Pagnoulle C, Jerome R. Noncovalent functionalization of multi-walled carbon nanotubes by pyrene containing polymers. Chemical Communications. 2003;**0**:2904-2905

[25] Behnam B, Shier WT, Nia AH, Abnous K, Ramezani M. Non-covalent functionalization of single-walled carbon nanotubes with modified polyethyleneimines for efficient gene delivery. International Journal of Pharmaceutics. 2013;**454**:204-215

[26] Cheng Q, Blais MO, Harris G, Jabbarzadeh E. PLGA carbon nanotube conjugates for intercellular delivery of caspase-3 into osteosarcoma cells. PLoS One. 2013;**8**:e81947

[27] Huang YP, Lin IJ, Chen CC, Hsu YC, Chang CC, Lee MJ. Delivery of small interfering RNAs in human cervical cancer cells by polyethylenimine-functionalized carbon nanotubes. Nanoscale Research Letters. 2013;**8**:267

[28] Gong H, Peng R, Liu Z. Carbon nanotubes for biomedical imaging: The recent advances. Advanced Drug Delivery Reviews. 2013;**65**:1951-1963

[29] Tong L, Liu Y, Dolash BD, Jung Y, Slipchenko MN, Bergstrom DE, et al. Label-free imaging of semiconducting and metallic carbon nanotubes in cells and mice using transient absorption microscopy. Nature Nanotechnology. 2012;**7**:56-61

[30] De La Zerda A, Zavaleta C, Keren S, Vaithilingam S, Bodapati S, Liu Z, et al. Carbon nanotubes as photoacoustic molecular imaging agents in living mice. Nature Nanotechnology. 2008;**3**:557-562

[31] Li J, Chang X, Chen X, Gu Z, Zhao F, Chai Z, et al. Toxicity of inorganic nanomaterials in biomedical imaging. Biotechnology Advances. 2014;**32**:727-743

[32] Yang N, Chen X, Ren T, Zhang P, Yang D. Carbon nanotube based biosensors. Sensors and Actuators, B: Chemical. 2015;**207**:690-715

[33] Vicentini FC, Janegitz BC, Brett CM, Fatibello-Filho O. Tyrosinase

biosensor based on a glassy carbon electrode modified with multi-walled carbon nanotubes and 1-butyl-3-methylimidazolium chloride within a dihexadecyl phosphate film. Sensors and Actuators, B: Chemical. 2013;**188**:1101-1108

[34] Pourasl AH, Ahmadi MT, Rahmani M, Chin HC, Lim CS, Ismail R, et al. Analytical modeling of glucose biosensors based on carbon nanotubes. Nanoscale Research Letters. 2014;**9**:33

[35] Tang X, Bansaruntip S, Nakayama N, Yenilmez E, Chang YL, Wang Q. Carbon nanotube DNA sensor and sensing mechanism. Nano Letters. 2006;**6**:1632-1636

[36] Alshehri R, Ilyas AM, Hasan A, Arnaout A, Ahmed F, Memic A. Carbon nanotubes in biomedical applications: Factors, mechanisms, and remedies of toxicity. Journal of Medicinal Chemistry. 2016;**59**:8149-8167

[37] Hasan A, Memic A, Annabi N, Hossain M, Paul A, Dokmeci MR, et al. Electrospun scaffolds for tissue engineering of vascular grafts. Acta Biomaterialia. 2014;**10**:11-25

[38] Hasan A, Paul A, Vrana NE, Zhao X, Memic A, Hwang Y-S, et al. Microfluidic techniques for development of 3D vascularized tissue. Biomaterials. 2014;**35**:7308-7325

[39] Memic A, Alhadrami HA, Hussain MA, Aldhahri M, Al Nowaiser F, Al-Hazmi F, et al. Hydrogels 2.0: Improved properties with nanomaterial composites for biomedical applications. Biomedical Materials. 2016;**11**:014104

[40] Jin GZ, Kim M, Shin US, Kim HW. Neurite outgrowth of dorsal root ganglia neurons is enhanced on aligned nanofibrous biopolymer scaffold with carbon nanotube coating. Neuroscience Letters. 2011;**501**:10-14

[41] Zhao B, Hu H, Mandal SK, Haddon RC. A bone mimic based on the self-assembly of hydroxyapatite on chemically functionalized single-walled carbon nanotubes. Chemistry of Materials. 2005;**17**:3235-3241

[42] Hirata E, Menard-Moyon C, Venturelli E, Takita H, Watari F, Bianco A, et al. Carbon nanotubes functionalized with fibroblast growth factor accelerate proliferation of bone marrow-derived stromal cells and bone formation. Nanotechnology. 2013;**24**:435101

[43] Ostrovidov S, Shi X, Zhang L, Liang X, Kim SB, Fujie T, et al. Myotube formation on gelatin nanofibers-multi-walled carbon nanotubes hybrid scaffolds. Biomaterials. 2014;**35**:6268-6277

[44] Cui HF, Vashist SK, Al-Rubeaan K, Luong JH, Sheu FS. Interfacing carbon nanotubes with living mammalian cells and cytotoxicity issues. Chemical Research in Toxicology. 2010;**23**:1131-1147

[45] Vittorio O, Raffa V, Cuschieri A. Influence of purity and surface oxidation on cytotoxicity of multiwalled carbon nanotubes with human neuroblastoma cells. Nanomedicine. 2009;**5**:424-431

[46] Koyama S, Kim YA, Hayashi T, Takeuchi K, Fujii C, Kuroiwa N, et al. In vivo immunological toxicity in mice of carbon nanotubes with impurities. Carbon. 2009;**47**:1365-1372

[47] Raffa V, Ciofani G, Nitodas S, Karachalios T, D'Alessandro D, Masini M, et al. Can the properties of carbon nanotubes influence their internalization by living cells? Carbon. 2008;**46**:1600-1610

[48] Murphy FA, Poland CA, Duffin R, Al-Jamal KT, Ali-Boucetta H, Nunes A, et al. Length-dependent retention of

carbon nanotubes in the pleural space of mice initiates sustained inflammation and progressive fibrosis on the parietal pleura. The American Journal of Pathology. 2011;**178**:2587-2600

[49] Kolosnjaj-Tabi J, Hartman KB, Boudjemaa S, Ananta JS, Morgant G, Szwarc H, et al. In vivo behavior of large doses of ultrashort and full-length single-walled carbon nanotubes after oral and intraperitoneal administration to Swiss mice. ACS Nano. 2010;**4**:1481-1492

[50] Fraczek A, Menaszek E, Paluszkiewicz C, Blazewicz M. Comparative in vivo biocompatibility study of single- and multi-wall carbon nanotubes. Acta Biomaterialia. 2008;**4**:1593-1602

[51] Di Giorgio ML, Di Bucchianico S, Ragnelli AM, Aimola P, Santucci S, Poma A. Effects of single and multi-walled carbon nanotubes on macrophages: Cyto and genotoxicity and electron microscopy. Mutation Research, Genetic Toxicology and Environmental Mutagenesis. 2011;**722**:20-31

[52] Coccini T, Roda E, Sarigiannis DA, Mustarelli P, Quartarone E, Profumo A, et al. Effects of water-soluble functionalized multi-walled carbon nanotubes examined by different cytotoxicity methods in human astrocyte D384 and lung A549 cells. Toxicology. 2010;**269**:41-53

[53] Montes-Fonseca SL, Orrantia-Borunda E, Aguilar-Elguezabal A, Gonzalez Horta C, Talamas-Rohana P, Sanchez-Ramirez B. Cytotoxicity of functionalized carbon nanotubes in J774A macrophages. Nanomedicine. 2012;**8**:853-859

[54] Gutierrez-Praena D, Pichardo S, Sanchez E, Grilo A, Camean AM, Jos A. Influence of carboxylic acid functionalization on the cytotoxic effects induced by single wall carbon nanotubes on human endothelial cells (HUVEC). Toxicology In Vitro. 2011;**25**:1883-1888

[55] Madani SY, Mandel A, Seifalian AM. A concise review of carbon nanotube's toxicology. Nanotechnology Reviews. 2013;**4**:21521

Chapter 13

Quality Control and Risk Management of Carbon Nanomaterials

Khalid Parwez and Suman V. Budihal

Abstract

Our atmosphere contains a substantial number of nanoparticles in which some are unintentionally produced, whereas others are intentionally produced engineered nanoparticle. Among all ENPs, the single-walled and multi-walled carbon nanotubes, spherical fullerenes, and dendrimers are attracting attention for biomedical applications, such as biosensor design, drug delivery, tumor therapy, and tissue engineering. Because of the inert nature of pristine carbon nanotubes (CNTs), it needs to be functionalized to make it reactive with other organic and inorganic materials. The functionalization leads to the addition of functional groups, e.g., C||O, C—O, —OH, and —COOH, to CNTs, which make them dispersible in solvents and suitable for numerous applications. Functionalized CNTs and their composite need to be tested for biocompatibility before real-time applications. Various toxicity mechanisms have been suggested for CNTs, including interference of transmembrane electron transfer, interruption/penetration of the cell envelope, oxidation of cell elements, and formulation of secondary products such as dissolved heavy metal ions or reactive oxygen species (ROS). Numerous studies have insinuated that well-functionalized CNTs are innoxious to animal cells, while raw CNTs or CNTs without functionalization manifest toxicity to cells at even modest dosage.

Keywords: nanotoxicology, carbon nanotubes, in vitro toxicity, in vivo toxicity

1. Introduction

In the past several years, a significant number of studies have been made to study the toxic effects of carbon nanotubes (CNTs). There are variations in the elucidations of these reports, and they mainly depend on the type of nanomaterials as well as functionalization methods. Properly functionalized carbon nanotubes were shown nontoxic to animals conducted by various groups [1–4], whereas raw carbon nanotubes were shown to be toxic to mice lungs in an in vivo study [5–8]. The latest research revealed that non-functionalized, long MWCNTs might be carcinogenic to mice [9]. Pristine nanotubes are indicated to cause oxidative stress and decrease cell viability [10, 11]; however, there is some sign that leftover catalyst particles also contribute to this effect [12]. The cytotoxicity can be decreased to zero via functionalization with a covalently attached polar functional group [13]. Likewise, the toxicity of noncovalently functionalized carbon nanotubes depends on the variety of the functional group. Cells were viable upon internalization of individually encapsulated DNA-wrapped SWCNT complex [14].

Therefore, the toxicity of carbon nanotube depends on the type of functionalization, aggregation behavior, and the presence of metal catalyst particle leftovers during synthesis.

2. An overview of carbon nanotube research

Manufacturing fundamental elements with great strength to weight ratio using carbon nanotube composite is the contemporary focus of the researchers. One of the likely utilizations of polymer nanocomposite is the CNT-augmented ultrafine fiber via electrospinning [15, 16], which has been known since the 1930s. Today, polymer fibers with nanometer diameter can be produced inexpensively using electrospinning technology. With <100 nm diameter, these fibers are being studied for drug delivery methods, energy storage, and improved functional garments [17–19]. These applications require improved (i) fiber strength, (ii) thermal conductivity, and (iii) electrical conductivity. Incorporating carbon nanotubes (CNTs) within electrospun fibers offers the probability of simultaneously improving all these three properties [20–22]. Xie et al. (2005) reviewed the dispersion and alignment of CNTs in the polymer matrix [23]. They found that the serious challenge is the development of means and ways to promote and increase the dispersion and alignment of CNTs in the matrix. Enhanced dispersion of CNTs in the polymer matrix will foster and extend the applications and developments of polymer/CNT nanocomposites.

Though the optical apprehension techniques are probably the most conventional in biology and life sciences, electrochemical or electronic detection techniques have also been adopted in biosensors/biochips due to their great sensitivity, high specificity, and low cost. Those techniques comprise of voltammetric techniques (cyclic voltammetry and differential pulse voltammetry), chronocoulometry, electrochemical impedance spectroscopy, and electronic detection based on electric field [24]. The sensors developed from CNTs have shown the ability to detect a range of analytes such as particular DNA sequences [25] as cancer biomarkers [26] and larger entities such as viruses [27]. These sensor devices have also been used to monitor enzymatic activities and study the behavior of potential drug molecules [28].

The apprehension of the analytes befalls with great specificity and sensitivity in a rationally precise time. Both SWCNT and MWCNT can be altered and conjugated to a bioactive unit and biological varieties including carbohydrates, amino acids and peptides, nucleic acid, and proteins, for various biological applications. Those biological applications are plausible only because the carbon nanotubes own some anomalous properties like the one-dimensional arrangement, large aspect ratio, outstanding mechanical characteristics, and chemical inertness [29]. The carbohydrate-functionalized carbon nanotubes have previously been used for the identification of pathogenic microorganisms, namely, [30, 31]. In the advancement of energy production and storage, nanotubes exhibit exceptional potential in supercapacitors [32], Li-ion batteries [33], solar cells [34], and fuel cells [35]. Energy applications could become the broadest application realm in the gross application of carbon nanotubes. For the advancement of Li-ion batteries' performance, MnO_2 and $LiFePO_4$ are being used as a cathode while MWCNTs and graphene as an anode. In the realm of the fuel cell, proton-exchange membrane fuel cells (PEMFCs), CNTs have been widely studied [36].

In a PEMFC, the conversion of chemical energy to electrical energy occurs via a direct electrochemical reaction, and its efficiency is directly dependent on the catalysts used [37]. The catalysts should have high endurance, low cost, and higher activities in oxygen reduction and/or fuel oxidation reaction [38]. Shortly, the most regularly used catalysts in the PEMFCs are metal NPs, mainly Pt and/or Pt-based alloys, because of high oxygen reduction and/or fuel oxidation reaction due to high Surface area : volume ratio and better Fermi levels for redox reactions [39]. Nonetheless, metal NPs are generally unstable and lose their catalytic activity due to their irreversible aggregation during the electrochemical processes.

Consequently, appropriate methods are obliged to fix and restrict these metal NPs from aggregation, e.g., carbon nanotubes (CNTs) are the most extensively adopted provision in modern development. Though the evolution of PEMFCs is under commercialization process, obstructions including how the CNTs influence the catalytic action of the metal/CNTs and high material cost continue. The evolution of numerous profoundly dynamic catalysts with the economical price for fuel cell commercialization would be one of the notable researches in this domain. Because of the enhanced production and intended use of CNTs in consumer commodities, there is a necessity for evaluation of the implied toxicity of these nanoparticles.

3. Toxicity studies of carbon nanotubes in vivo

In vivo toxicity knowledge impersonated a vital role in risk evaluation. Those techniques can be applied to determine acute toxicity, chronic toxicity, developmental toxicity, genotoxicity, and reproductive toxicity. In vivo study is indispensable in the fields of medicine including cancer therapy. Several animal trials are performed to highlight the possible serious impressions of newly formed medicines and chemical substances on the human. In some trials, researchers attempt to simulate situations concerning humans (e.g., arthritis, cystic fibrosis, and cancer) in animals, to assess the capabilities of new medicines in treatment. To inscribe the potential side effects of CNTs on human health and environment, animal models have been used to investigate the toxicity of CNTs. Non-functionalized CNTs were instilled intratracheally (IT) into animals, exhibited as pulmonary toxicity including inflammation and fibrotic responses due to the collection of raw CNTs in the lung airways [5].

These outcomes suggest that aerosol vulnerability of untreated CNTs in the workplace should be shunned to preserve human health. Notwithstanding, intratracheal instillation of functionalized soluble CNTs has little inference to the toxicology profile. In the latest pilot study, asbestos-like pathogenicity was observed by Poland et al. [9] when the mesothelial lining of the body cavity of mice was exposed to large MWCNTs of 80–160 nm diameter and 10–50 nm length [9]. Yet the assumption of this finding for probable negative effects of CNTs on human health is inadequate. It should be heeded that the MWCNT materials utilized in this research were just sonicated in bovine serum albumin (BSA) without surface functionalization.

Furthermore, no noticeable toxic result was observed for smaller and tinier MWCNTs of 1–20 nm length and 10–14 nm diameter, appreciating that the toxicology characterizations of CNTs may vary between CNTs of varying sizes. It is deserving asserting that functionalized SWCNTs utilized in biomedical research have a length of 50–300 nm and diameter of 1–2 nm, which is completely distinct from the geometry of MWCNTs adopted by Poland et al. [9]. Gambhir and colleague applied covalently and noncovalently PEGylated SWCNTs to investigate the in vivo toxicity [3]. The PEGylated SWCNTs (-3 mg kg^{-1}) were intravenously infused into mice and inspected over 4 months. Systolic blood pressure, total blood counts, and serum chemistry are registered every month. Necropsy and tissue histology analyses were executed at the completion of 4 months. The blood chemistry and histological investigations were normal. Those experiments insinuate that functionalized biocompatible SWCNTs may be secured for in vivo biological reinforcements. An added investigation revealed related outcomes, confirming that PEGylated SWCNTs are gradually eliminated from the body after systemic administration in mouse models, without manifesting apparent toxicity in the system [40]. Yang et al. acknowledged that SWCNTs dangled in Tween-80 manifested lesser toxicities to the experimented mice at a high dose of −40 mg kg^{-1}, following intravenous inoculation for 3 months [41]. Toxicity may be due to the oxidative stress engendered by SWCNTs assembled in the liver and lungs of mice [42]. The toxicity published was dose-dependent and appeared to be less acceptable at lower doses. A current article by the same group unveiled that covalently PEGylated SWCNTs displayed an ultra-long blood dissemination half-life in rodents. Though the long-term toxicity of altered SWCNTs is still to be investigated, no critical toxicity has been recorded too at a higher dose of 24 mg kg^{-1}.

4. Respiratory toxicity

A guinea pig was inoculated intratracheally with the soot of CNT. Breathing rate, tidal capacity, pulmonary obstruction, bronchoalveolar fluid, and protein content were estimated. The authors admitted that working with soot-carrying CNT was probably not a health jeopardy, but they did not present their pathological investigation [43]. Research in mice is conducted by Lam et al. [5], and they authenticated that SWCNT could be toxic if they entered the lungs; Warheit et al. [6] conveyed a related investigation in rats, reporting the granuloma development apparently due to the collection of CNT. Muller et al. analyzed carbon black, MWCNT, and asbestos influences, implanted in the trachea of rodents. Scholars demonstrated dose-dependent inflammation, and granuloma production, increased considerably with MWCNT than with carbon black than asbestos. The early granulomatous reaction, abnormal acute inflammatory response, and progressive fibrosis were observed upon exposure of SWCNT in mice. Pharyngeal aspiration was used alternately of the intratracheal instillation used in the earlier investigations and rendered aerosolization of fine SWNCT particles. Another contemporary study insinuates shifts in deposition prototype and pulmonary response when SWCNT is uniformly dispersed in the suspension antecedent to pharyngeal aspiration [44]. Current research insinuates MWCNT immigration to the subpleural and associated pleural mononuclear cells and subpleural fibrosis in mice upon inhalation [45] and further admonition, and decent security models are prescribed when manipulating CNT. Research by [46] confirms the earlier reports; they characterized in vitro and in vivo stimulation of collagen deposition, lung fibroblast propagation, and metalloproteinase intensified expression without inflammation when dispersed SWCNT was applied. Following inhalation, the different variety of nanoparticles may enter the central nervous system (CNS) [47] by a method called transcytosis [48]. The investigation unveiled that sniffed gold nanoparticles aggregate in the olfactory tubercle of rats and enter the cerebral cortex, lung, and the distinct organs such as the tongue, esophagus, kidney, spleen, aorta, septum, heart, and blood [49]. Those remarks vindicate that nanoparticles can infiltrate into the CNS via the olfactory venation if they are being in high doses in the air. Those nanoparticles may impact not only on the respiratory tract and neighboring organs but disseminated to remote organs.

5. Bio-distribution of carbon nanotubes

Knowledge of bio-distribution of CNTs following systemic inoculation inside animals is a pretty serious concern. Numerous investigators investigated in vivo bio-distribution and pharmacokinetic investigations in the preceding several years. Scientist adopted various CNT materials, different surface functionalization methods, and various tracking methodologies. Consequently, they got unsteady and seldom ambiguous results. Singh et al. and Lacerda et al. utilized radiolabeled (1n-DTPA) SWCNTs and MWCNTs to describe bio-distribution [50, 51]. Exceptionally, following intravenous inoculation of CNTs into mice, no uptake in the reticuloendothelial system (RES) such as the liver and spleen was witnessed. But, quick urinal removal of CNTs was witnessed. More than ninety five percent of CNTs were removed within 3 hours. Those results are comparable to the in vivo response of minute particles yet distinct from that prognosticated of maximum nanoparticles with sizes exceeding the glomerular filtration threshold. To defend their conclusions, the researchers stated that the short diameters of CNTs were eliminated in urine notwithstanding they were large in length. However, this theory is unsettled. For example, for the protein bio-distribution and elimination function of quantum dots (QDs), published by Choi et al., it is observed that the 6 nm maximum size of spherical QDs including coatings was obliged to fast urinal elimination. Nevertheless, the QDs are much shorter than the diameter of SWCNT bundles (10–40 nm) or MWCNTs (20–80 nm) [51] practiced in those bio-distribution investigations. Therefore, the inscribed fast urinal excretion of CNTs requires validation. Various other labs have also assessed the bio-distribution of radiolabeled CNTs in rodents. Wang et al. noticed delayed urinal elimination and weak RES uptake in their primary research. But, consecutive articles by the same association utilizing ^{14}C-taurine-functionalized CNTs recorded steadfast liver accumulation of CNTs following intravenous inoculation [52]. Research carried out by McDevitt et al., utilizing antibody-conjugated radiolabeled CNTs functionalized by 1,3-dipolar cycloaddition, also confirmed delayed urinal excretion and high CNT uptake in the liver and spleen [53]. The bio-distribution investigations of radiolabeled, PEGylated SWCNTs unveiled uptake of SWCNT in RES organs without active clearance [54]. A substantial quantity of CNTs is persisting, even after 15 days. The radiolabel system is a proper technique to identify the bio-distribution of material but may commence to inaccurate outcomes, if excess-free radioisotopes in the radiolabeled CNT specimens are not separated effectively. The free radioisotopes are tiny particles that could be quickly excreted in urine following intravenous inoculation. Moreover, radiolabels could be undeviatingly released from CNTs in vivo and be regularly eliminated in the free form. Consequently, radiolabeling is not an excellent approach to investigate the elimination and long-term predestination of CNTs. The expert has discovered that photoluminescence is the inherent characteristics of CNTs. Cherukuri et al. used single semiconducting SWCNTs which show NIR photoluminescence, to trace nanotubes in rabbits [55]. Without obtaining complete bio-distribution data, the expert could not testify SWCNT photoluminescence signals in every organ besides the liver. Yang et al. carried out the research to comprehend the bio-distribution of ^{13}C-fortified unfunctionalized SWCNTs over a month utilizing isotope ratio mass spectroscopy [56]. The event conferred unusual nanotube uptake in the liver, lung, and spleen without notable elimination within 28 days. Raman spectroscopy has been applied to analyze the long-term predestination of PEGylated nanotubes in rodents. It was reported that most of the PEGylated SWCNTs were assembled in the liver and spleen following intravenous inoculation but gradually eliminated through the biliary pathway toward the feces within months. A low SWCNT Raman signal was also identified in the mouse kidney and bladder. It is unveiled that little portion of SWCNTs with short lengths was eliminated into the urine.

6. Toxicity of carbon nanotubes in vitro

In vitro, toxicological investigations are a highly significant means for nanotoxicology, corresponding to in vivo investigations because of moderate expense, lessening ethical anxieties, and diminishing the number of laboratory animals needed for trial.

7. CNT toxicity investigations in animal cell lines

The subject of carbon nanotube toxicity is still unresolved even in cell culture experiments. Inhibition of HEK 293 cell proliferation following exposure to SWCNTs [10], MWCNTs, inducing cell cycle arrest and increasing apoptosis/necrosis of human skin fibroblasts were examined by different research groups [57]. Nevertheless, it is worth stating that functionalized CNTs were not used in those investigations. Bottini et al. observed T-lymphocyte apoptosis evoked by oxidized MWCNTs [58]. Because simple oxidation, used in these studies, is not enough to disperse carbon nanotubes in saline and cell culture media since it is not a kind of biocompatible functionalization. Sayes et al. indicated that toxicity of CNTs was also dependent on the density of functionalization. Inconsiderable toxicity was observed for those functionalized with the high density of phenyl-SO3X groups [13]. These results are understandable because CNTs without proper functionalization carry a highly hydrophobic surface. Consequently, they may aggregate in the cell culture medium. The aggregation of CNT channels to binding of several biological species, including proteins, via hydrophobic interactions, provokes cell toxicity. Khalid et al. reported no toxicity of functionalized MWCNT to Saos cell lines up to the tested concentration of 1000 μg/mL [59]. Other factors like surfactants may also play a role in the noted toxicity of CNT in vitro. Extra surfactants, present in the CNT suspensions, are known to be highly toxic to cells [60]. The metal catalyst, used during the synthesis of CNTs, should also be examined as an important factor when the toxicity of carbon nanotubes is analyzed [61]. Furthermore, proper analytical methods must be hired in toxicity analysis to prevent interference of carbon nanotubes with the test reagents [62]. Davoren et al. reported concentration-dependent cytotoxicity of SWCNT on a lung carcinoma cell line (A549) [63]. Another study, led by Sharma, unveiled that SWCNT induced oxidative stress in rat lung cells [64]. Herzog et al. reported the same oxidative stress linked to alterations in primary bronchial epithelial cells and A549 cells, but the study also revealed that the reaction is strongly dependent on the dispersion medium used [65]. Pulskamp used two cells lines (human A549 and rat macrophages NR8383) and tested with CNTs and revealed, as oxidative stress was provoked to these cell lines. However, when purified SWCNT corresponded with commercial CNT, it is unveiled that all the biological consequences are associated with the metal traces. There is a complicated result between WST (2-(4-iodopheny1)-3-(4-nitropheny1)-5-(2,4-disulfopheny1)-2H-tetrazolium, monosodium salt) and MTT (344, 5-dimethylthiazol-2-yl)-2,5-diphenyltetrazolium bromide) viability assays. These dyes depend on the mitochondrial dehydrogenase activity [66]. The modifications can only be described based on associations of CNT with non-soluble formazan crystals in MTT. That is why suitable assay methods and well-characterized materials are the most important requirements for in vitro toxicity assays of carbon nanotubes.

8. CNT toxicity investigations in bacteria and yeast cells

As an option to animal cell lines, bacteria and yeast can be a relevant model for studying how single-celled microorganisms react to the environmental stressors such as CNTs [67]. Copious toxicity mechanisms have been suggested for CNT including interruption/penetration of the cell envelope, oxidation of cell ingredients, the arrest of transmembrane electron transfer, and generation of secondary products such as reactive oxygen species (ROS) or dissolved heavy metal ions [68]. Toxicity of a CNT is depending on its structure along with its geometry and surface functionalization. Various researches have shown that adequately functionalized, serum-stable CNTs are innoxious to animal cells, whereas CNTs without functionalization seemed critically toxic to human or animal cell lines at the moderate dosage [2]. The SWCNT displays a potent antimicrobial response for both suspended and deposited bacteria and interrupts the accumulation of bacterial films. The immediate interaction among the SWCNT and bacteria is apparently the central cause to induce cell death [68]. Well-dispersed individual SWCNT is more toxic than agglomerates due to greater physical puncturing of bacterial membranes and impairs the cell integrity [69]. The CNT bacteria interplay is determined by surface functionalization and length of CNT. It may govern the toxic effect also. A negatively charged or neutral SWCNT functionalized with -OH or -COOH aggregates more efficiently with bacteria and diminishes bacteria viability as contrasted to the positively charged SWCNTs functionalized with -NH_2 [70]. Likewise, longer SWCNTs exhibited concentration and time-dependent toxicity to bacteria, whereas short SWCNTs were limited toxically as they aggregate themselves [71]. The purity of SWCNTs may also influence bacterial toxicity. Pure SWCNTs were observed to be less toxic than SWCNTs with higher metal content due to glutathione oxidation following contact [72]. Additionally, greater ionic strength suspensions, such as phosphate buffered saline (PBS) or brain heart infusion broth, also lessen SWCNT toxicity due to decreased intensity of interactions between SWCNT and cells, compared to low ionic strength suspensions (deionized water or saline). Likewise, a film with natural organic matter (NOM) limits SWCNT toxicity, notwithstanding diminished aggregation [73]. Other studies unveiled that SWCNT reduces enzyme activity and microbial biomass at concentration 300 mg kg^{-1} and above [74]. As it is clear that SWCNT provokes bacterial death, a surface coating with SWCNT would decrease biofilm expansion in both real and industrial settings [75]. The MWCNT appears to be runty toxic to bacteria as contrasted to SWCNT [76]. The decreased toxicity may be due to minor interactions among bacteria and MWCNT. The limited interaction might be due to the greater rigidity and presumably inferior van der Waal's forces at the MWCNT surface. Thin MWCNT with less diameter exhibits greater toxicity to bacteria corresponding to larger ones [77]. When the consequence of the length of MWCNT was estimated, shorter MWCNTs were extra toxic to *Pseudomonas fluorescens* compared to long MWCNT [78]. When MWCNTs are uncapped, debundled, and dispersed in solution, the toxicity to bacteria raised [79]. The purity of CNT has also been vindicated to influence the toxicity in microorganisms. Furthermore, when the toxicity within pristine and purified MWCNT was studied in two bacterial strains (*Escherichia coli* and *Cupriavidus metallidurans*), no variation in toxicity of MWCNT was perceived between the two forms [80]. Heating refinement of CNTs presumably has the inadequate capability to modify the surface corresponding to acid processing, consequently sustaining toxicity of the raw form. However, in both the investigations, gum arabic (GA, 0.25 wt%) was used to suspend CNTs, which might have altered the surface, influencing toxicity. Meanwhile in soil toxicity assay, MWCNT, reduced microbial biomass and enzyme activity at concentration 5000 mg kg^{-1} [81]. In a separate research, the conidia of the fungi *Paecilomyces fumosoroseus* were incubated for 865 hours with 0.2 mg L^{-1} raw and/or carboxylated MWCNT. Mycelium growth on solid medium was witnessed following incubation. Association among the fungi and CNTs had no notable effect on germination and biomass production, but the loss of biomass was witnessed following exposure to raw MWCNT for 865 hours [82]. Mechanical impacts of CNT, as observed in bacteria, might have caused the effects.

9. Ecotoxicity of carbon nanotubes

As the production and widespread application of CNTs in industrial and customer products are progressing, the release of this nanomaterial into the environment too will scale up. Many scientific reviews have evaluated the sources, behavior, fate, and the mechanisms of toxicity of carbon nanomaterial. Maximum of these assessments apprehended that additional research is obligatory in the field of nano-ecotoxicology (**Table 1**).

Organism tested	Types of CNTs	LOEC	EC 50	Mechanism of toxicity	References
Chlorella vulgaris	Pristine CNT	0.053 mg L^{-1}	1.8 mg L^{-1}	Oxidative stress, agglomeration and physical interactions	[83]
	MWCNT of diameter 10, 20-40, and 60-100 IIIII	NA	41.0,12.7, and 12.4 mg L^{-1}, respectively	Oxidative stress, agglomeration and physical interactions	[84]
Pseudokirchneriella subcapitata	Pristine CNT	0.053 mg L^{-1}	2.5 mg L^{-1}	Oxidative stress, agglomeration and physical interactions	[83]
	SWCNT	0.25 mg L	NA	Oxidative stress, agglomeration and physical interactions	[85]
Thalassiosira pseudonanas	DWCNTs	0.1 mg L^{-1}	1.86 mg L^{-1}	Oxidative stress, agglomeration and physical interactions	[86]
Dunaliella tertiolecta	MWCNT	NA	0.8 mg L^{-1}	Oxidative stress and photosynthesis inhibition	[87]
Tetrahymena thermophila	SWCNT	1.6 mg L^{-1}	NA	Physical interactions	[88]
Stylonychia mytilus	Functionalized MWCNT	1 mg L^{-1}	NA	Physical interactions	[89]
attenuata	SWCNT	1.10 mg L^{-1}	*H y d r a* NA	Physical interactions	[89]
Daphnia magna	SWCNT (60% pure)	NA	1.3 mg L^{-1}	Physical interactions	[90]
	MWCNT resuspended in NOM	NOEC 20 mg L-1	NA	No toxicity	[91]
	MWCNT grafted with polyethylenimine	NA	25 mg L^{-1}	Increased size of the surface coating	[92]
Ceriodaphnia . dubia	MWCNT resuspended in NOM	0.25 mg L^{-1}	NA	Agglomeration	[93]
Danio rerio embryo	SWCNT	120 mg L^{-1}	NA	Agglomeration	[94]
Oryzias melastigma	DWCNT	10 mg L^{-1}	NA	Agglomeration	[86]
Xenopus leavis larvae	DWCNT	10 mg L^{-1}	NA	Physical interactions	[95]
Drosophila melanogaster	SWCNT and MWCNT in 1 g kg^{-1} food			No toxicity	[96]
Female Fisher rats	Oral gavage of 0.64 mg ka^{1} SWCNT	NA	NA	Increased levels of oxidative damage to DNA in liver and lung tissue	[97]
Sprague—Dawleyrat	1000mg kg^{-1} of SWCNT from gestation day 6 to 19	NA	NA	No teratogenicity	[98]

Abbreviations: LOEC: Least observable effect concentration, EC 50: Effective concentration 50, NOEC: No observed effect concentration, NOM: Natural organic matter.

Table 1.

Summary of the studies related to eco-toxicity of the CNTs on different organisms [82–98].

10. Conclusions

Toxicity of carbon nanomaterials is an essential concern in the modern world for the scientific community, environmentalists, and governments. The chances of exposure to the environment are more like the application of carbon nanomaterial which is increasing every day. Some research shows the different toxicity patterns for the materials when it is exposed to living cells in vitro or in vivo, whereas other studies say that the adequately functionalized bearing carboxylic or hydroxyl group and serum-stable CNTs are safe for living cells. I want to conclude my discussion by highlighting the factors involved in toxicity and the toxicity mechanism. The toxicity of the nanomaterials depends on many factors including functionalization, catalyst, size, shape, dimensions, dispersion, and methods used for detecting toxicity. The pristine carbon nanotubes are more damaging to the cells than the functionalized one. The covalently functionalized CNTs are more compatible for the cells than non-covalent functionalization. The catalyst used during the production of the nanotubes like platinum or iron also contributes to the toxicity of the cells. Hence it is imperative to differentiate the toxicity of carbon nanotubes and catalyst. Dispersion in the high ionic strength solvent like PBS makes the CNTs more compatible with living cells compared to the less ionic strength solvent like deionized water. Hence it is always recommended to prepare the solution in PBS or other high ionic strength solvents for better compatibility and less toxicity. The short and broken CNTs with a small diameter are observed to be damaging to bacterial cells because of physical puncturing. In vivo studies help us to understand the acute toxicity, chronic toxicity, developmental toxicity, genotoxicity, and reproductive toxicity of CNTs in laboratory animals. No critical acute toxicity, chronic toxicity, developmental toxicity, genotoxicity, and reproductive toxicity are observed following intravenous or intratracheal instillation of CNTs. Adequately functionalized CNTs are biocompatible and promptly eliminated through urine or biliary pathway following intravenous inoculation. Pharmacokinetic studies of CNTs show very less or no uptake of CNTs to the reticuloendothelial systems including the liver, lung, and spleen. Various mechanisms are also listed to study the toxicity of the CNTs to the living cells which includes oxidation of cell components, arrest of electron transport chain, reactive oxygen species, and physical puncturing of the cell. Further studies need to be conducted in the field of eco-toxicity of CNTs and validation of the toxicological data for the safety of aquatic and aerial animals. These studies shall help the public regulatory organization to frame a rule for ensuring the safety of this modern engineered nanoparticle.

Author details

Khalid Parwez[1*] and Suman V. Budihal[2]

1 Department of Clinical Laboratory Science, Faculty of Applied Medical Science, Shaqra University (SU), Al-Dawadmi, Kingdom of Saudi Arabia

2 Department of Physiology, Kasturba Medical College, Manipal Academy of Higher Education, Manipal, Karnataka, India

*Address all correspondence to: drkhalidpm@gmail.com

IntechOpen

© 2019 The Author(s). Licensee IntechOpen. This chapter is distributed under the terms of the Creative Commons Attribution License (http://creativecommons.org/licenses/by/3.0), which permits unrestricted use, distribution, and reproduction in any medium, provided the original work is properly cited.

References

[1] Shi Kam NW, Jessop TC, Wender PA, Dai H. Nanotube molecular transporters: Internalization of carbon nanotube–protein conjugates into mammalian cells. Journal of the American Chemical Society. 2004;**126**(22):6850-6851

[2] Dumortier H, Lacotte S, Pastorin G, Marega R, Wu W, Bonifazi D, et al. Functionalized carbon nanotubes are non-cytotoxic and preserve the functionality of primary immune cells. Nano Letters. 2006;**6**(7):1522-1528

[3] Schipper ML, Nakayama-Ratchford N, Davis CR, Kam NW, Chu P, Liu Z, et al. A pilot toxicology study of single-walled carbon nanotubes in a small sample of mice. Nature Nanotechnology. 2008;**3**(4):216

[4] Wu P, Chen X, Hu N, Tam UC, Blixt O, Zettl A, et al. Biocompatible carbon nanotubes generated by functionalization with glycodendrimers. Angewandte Chemie. 2008;**120**(27):5100-5103

[5] Lam CW, James JT, McCluskey R, Hunter RL. Pulmonary toxicity of single-wall carbon nanotubes in mice 7 and 90 days after intratracheal instillation. Toxicological Sciences. 2004;**77**(1):126-134

[6] Warheit DB, Laurence BR, Reed KL, Roach DH, Reynolds GA, Webb TR. Comparative pulmonary toxicity assessment of single-wall carbon nanotubes in rats. Toxicological Sciences. 2004;**77**(1):117-125

[7] Shvedova AA, Kisin ER, Mercer R, Murray AR, Johnson VJ, Potapovich AI, et al. Unusual inflammatory and fibrogenic pulmonary responses to single walled carbon nanotubes in mice. American Journal of Physiology. Lung Cellular and Molecular Physiology. 2005;**289**(5):698-708

[8] Muller J, Huaux F, Moreau N, Misson P, Heilier JF, Delos M, et al. Respiratory toxicity of multi-wall carbon nanotubes. Toxicology and Applied Pharmacology. 2005;**207**(3):221-231

[9] Poland CA, Duffin R, Kinloch I, Maynard A, Wallace WA, Seaton A, et al. Carbon nanotubes introduced into the abdominal cavity of mice show asbestos-like pathogenicity in a pilot study. Nature Nanotechnology. 2008;**3**(7):423

[10] Cui D, Tian F, Ozkan CS, Wang M, Gao H. Effect of single wall carbon nanotubes on human HEK293 cells. Toxicology Letters. 2005;**155**(1):73-85

[11] Manna SK, Sarkar S, Barr J, Wise K, Barrera EV, Jejelowo O, et al. Single-walled carbon nanotube induces oxidative stress and activates nuclear transcription factor-κB in human keratinocytes. Nano Letters. 2005;**5**(9):1676-1684

[12] Kagan VE, Tyurina YY, Tyurin VA, Konduru NV, Potapovich AI, Osipov AN, et al. Direct and indirect effects of single walled carbon nanotubes on RAW 264.7 macrophages: Role of iron. Toxicology Letters. 2006;**165**(1):88-100

[13] Sayes CM, Liang F, Hudson JL, Mendez J, Guo W, Beach JM, et al. Functionalization density dependence of single-walled carbon nanotubes cytotoxicity in vitro. Toxicology Letters. 2006;**161**(2):135-142

[14] Kam NW, O'Connell M, Wisdom JA, Dai H. Carbon nanotubes as multifunctional biological transporters and near-infrared agents for selective cancer cell destruction. Proceedings of the National Academy of Sciences of the United States of America. 2005;**102**(33):11600-11605

[15] Khalid P, Hussain MA, Suman VB, Arun AB. Toxicology of carbon

nanotubes—A review. International Journal of Applied Engineering Research. 2016;**11**(1):148-157

[16] Formhals A. Process and apparatus for preparing artificial threads. US patent. 2 October 1934;**1**(975): 504

[17] Reneker DH, Chun I. Nanometre diameter fibres of polymer, produced by electrospinning. Nanotechnology. 1996;**7**(3):216

[18] Kenawy ER, Bowlin GL, Mansfield K, Layman J, Simpson DG, Sanders EH, et al. Release of tetracycline hydrochloride from electrospun poly(ethylene-co-vinylacetate), poly(lactic acid), and a blend. Journal of Controlled Release. 2002;**81**(1-2):57-64

[19] Laxminarayana K, Jalili N. Functional nanotube-based textiles: Pathway to next generation fabrics with enhanced sensing capabilities. Textile Research Journal. 2005;**75**(9):670-680

[20] Ko F, Gogotsi Y, Ali A, Naguib N, Ye H, Yang GL, et al. Electrospinning of continuous carbon nanotube-filled nanofiber yarns. Advanced Materials. 2003;**15**(14):1161-1165

[21] Dror Y, Salalha W, Khalfin RL, Cohen Y, Yarin AL, Zussman E. Carbon nanotubes embedded in oriented polymer nanofibers by electrospinning. Langmuir. 2003;**19**(17):7012-7020

[22] Lim JY, Lee CK, Kim SJ, Kim IY, Kim SI. Controlled nanofiber composed of multi-wall carbon nanotube/poly(ethylene oxide). Journal of Macromolecular Science, Part A: Pure and Applied Chemistry. 2006;**43**(4-5):785-796

[23] Xie XL, Mai YW, Zhou XP. Dispersion and alignment of carbon nanotubes in polymer matrix: A review. Materials Science & Engineering R: Reports. 2005;**49**(4):89-112

[24] Cai H, Cao X, Jiang Y, He P, Fang Y. Carbon nanotube-enhanced electrochemical DNA biosensor for DNA hybridization detection. Analytical and Bioanalytical Chemistry. 2003;**375**(2):287-293

[25] Tu X, Manohar S, Jagota A, Zheng M. DNA sequence motifs for structure-specific recognition and separation of carbon nanotubes. Nature. 2009;**460**(7252):250

[26] Thakare VS, Das M, Jain AK, Patil S, Jain S. Carbon nanotubes in cancer theragnosis. Nanomedicine. 2010;**5**(8):1277-1301

[27] Patolsky F, Zheng G, Hayden O, Lakadamyali M, Zhuang X, Lieber CM. Electrical detection of single viruses. Proceedings of the National Academy of Sciences of the United States of America. 2004;**101**(39):14017-14022

[28] Prato M, Kostarelos K, Bianco A. Functionalized carbon nanotubes in drug design and discovery. Accounts of Chemical Research. 2007;**41**(1):60-68

[29] Endo M, Strano MS, Ajayan PM. Potential applications of carbon nanotubes. In: Carbon Nanotubes. Berlin, Heidelberg: Springer; 2007. pp. 13-62

[30] Elkin T, Jiang X, Taylor S, Lin Y, Gu L, Yang H, et al. Immuno-carbon nanotubes and recognition of pathogens. ChemBioChem. 2005;**6**(4):640-643

[31] Wang H, Gu L, Lin Y, Lu F, Meziani MJ, Luo PG, et al. Unique aggregation of anthrax (*Bacillus anthracis*) spores by sugar-coated single-walled carbon nanotubes. Journal of the American Chemical Society. 2006;**128**(41):13364-13365

[32] Kim YJ, Kim YA, Chino T, Suezaki H, Endo M, Dresselhaus MS. Chemically

modified multiwalled carbon nanotubes as an additive for supercapacitors. Small. 2006;**2**(3):339-345

[33] Leroux F, Metenier K, Gautier S, Frackowiak E, Bonnamy S, Beguin F. Electrochemical insertion of lithium in catalytic multi-walled carbon nanotubes. Journal of Power Sources. 1999;**81**:317-322

[34] Landi BJ, Castro SL, Ruf HJ, Evans CM, Bailey SG, Raffaelle RP. CdSe quantum dot-single wall carbon nanotube complexes for polymeric solar cells. Solar Energy Materials and Solar Cells. 2005;**87**(1-4):733-746

[35] Wang C, Waje M, Wang X, Tang JM, Haddon RC, Yan Y. Proton exchange membrane fuel cells with carbon nanotube based electrodes. Nano Letters. 2004;**4**(2):345-348

[36] Kamavaram V, Veedu V, Kannan AM. Synthesis and characterization of platinum nanoparticles on in situ grown carbon nanotubes based carbon paper for proton exchange membrane fuel cell cathode. Journal of Power Sources. 2009;**188**(1):51-56

[37] Wang X, Waje M, Yan Y. CNT-based electrodes with high efficiency for PEMFCs. Electrochemical and Solid-State Letters. 2005;**8**(1):A42-A44

[38] Kundu S, Nagaiah TC, Xia W, Wang Y, Dommele SV, Bitter JH, et al. Electrocatalytic activity and stability of nitrogen-containing carbon nanotubes in the oxygen reduction reaction. The Journal of Physical Chemistry C. 2009;**113**(32):14302-14310

[39] Li W, Liang C, Qiu J, Zhou W, Han H, Wei Z, et al. Carbon nanotubes as support for cathode catalyst of a direct methanol fuel cell. Carbon. 2002;**40**(5):787-790

[40] Liu Z, Davis C, Cai W, He L, Chen X, Dai H. Circulation and long-term fate of functionalized, biocompatible single-walled carbon nanotubes in mice probed by Raman spectroscopy. Proceedings of the National Academy of Sciences of the United States of America. 2008;**105**(5):1410-1415

[41] Yang ST, Wang X, Jia G, Gu Y, Wang T, Nie H, et al. Long-term accumulation and low toxicity of single-walled carbon nanotubes in intravenously exposed mice. Toxicology Letters. 2008;**181**(3):182-189

[42] Yang ST, Fernando KS, Liu JH, Wang J, Sun HF, Liu Y, et al. Covalently PEGylated carbon nanotubes with stealth character in vivo. Small. 2008;**4**(7):940-944

[43] Huczko A, Lange H, Całko E, Grubek-Jaworska H, Droszcz P. Physiological testing of carbon nanotubes: Are they asbestos-like? Fullerene Science and Technology. 2001;**9**(2):251-254

[44] Mercer RR, Scabilloni JF, Wang L, Kisin ER, Murray AR, Schwegler-Berry D, et al. Alteration of deposition pattern and pulmonary response as a result of improved dispersion of aspirated single walled carbon nanotubes in a mouse model. American Journal of Physiology. Lung Cellular and Molecular Physiology. 2008;**294**(1):87-97

[45] Ryman-Rasmussen JP, Cesta MF, Brody AR, Shipley-Phillips JK, Everitt JI, Tewksbury EW, et al. Inhaled carbon nanotubes reach the subpleural tissue in mice. Nature Nanotechnology. 2009;**4**(11):747

[46] Wang L, Mercer RR, Rojanasakul Y, Qiu A, Lu Y, Scabilloni JF, et al. Direct fibrogenic effects of dispersed single-walled carbon nanotubes on human lung fibroblasts. Journal of Toxicology and Environmental Health, Part A. 2010;**73**(5-6):410-422

[47] Elder A, Gelein R, Silva V, Feikert T, Opanashuk L, Carter J, et al. Translocation of inhaled ultrafine manganese oxide particles to the central nervous system. Environmental Health Perspectives. 2006;**114**(8):1172-1178

[48] Zensi A, Begley D, Pontikis C, Legros C, Mihoreanu L, Wagner S, et al. Albumin nanoparticles targeted with Apo E enter the CNS by transcytosis and are delivered to neurones. Journal of Controlled Release: Official Journal of the Controlled Release Society. 2009;**137**(1):78-86

[49] Yu LE, Lanry Yung LY, Ong CN, Tan YL, Suresh Balasubramaniam K, Hartono D, et al. Translocation and effects of gold nanoparticles after inhalation exposure in rats. Nanotoxicology. 2007;**1**(3):235-242

[50] Singh R, Pantarotto D, Lacerda L, Pastorin G, Klumpp C, Prato M, et al. Tissue biodistribution and blood clearance rates of intravenously administered carbon nanotube radiotracers. Proceedings of the National Academy of Sciences of the United States of America. 2006;**103**(9):3357-3362

[51] Lacerda L, Soundararajan A, Singh R, Pastorin G, Al-Jamal KT, Turton J, et al. Dynamic imaging of functionalized multi-walled carbon nanotube systemic circulation and urinary excretion. Advanced Materials. 2008;**20**(2):225-230

[52] Deng XY, Yang ST, Nie HY, Wang HF, Liu YF. A generally adoptable radiotracing method for tracking carbon nanotubes in animals. Nanotechnology. 2008;**19**(7):75101

[53] McDevitt MR, Chattopadhyay D, Kappel BJ, Jaggi JS, Schiffman SR, Antczak C, et al. Tumor targeting with antibody functionalized, radiolabeled carbon nanotubes. Journal of Nuclear Medicine. 2007;**48**:1180-1189

[54] Liu Z, Cai WB, He LN, Nakayama N, Chen K, Sun XM, et al. In vivo biodistribution and highly efficient tumour targeting of carbon nanotubes in mice. Nature Nanotechnology. 2007;**2**(1):47-52

[55] Cherukuri P, Gannon CJ, Leeuw TK, Schmidt HK, Smalley RE, Curley SA, et al. Mammalian pharmacokinetics of carbon nanotubes using intrinsic near infrared fluorescence. Proceedings of the National Academy of Sciences of the United States of America. 2006;**103**(50):18882-18886

[56] Yang ST, Guo W, Lin Y, Deng XY, Wang HF, Sun HF, et al. Biodistribution of pristine single-walled carbon nanotubes in vivo. Journal of Physical Chemistry C. 2007;**111**(48):17761-17764

[57] Ding LH, Stilwell J, Zhang TT, Elboudwarej O, Jiang HJ, Selegue JP, et al. Molecular characterization of the cytotoxic mechanism of multiwall carbon nanotubes and nano-onions on human skin fibroblast. Nano Letters. 2005;**5**(12):2448-2464

[58] Bottini M, Bruckner S, Nika K, Bottini N, Bellucci S, Magrini A, et al. Multi-walled carbon nanotubes induce T lymphocyte apoptosis. Toxicology Letters. 2006;**160**(2):121-126

[59] Khalid P, Hussain MA, Rekha PD, Arun AB. Carbon nanotube-reinforced hydroxyapatite composite and their interaction with human osteoblast in vitro. Human & Experimental Toxicology. 2014;**33**(9):897-990

[60] Dong L, Joseph KL, Witkowski CM, Craig MM. Cytotoxicity of single-walled carbon nanotubes suspended in various surfactants. Nanotechnology. 2008;**19**(25):255702

[61] Plata DL, Gschwend PM, Reddy CM. Industrially synthesized single walled carbon nanotubes: Compositional data for users, environmental risk assessments and

source apportionment. Nanotechnology. 2008;**19**(18):185706

[62] Casey A, Herzog E, Davoren M, Lyng FM, Byrne HJ, Chambers G. Spectroscopic analysis confirms the interactions between single walled carbon nanotubes and various dyes commonly used to assess cytotoxicity. Carbon. 2007;**45**(7):1425-1432

[63] Davoren M, Herzog E, Casey A, Cottineau B, Chambers G, Byrne HJ, et al. In vitro toxicity evaluation of single walled carbon nanotubes on human A549 lung cells. Toxicology In Vitro. 2007;**21**(3):438-448

[64] Sharma CS, Sarkar S, Periyakaruppan A, Barr J, Wise K, Thomas R, et al. Single walled carbon nanotubes induces oxidative stress in rat lung epithelial cells. Journal of Nanoscience and Nanotechnology. 2007;**7**(7):2466-2472

[65] Herzog E, Casey A, Lyng FM, Chambers G, Byrne HJ, Davoren M. A new approach to the toxicity testing of carbon-based nanomaterials—The clonogenic assay. Toxicology Letters. 2007;**174**(1-3):49-60

[66] Pulskamp K, Diabate S, Krug HF. Carbon nanotubes show no sign of acute toxicity but induce intracellular reactive oxygen species in dependence on contaminants. Toxicology Letters. 2007;**168**(1):58-74

[67] Boor KJ. Bacterial stress responses: What doesn't kill them can make them stronger. PLoS Biology. 2006;**4**(1):e23

[68] Kang S, Pinault M, Pfefferle LD, Elimelech M. Single walled carbon nanotubes exhibit strong antimicrobial activity. Langmuir. 2007;**23**(17):8670-8673

[69] Liu S, Wei L, Hao L, Fang N, Chang MW, Xu R, et al. Sharper and faster nano darts kill more bacteria: A study of antibacterial activity of individually dispersed pristine single-walled carbon nanotube. ACS Nano. 2009;**3**(12):3891-3902

[70] Arias LR, Yang L. Inactivation of bacterial pathogens by carbon nanotubes in suspensions. Langmuir. 2009;**25**(5):3003-3012

[71] Yang C, Mamouni J, Tang Y, Yang L. Antimicrobial activity of single walled carbon nanotubes length effect. Langmuir. 2010;**26**(20):16013-16019

[72] Vecitis CD, Zodrow KR, Kang S, Elimelech M. Electronic structure dependent bacterial cytotoxicity of single walled carbon nanotubes. ACS Nano. 2010;**4**(9):5471-5479

[73] Kang S, Mauter MS, Elimelech M. Microbial cytotoxicity of carbon based nanomaterials: Implications for river water and wastewater effluent. Environmental Science & Technology. 2009;**43**(7):2648-2653

[74] Jin L, Son Y, Yoon TK, Kang YJ, Kim W, Chung H. High concentrations of single walled carbon nanotubes lower soil enzyme activity and microbial biomass. Ecotoxicology and Environmental Safety. 2013;**88**:9-15

[75] Rodrigues DF, Elimelech M. Toxic effects of single walled carbon nanotubes in the development of *E. coli* biofilm. Environmental Science & Technology. 2010;**44**(12):4583-4589

[76] Kang S, Mauter MS, Elimelech M. Physicochemical determinants of multiwalled carbon nanotube bacterial cytotoxicity. Environmental Science & Technology. 2008;**42**(19):7528-7534

[77] Zheng H, Liu L, Lu Y, Long Y, Wang L, Ho KP. Rapid determination of nanotoxicity using luminous bacteria. Analytical Sciences. 2010;**26**(1):125-128

[78] Riding MJ, Martin FL, Trevisan J, Llabjani V, Patel II, Jones KC, et al.

Concentration dependent effects of carbon nanoparticles in gram negative bacteria determined by infrared spectroscopy with multivariate analysis. Environmental Pollution. 2012;**163**:226-234

[79] Kang S, Herzberg M, Rodrigues DF, Elimelech M. Antibacterial effects of carbon nanotubes: Size does matter. Langmuir. 2008;**24**(13):6409-6413

[80] Simon-Deckers A, Loo S, Mayne-L'hermite M, Herlin-Boime N, Menguy N, Reynaud C, et al. Size composition and shape dependent toxicological impact of metal oxide nanoparticles and carbon nanotubes toward bacteria. Environmental Science & Technology. 2009;**43**(21):8423-8429

[81] Chung H, Son Y, Yoon TK, Kim S, Kim W. The effect of multi walled carbon nanotubes on soil microbial activity, Ecotoxicology and Environmental Safety, 2011;**74**(4):569-575

[82] Gorczyca A, Kasprowicz MJ, Lemek T. Physiological effect of multi walled carbon nanotubes (MWCNTs) on conidia of the entomopathogenic fungus, *Paecilomyces fumosoroseus* Deuteromycotina: Hyphomycetes. Journal of Environmental Science and Health, Part A. 2009;**44**(14):1592-1597

[83] Schwab F, Bucheli TD, Lukhele LP, Magrez A, Nowack B, Sigg L, et al. Are carbon nanotube effects on green algae caused by shading and agglomeration? Environmental Science & Technology. 2011;**45**(14):6136-6144

[84] Long Z, Ji J, Yang K, Lin D, Wu F. Systematic and quantitative investigation of the mechanism of carbon nanotubes' toxicity toward algae. Environmental Science & Technology. 2012;**46**(15):8458-8466

[85] Youn S, Wang R, Gao J, Hovespyan A, Ziegler KJ, Bonzongo JC, et al. Mitigation of the impact of single-walled carbon nanotubes on a freshwater green algae: *Pseudokirchneriella subcapitata*. Nanotoxicology. 2011;**6**(2):161-172

[86] Kwok KW, Leung KM, Flahaut E, Cheng J, Cheng SH. Chronic toxicity of double walled carbon nanotubes to three marine organisms: Influence of different dispersion methods. Nanomedicine. 2010;**5**(6):951-961

[87] Wei LP, Thakkar M, Chen YH, Ntim SA, Mitra S, Zheng XY. Cytotoxicity effects of water dispersible oxidized multiwalled carbon nanotubes on marine alga, *Dunaliella tertiolecta*. Aquatic Toxicology. 2010;**100**(2):194-201

[88] Ghafari P, St-Denis CH, Power ME, Jin X, Tsou V, Mandal HS, et al. Impact of carbon nanotubes on the ingestion and digestion of bacteria by ciliated protozoa. Nature Nanotechnology. 2008;**3**(6):347-351

[89] Zhu Y, Zhao Q, Li Y, Cai X, Li W. The interaction and toxicity of multi walled carbon nanotubes with *Stylonychia mytilus*. Journal of Nanoscience and Nanotechnology. 2006;**6**(5):1357-1364

[90] Zhu X, Zhu L, Chen Y, Tian S. Acute toxicities of six manufactured nanomaterial suspensions to *Daphnia magna*. Journal of Nanoparticle Research. 2009;**11**(1):67-75

[91] Kim KT, Edgington AJ, Klaine SJ, Cho JW, Kim SD. Influence of multi-walled carbon nanotubes dispersed in natural organic matter on speciation and bioavailability of copper. Environmental Science & Technology. 2009;**43**(23):8979-8984

[92] Petersen EJ, Pinto RA, Mai DJ, Landrum PF, Weber WJ Jr. Influence of polyethyleneimine graftings of multi walled carbon nanotubes on their accumulation and elimination by and toxicity to *Daphnia magna*.

Environmental Science & Technology. 2011;**45**(3):1133-1138

[93] Li M, Huang CP. The responses of *Ceriodaphnia dubia* toward multi-walled carbon nanotubes: Effect of physical–chemical treatment. Carbon. 2011;**49**(5):1672-1679

[94] Cheng J, Flahaut E, Cheng SH. Effect of carbon nanotubes on developing zebrafish (*Danio rerio*) embryos. Environmental Toxicology and Chemistry. 2007;**26**(4):708-716

[95] Bourdiol F, Mouchet F, Perrault A, Fourquaux I, Datas L, Gancet C, et al. Biocompatible polymer-assisted dispersion of multi walled carbon nanotubes in water, application to the investigation of their ecotoxicity using *Xenopus laevis* amphibian larvae. Carbon. 2013;**54**:175-191

[96] Leeuw TK, Reith RM, Simonette RA, Harden ME, Cherukuri P, Tsyboulski DA, et al. Single-walled carbon nanotubes in the intact organism: Near-TR imaging and biocompatibility studies in Drosophila. Nano Letters. 2007;**7**(9):2650-2654

[97] Aschberger K, Johnston HJ, Stone V, Aitken RJ, Hankin SM, Peters SA, et al. Review of carbon nanotubes toxicity and exposure appraisal of human health risk assessment based on open literature. Critical Reviews in Toxicology. 2010;**40**(9):759-790

[98] Lim JH, Kim SH, Lee IC, Moon C, Kim SH, Shin DH, et al. Evaluation of maternal toxicity in rats exposed to multi-wall carbon nanotubes during pregnancy. Environmental Health and Toxicology. 2011;**26**:e2011006

Chapter 14

Advantages and Limitations of CNT-Polymer Composites in Medicine and Dentistry

Neeraja Turagam and Durga Prasad Mudrakola

Abstract

The past two decades have seen great technological advancements in the fields of optics, biochemistry, and physics allowing the fundamentals of our own human biology to be understood and controlled. At the forefront of this great understanding lies a tiny structure made of carbon called nanotube. Many studies have demonstrated that peptides, medicinal molecules, and nucleic acids, when bonded to carbon nanotubes, are delivered considerably more safely and effectively into cells than by traditional methods. Two types of carbon nanotubes have been researched for use in biomedical applications. The first is SWNT, single walled and second MWNT, multi-walled nanotube. Shell structures can be used for delivering anticancer drugs to tumors in various parts of the human body. In dentistry, the carbon nanotubes along with polymers prevent shrinkage and dimensional changes in resin and help in better fit at bone implant interface as well as in delivering well-fitting dentures. Evolution of gene therapy, cancer treatments, and innovative new answers for life-threatening diseases on the horizon, the science of nanomedicine has become an ever growing field that has an incredible ability to bypass barriers previously thought unavoidable.

Keywords: MWNT, SWNT, peptides, nanodentistry, nanomedicine, nanocomposites

1. Introduction

Carbon is an important element to various sciences, from physics, chemistry, and materials science to life science, but conventional carbon formulation in the micron scale may not be the optimal implant material [1]. Then the nanomaterial's such as the carbon nanotubes (CNTs), with unique electrical, mechanical, and surface properties, have captured the attention and aroused the interest of many scientists, since CNTs were discovered by Iijima in 1991 and up to now appear well suited as a biomaterial [2–7]. CNTs are substances with cylindrical structure of about 1 nm diameter and 1–10 m length, consisting of only carbon atoms. In general, CNTs contain single-wall carbon nanotubes (SWCNTs) and multiwall carbon nanotubes (MWCNTs).

SWCNTs are viewed microscopically as rolled-up structures of single sheets of graphene and individual carbon structures, approximately 1 nm in diameter and up to a millimeter or more in length, and MWCNTs are similar to hollow graphite fibers, except that they have a much higher degree of structural perfection, which are having a diameter of 10–200 nm [8–11]. Lu and Tsai investigated the load transfer efficiency in double-walled carbon nanotubes (DWCNTs, a hollow cylindrical structure, which contains two concentric graphene layers) using multiscale

 IntechOpen

finite element modeling, and the results showed that increasing of CNTs' length can effectively improve the load transfer efficiency in the outermost layers, while the DWCNTs with incremental covalent exhibit increasing load transfer efficiency in the inner layer. Besides, compared with single walled, the double walled nanotubes have decreased potential of load transfer efficiency [12].

Several studies proved increase mechanical properties of CNTs-based reinforced composites by the adding of carbon nanotubes (**Figure 1**). CNTs reinforced composites have been investigated thoroughly for numerous aspects of life and biomedical applications. The review introduced fabrication of CNTs reinforced composites, CNTs reinforced with ceramic and metal matrix composites their biocompatibility (in vivo), cell experiments (in vitro) and mechanical properties.

1.1 Early thinking

The late Nobel prize winning physicist Richard P. Feynman in 1959 speculated the potential of nano size devices as early as 1959. In his historic lecture in 1959, he concluded saying, "this is a development which I think cannot be avoided" [13].

1.2 Nanomaterials in dentistry

Inspite of the better understanding and use of chemistry and materials, recent developments in physical properties, no material has been found to be ideal for any kind of dental application [14]. Silver amalgam, as a dental restorative material has been used for more than a century, but for the toxicity and esthetics which has been of major concern for many many years [15–20]. In contrast the composite restorative materials have very good esthetics, and are very technique sensitive [21]. Nature has arranged complex biominerals in the best possible way from the micro to the nano-scale and no one can yet combine biological and physical properties to get ideal structures [22].

1.3 Access to nanodentistry

The practical applications in dentistry has various approaches [23, 24]. Broadly, two key approaches in nanotechnology are present for creating smaller or better materials. One being the top-down and the other is bottom-up. Top-down approach is based on solid-state processing of materials. The "top-down" approaches are used to fabricate functional structures at micro and nanoscales such as chemical vapor

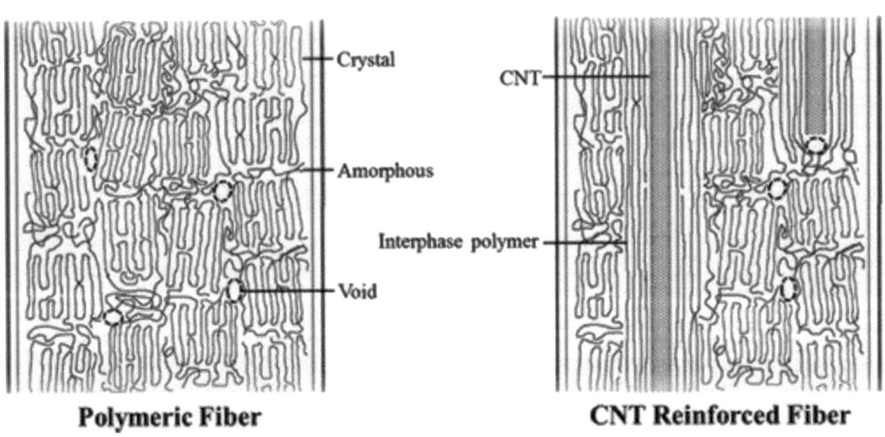

Figure 1.
CNTs-based reinforced composites

deposition (CVD), monolithic processing, wet and plasma etching [25]. These approaches are used in electronics industry as well as for coatings of medical implants and stent using chemical vapour deposition technology for increased blood flow [26].

The "bottom-up" approach entangles the fabrication of materials via edifice up particles by harvesting atomic elements. Bottom-up processing is based on extremely organized chemical synthesis and growth of materials [27] which occurs in repairing of cells, tissues or organ systems and protein synthesis as well.

Nanodentistry will make possible the maintenance of near-perfect oral health through the use of nanomaterials, biotechnology, including tissue engineering, and nanorobotics. Oral health and disease trends may change the focus on specific diagnostic and treatment modalities [28, 29].

1.3.1 Hypersensitivity cure

Dentin hypersensitivity is due to changes in pressure transmitted hydrodynamically to the pulp. Hence, teeth having hypersensitivity have eight times increased surface density of dentinal tubules and tubules with diameters twice as large than nonsensitive teeth. Dental nanorobots could precisely and selectively obstruct selected tubules in minutes using native biological materials.

1.3.2 Local anesthesia

A colloidal suspension with millions of active analgesic micron-size dental robots will be introduced in the gums of the patient. On coming in contact with the surface of tooth or mucosa, the ambulating nanorobots enter the pulp via the gingival sulcus, lamina propria, and dentinal tubules. Once introduced in the pulp, the dentist commands analgesic dental robots to stop all sensitivity and reactions in any specific tooth that needs treatment. The dentist orders the nanorobots to restore all sensation, after finishing all the oral treatments to relinquish control of nerve traffic, and to egress from the tooth by similar pathways used for ingress.

1.3.3 Orthodontic treatment

Orthodontic nanorobots could directly stimulate and manipulate the periodontal tissues, leading to rapid and painless tooth straightening, rotating, and vertical repositioning in few hours. Nanotechnology derived orthodontic wire is a new and advanced stainless steel wire which has the following properties (a) ultra-high strength (b) good deformability (c) corrosion resistance (d) good surface finish.

1.3.4 Nanoimpression

The introduction of Nanofillers into Polyvinylsiloxanes yields a siloxane impression material with properties superior to conventional impression materials.

Advantages (a) Better flow (b) Improved hydrophilic properties leading to fewer voids at margin and better model pouring (c) Enhanced detail precision.

- Nanosolutions: These are unique, dispersible nanoparticles with superior properties that can be produced from nanosolutions. This can be made use of dentin bonding agents (AdperTM) because of better dentin bond strength and better performance.

- Nanorobotic dentifrice [dentifrobots]: subocclusal nanorobotic dentifrice present in tooth paste or mouthwash could monitor all supragingival and subgingival surfaces, metabolizing the organic matter which is trapped into odorless and harmless vapors required for continuous calculus debridement. These invisibly small dentifrobots [1–10 μm], crawling at 1–10 μm/s are purely mechanical devices which are inexpensive. They would safely get deactivated themselves when swallowed and would be programmed with strict occlusal avoidance protocol.

- Dental durability and cosmetics: durability of the tooth along with aesthetics may be improved by replacing layers of upper enamel with pure sapphire and diamond embedded carbon nanotubes as they are more fracture resistant as nanostructured composites.

- Photosensitizers and carriers: quantum dots can be used as photosensitizers and carriers as they are bound to bind to the antibody present on the surface of the target cell. They can give rise to reactive oxygen species and when stimulated by UV light and thus will be lethal to the target cell.

- Diagnosis of oral cancer.

2. Nanoelectromechanical systems (NEMS)

They transform biochemical to electrical signals. NEMS biosensors exhibit specificity and sensitivity to detect the presence of abnormal cells at molecular level.

Oral fluid nanosensor test (OFNASET) used for multiplex detection of salivary biomarkers for oral cancer.

Optical Nano Biosensor - The nanobiosensor is a unique fiberoptics-based tool which allows the minimally invasive analysis of intracellular components (Cytochrome C1).

2.1 Treatment of oral cancer

Nanotechnology in field of cancer therapeutics has offered highly specific tools in the form of multifunctional Dendrimers. Nanoshells are miniscule beads with metallic outer layers designed to produce intense heat by absorbing specific wavelengths of radiations that can be used for selective destruction of cancer cells leaving aside intact, adjacent normal cell [30].

2.2 Nanocomposites

Nanocomposites are produced by homogeneously distributed nanoparticles in resins or coatings. Nanofillers used includes an aluminosilicate powder with mean particle size of 80 nm and a 1:4 M ratio of alumina to silica and a refractive index of 1.508 [31].

2.3 Advantages

- Increased hardness.

- Increased flexural strength, translucency.

- 50% reduction in filling shrinkage.

- excellent handling properties.

2.4 Challenges faced by nanodentistry

- Precise positioning and assembly of molecular scale virus in humans [31].
- Economical nanorobot mass production technique.
- Biocompatibility.
- Simultaneous coordination of activities of large numbers of independent micron-scale robots.
- Social issues of public acceptance, ethics, regulation.

2.5 Nanomaterials used for dental tissue regeneration

Pulp stem cells are purified in the lab and grown in sheets on scaffolds composed of nanofibers of biodegradable collagen type I or fibronectin used for pulp regeneration [32, 33]. Self-assembling polypeptide hydrogels have been used for pulp tissue regeneration with the formation of a nanofiber mesh for supporting the growing cells [34]. Puramatrix proven to enhance cell growth contains amino acids repeats of alanine, arginine and aspartate [35]. Natural silk based nanomaterials are being used for various tissue regeneration applications [36]. Injectable self-assembly collagen I scaffold containing exfoliated teeth stem cells led to the formation of pulp like tissue and functional odontoblasts [37]. Collagen type I is found in the form of nanofibers in dentin (~80–90% of organic matrix) and bone with abundant fibrous protein [38]. Odontogenic differentiation and mineralization was promoted in the presence of type I collagen scaffolds [39, 40].

2.6 Nanomedicine

Nanomedicine is the application of nanotechnology (the engineering of tiny machines) to the prevention and treatment of disease in the human body. This evolving discipline has the potential to dramatically change medical science.

2.7 Current status of nanomedicine

2.7.1 Diagnostics

Nanorobots are expected to circulate in the vascular system and send out signals when imbalances appear in the circulatory and lymphatic system. To monitor brain activity fixed nanomachines could be inserted in the nervous system of the human body. Latest nanomedical heart trackers are present in the major hospitals to accurately track and treat the heart beat and its downfalls as needed in the body [41]. The present and potential diagnostic uses is large being fullerene-based sensors, imaging (cellular, etc.), monitoring, lab on a chip, nanosensors, scanning probe microscopy, protein microarrays intracellular devices, intracellular biocomputers and intracellular sensors/reporters, endoscopic robots and microscopes.

2.7.2 Protein and peptide delivery

Protein and peptide molecules form the functional units of cells. Their molecular derangements lead to many illnesses. Targeted or controlled delivery of these molecules using nano particles and dendrimers is an emerging field called nano bio pharmaceutics.

2.7.3 Drugs dispersion and drug delivery

Drug delivery is based on developing nanoscale molecules to improve drug bioavailability. Nanomedicine based tools and devices are being developed for imaging. By the use of nanoparticle contrast agents, images such as ultrasound and magnetic resonance imaging (MRI) have improved distribution and contrast [42]. Triggered response is one way for drug molecules to be used more efficiently. The strength of drug delivery systems is their ability to alter the bio distribution and pharmacokinetics of the drug. Drugs are placed in the body and only activate on encountering a particular signal. For example, a drug with poor solubility will be replaced by a drug delivery system where both hydrophilic and hydrophobic environments exist thus improving its solubility.

2.7.4 Oncology

The small size of nanoparticles enhances their use in oncology. Quantum dots (nanoparticles with quantum confinement properties, such as size-tunable light emission), when used in conjunction with MRI, produces exceptional images of tumor sites [43]. Diagnosis of cancer at early stages can be detected from a few drops of the patient's blood by using sensor test chips containing thousands of nanowires, able to detect proteins and other biomarkers left behind by cancer cells [44]. Prof. Jennifer West has demonstrated the use of 120 nm diameter nanoshells coated with gold to kill cancer tumors in mice. By irradiating the area of the tumor with an infrared laser, which passes through flesh without heating it, the gold is heated sufficiently to cause death of the cancer cells [45].

2.7.5 Surgery

With the help of gold-coated nano shells, infrared laser and flesh welder bloodless surgery can be done with greater efficiency [46].

2.7.6 Nanomaterials for brachytherapy

BrachySil™ (Sivida, Australia) delivers 32P, clinical trial.

2.7.7 Drug delivery across

The blood-brain barrier/more effective treatment of brain tumors, Alzheimer's, Parkinson's in development.

2.7.8 Nanovectors for gene therapy

Non-viral gene delivery systems.

2.7.9 Cell repair machines

Direct cell and tissue repair can be done using molecular machines, however by using drugs and surgery only tissues can repair themselves. Access to cells by inserting needles into cells by molecular machines without killing them is possible [47].

2.7.10 Ethics and nanomedicine

Currently the most significant concerns involve risk assessment, risk management and risk communication of ENMs in clinical trials [48]. Implanting a computing chip in humans raises many ethical concerns. The chip can diagnose diseases and can also analyze our DNA to determine the diseases to which one may be susceptible to in later stages. Ethical issues concerning a patient's right-to know, right-not-to-know and the duty-to-know arise [49]. Increase in the current level of accuracy and efficiency of diagnostic and therapeutic procedures by augmenting the targeting and distribution by nanoparticles, the dangers of nanotoxicity becomes a paramount next step in better understanding of their medical needs [50].

2.7.11 Adverse reactions

Multiwalled carbon nanotubes led to asbestos like effects on the mesothelium due to high doses of intracavitary injection in rodents. Whether the inhalation of MWCNT will translocate to sensitive mesothelial sites has not been answered yet [51]. It will also be important to know their adverse effects, if any, in pediatric, geriatric and differing pathophysiological conditions like pregnancy, lactation, congestive heart failure, uremia, etc. (**Figure 1**).

3. Conclusions

Nanomedicine and nanodentistry will have an impact on many medical applications. The usefulness of these are not only therapeutic but also diagnostic. Development of applications of nanomedicine and nanodentistry is very complex and needs an integrated approach of all stakeholders. Future applications of nanodentistry will include nanorobotics, carbon nanotubes, nanocomposites whereas nanomedicine will include activity monitors, biochips, insulin pumps, needle less injectors, medical flow sensors and blood pressure, glucose monitoring devices and drug injecting systems. What nanomedicine and nanodentistry will be able to achieve in the future is beyond current imagination. However, it will be a tough task to handle the ethical issues which will be arising with the same pace.

Conflict of interest

Authors have no 'conflict of interest' declaration.

Author details

Neeraja Turagam* and Durga Prasad Mudrakola
Faculty of Dentistry, AIMST University, Kedah, Malaysia

*Address all correspondence to: neer222@gmail.com

IntechOpen

© 2019 The Author(s). Licensee IntechOpen. This chapter is distributed under the terms of the Creative Commons Attribution License (http://creativecommons.org/licenses/by/3.0), which permits unrestricted use, distribution, and reproduction in any medium, provided the original work is properly cited.

References

[1] Mortier J, Engelhardt M. Foreign body reaction to a carbon fiber implant in the knee: Case report and literature survey. Zeitschrift für Orthopädie und Ihre Grenzgebiete. 2000;**138**(5):390-394

[2] Iijima S. Helical microtubules of graphitic carbon. Nature. 1991;**354**(6348):56-58

[3] Li X, Liu X, Huang J, Fan Y, Cui F-Z. Biomedical investigation of CNT based coatings. Surface and Coatings Technology. 2011;**206**(4):759-766

[4] Firkowska I, Olek M, Pazos-Peréz N, Rojas-Chapana J, Giersig M. Highly ordered MWNT-based matrixes: Topography at the nanoscale conceived for tissue engineering. Langmuir. 2006;**22**(12):5427-5434

[5] Li XM, Feng Q, Liu X, Dong W, Cui F. The use of nanoscaled fibers or tubes to improve biocompatibility and bioactivity of biomedical materials. Journal of Nanomaterials. 2013;**3**:1-16

[6] Kam NWS, Jessop TC, Wender PA, Dai H. Nanotube molecular transporters: Internalization of carbon nanotube-protein conjugates into mammalian cells. Journal of the American Chemical Society. 2004;**126**(22):6850-6851

[7] Kam NWS, Liu Z, Dai H. Functionalization of carbon nanotubes via cleavable disulfide bonds for efficient intracellular delivery of siRNA and potent gene silencing. Journal of the American Chemical Society. 2005;**127**(36):12492-12493

[8] Ebbesen TW. Carbon nanotubes: prepartion and properties. From carbon fibers to nanotubes. 1997:42-49

[9] Ajayan PM. Nanotubes from carbon. Chemical Reviews. 1999;**99**(7):1787-1799

[10] De Jong KP, Geus JW. Carbon nanofibers: Catalytic synthesis and applications. Catalysis Reviews. 2000;**42**(4):481-510

[11] Ono-Ogasawara M, Myojo T. Characteristics of multi-walled carbon nanotubes and background aerosols by carbon analysis, particle size and oxidation temperature. Advanced Powder Technology. 2012;**24**(1):263-269

[12] Lu TC, Tsai JL. Characterizing load transfer efficiency in double-walled carbon nanotubes using multiscale finite element modeling. Composites: Part B. 2013;**44**(1):394-402

[13] Feynman R. There's plenty of room at the bottom. In: Gilbert HD, editor. Miniaturization. New York: Reinhold; 1961. pp. 282-296

[14] Mitra SB, Wu D, Holmes BN. An application of nanotechnology in advanced dental materials. Journal of the American Dental Association (Chicago, IL). 2003;**134**:1382-1390

[15] Jones DW. A Canadian perspective on the dental amalgam issue. British Dental Journal. 1998;**184**:581-586

[16] Warfvinge K. Mercury exposure of a female dentist before pregnancy. British Dental Journal. 1995;**178**:149-152

[17] Smart ER, Macleod RI, Lawrence CM. Resolution of lichen-planus following removal of amalgam restorations in patients with proven allergy to mercury salts—A pilot-study. British Dental Journal. 1995;**178**:108-112

[18] Eley BM. The future of dental amalgam: A review of the literature. 7: Possible alternative materials to amalgam for the restoration of posterior teeth. British Dental Journal. 1997;**183**:11-14

[19] Mclean JW. Alternatives to amalgam alloys: 1. British Dental Journal. 1984;**157**:432-433

[20] Yardley RM. Alternatives to amalgam alloys: 2. British Dental Journal. 1984;**157**:434-435

[21] Saunders SA. Current practicality of nanotechnology in dentistry. Part 1: Focus on nanocomposite restoratives and biomimetics. Clinical, Cosmetic and Investigational Dentistry. 2009;**1**:47-61

[22] Kanaparthy R, Kanaparthy A. The changing face of dentistry: Nanotechnology. International Journal of Nanomedicine. 2011;**6**:2799-2804

[23] Subramani K, Ahmed W. Emerging Nanotechnologies in Dentistry: Processes, Materials and Applications. Amsterdam, The Netherlands: William Andrew; 2011

[24] Mikkilineni M, Rao A, Tummala M, Elkanti S. Nanodentistry: New buzz in dentistry. European Journal of General Dentistry. 2013;**2**:109

[25] Zhang L, Webster TJ. Nanotechnology and nanomaterials: Promises for improved tissue regeneration. Nano Today. 2009;**4**:66-80

[26] Roszek B, De Jong W, Geertsma R. Nanotechnology in Medical Applications: State-of-the-Art in Materials and Devices. Bilthoven, The Netherlands: Rijksinstituut voor Volksgezondheid en Milieu RIVM; 2005

[27] Wickson F. Narratives of nature and nanotechnology. Nature Nanotechnology. 2008;**3**:313-315

[28] Whitesides GM, Love JC. The art of building small. Scientific American. 2001;**285**(3):33-41

[29] Freitas RA Jr. Nanodentistry. Journal of the American Dental Association. 2000;**131**(11):1559-1565

[30] Jhaver HM, Balaji PR. Nanotechnology: The future of dentistry. Journal of Indian Society of Periodontology 2005;**5**:15-17

[31] Pratap R. Engaging private enterprise in nanotech research in india: ICS. Trieste. 2005. pp. 675-680

[32] Venugopal J, Ramakrishna S. Applications of polymer nanofibers in biomedicine and biotechnology. Applied Biochemistry and Biotechnology. 2005;**125**:147-157

[33] Fukuda J, Khademhosseini A, Yeh J, Eng G, Cheng J, Farokhzad OC, et al. Micropatterned cell co-cultures using layer-by-layer deposition of extracellular matrix components. Biomaterials. 2006;**27**:1479-1486

[34] Galler KM, Cavender A, Yuwono V, Dong H, Shi S, Schmalz G, et al. Self-assembling peptide amphiphile nanofibers as a scaffold for dental stem cells. Tissue Engineering. Part A. 2008;**14**:2051-2058

[35] Misawa H, Kobayashi N, Soto-Gutierrez A, Chen Y, Yoshida A, Rivas-Carrillo JD, et al. Pura matrix facilitates bone regeneration in bone defects of calvaria in mice. Cell Transplantation. 2006;**15**:903-910

[36] Zafar MS, Al-Samadani KH. Potential use of natural silk for bio-dental applications. Journal of Taibah University Medical Sciences. 2014;**9**:171-177

[37] Demarco FF, Conde M, Cavalcanti BN, Casagrande L, Sakai VT, Nör JE. Dental pulp tissue engineering. Brazilian Dental Journal. 2011;**22**:3-14

[38] Wiesmann H, Meyer U, Plate U, Höhling H. Aspects of collagen mineralization in hard tissue formation. International Review of Cytology. 2004;**242**:121-156

[39] Kim NR, Lee DH, Chung P, Yang H. Distinct differentiation properties of human dental pulp cells on collagen, gelatin, and chitosan scaffolds. Oral Surgery, Oral Medicine, Oral Pathology, Oral Radiology, and Endodontics. 2009;**108**:e94-e100

[40] Mizuno M, Miyamoto T, Wada K, Watatani S, Zhang GX. Type I collagen regulated dentin matrix protein-1 (Dmp-1) and osteocalcin (OCN) gene expression of rat dental pulp cells. Journal of Cellular Biochemistry. 2003;**88**:1112-1119

[41] Saha M. Nanomedicine: Promising tiny machine for the healthcare in future—A review. Oman Medical Journal. 2009;**24**(4):242-247

[42] LaVan DA, McGuire T, Langer R. Small-scale systems for in vivo drug delivery. Nature Biotechnology. 2003;**21**(10):1184-1191

[43] Nie S, Xing Y, Kim GJ, Simons JW. Nanotechnology applications in cancer. Annual Review of Biomedical Engineering. 2007;**9**:257-288

[44] Zheng G, Patolsky F, Cui Y, Wang WU, Lieber CM. Multiplexed electrical detection of cancer markers with nanowire sensor arrays. Nature Biotechnology. 2005;**23**(10):1294-1301

[45] Loo C, Lin A, Hirsch L, Lee MH, Barton J, Halas N, et al. Nanoshell-enabled photonics-based imaging and therapy of cancer. Technology in Cancer Research & Treatment. 2004;**3**(1):33-40

[46] Gobin AM, O'Neal DP, Watkins DM, Halas NJ, Drezek RA, West JL. Near infrared laser-tissue welding using nanoshells as an exogenous absorber. Lasers in Surgery and Medicine. 2005;**37**(2):123-129

[47] Cathy G. The potential and the pitfalls of nanomedicine. 2007. Available from: http://www.nanowerk.com/spotlight/spotid=1891.php [Accessed: 24 December 2011]

[48] Resnik DB, Tinkle SS. Ethics in nanomedicine. Nanomedicine (London, England). 2007;**2**(3):345-350

[49] Evers J, Aerts S, De Tavernier J. An ethical argument in favor of nano-enabled diagnostics in livestock disease control. NanoEthics. 2008;**2**:163-178

[50] Minchin R. Nanomedicine: Sizing up targets with nanoparticles. Nature Nanotechnology. 2008;**3**(1):12-13

[51] Oberdörster G. Safety assessment for nanotechnology and nanomedicine: Concepts of nanotoxicology. Journal of Internal Medicine. 2010;**267**(1):89-105